Lecture Notes in Computer Science   14466

Founding Editors

Gerhard Goos
Juris Hartmanis

Editorial Board Members

The series Lecture Notes in Computer Science (LNCS), including its subseries Lecture Notes in Artificial Intelligence (LNAI) and Lecture Notes in Bioinformatics (LNBI), has established itself as a medium for the publication of new developments in computer science and information technology research, teaching, and education.

LNCS enjoys close cooperation with the computer science R & D community, the series counts many renowned academics among its volume editors and paper authors, and collaborates with prestigious societies. Its mission is to serve this international community by providing an invaluable service, mainly focused on the publication of conference and workshop proceedings and postproceedings. LNCS commenced publication in 1973.

Michael A. Bekos · Markus Chimani
Editors

Graph Drawing and Network Visualization

31st International Symposium, GD 2023
Isola delle Femmine, Palermo, Italy, September 20–22, 2023
Revised Selected Papers, Part II

Springer

Editors
Michael A. Bekos 🄳
University of Ioannina
Ioannina, Greece

Markus Chimani 🄳
Osnabrück University
Osnabrück, Germany

ISSN 0302-9743 ISSN 1611-3349 (electronic)
Lecture Notes in Computer Science
ISBN 978-3-031-49274-7 ISBN 978-3-031-49275-4 (eBook)
https://doi.org/10.1007/978-3-031-49275-4

This Springer imprint is published by the registered company Springer Nature Switzerland AG
The registered company address is: Gewerbestrasse 11, 6330 Cham, Switzerland

Paper in this product is recyclable.

Preface

This volume contains the papers presented at GD 2023, the 31st International Symposium on Graph Drawing and Network Visualization, held on September 20–22, 2023 in Isola delle Femmine (Palermo), Italy. Graph drawing is concerned with the geometric representation of graphs and constitutes the algorithmic core of network visualization. Graph drawing and network visualization are motivated by applications where it is crucial to visually analyze and interact with relational datasets. Information about the conference series and past symposia is maintained at http://www.graphdrawing.org.

A total of 122 participants from 18 different countries attended the conference. With regards to the program itself, regular papers could be submitted to one of two distinct tracks: Track 1 for papers on combinatorial and algorithmic aspects of graph drawing and Track 2 for papers on experimental, applied, and network visualization aspects. Short papers were given a separate category, which welcomed both theoretical and applied contributions. An additional track was devoted to poster submissions. All the tracks were handled by a single Program Committee. As in previous editions of GD, the papers in the different tracks did not compete with each other, but all program committee members were invited to review papers from either track in a "light-weight double-blind" process.

In response to the call for papers, the Program Committee received a total of 114 submissions, consisting of 100 papers (52 in Track 1, 23 in Track 2, and 25 in the short paper category) and 14 posters. More than 300 reviews were provided, about a third having been contributed by external sub-reviewers. After extensive electronic discussions by the Program Committee via EasyChair, interspersed with virtual meetings of the Program Chairs producing incremental accept/reject proposals, 31 long papers, 7 short papers, and 11 posters were selected for inclusion in the scientific program of GD 2023. This resulted in an overall paper acceptance rate (not considering posters) of 38% (46% in Track 1, 30% in Track 2, and 28% in the short paper category). As is common in GD, some hard choices had to be made in particular during the final acceptance/rejection round, where several papers that clearly had merit still did not make the cut. However, the number of submitted high-quality papers speaks for the community. Authors published an electronic version of their accepted papers on the arXiv e-print repository; a conference index with links to these contributions was made available before the conference.

There were two invited lectures at GD 2023. Monique Teillaud from INRIA Nancy - Grand Est, LORIA (France) described the intrinsics of *"The CGAL Project"*, while Michael Kaufmann from Universität Tübingen (Germany) focused *"On Orthogonal Drawings of Plane and Not So Plane Graphs"*. Abstracts of both invited lectures are included in these proceedings.

The conference gave out best paper awards in Track 1 and Track 2, as well as a best presentation award and a best poster award. Based on a majority vote of the Program Committee, "On the Biplanarity of Blowups" by David Eppstein was chosen as the best paper in Track 1. In Track 2, the best paper was chosen to be "Celtic-Graph: Drawing Graphs as Celtic Knots and Links" by Peter Eades, Niklas Gröne, Karsten Klein, Patrick Eades, Leo Schreiber, Ulf Hailer, and Falk Schreiber. Based on a majority vote of conference participants, the best presentation award was given to Julia Katheder for her presentation of the paper "Weakly and Strongly Fan-Planar Graphs". Also chosen by the conference audience, the best poster award was given to "What Happens at Dagstuhl? Uncovering Patterns through Visualization" by Felix Klesen, Jacob Miller, Fabrizio Montecchiani, Martin Nöllenburg, and Markus Wallinger. Many thanks to Springer whose sponsorship funded the prize money for these awards.

A PhD School on graph databases was held on the day prior to the conference. The lectures were led by Nikolay Yakovets (TU Eindhoven), Andreas Kollegger (Neo4J), and Fouli Argyriou (yWorks). This year, we had the special treat of celebrating Giuseppe Liotta's 60th birthday on the day after the conference – an opportunity most of the conference participants happily took. Congratulations, Beppe, for so many successful and graph drawing inspiring years!

As is traditional, the 31st Annual Graph Drawing Contest was held during the conference. The contest was divided into two parts, creative topics and the live challenge. The creative topics task featured a single graph: a board-game recommendations network (a data set containing the top-100 games found on www.boardgamegeek.com). The live challenge focused on minimizing the number of crossings on point set embeddings. There were two categories: manual and automatic. We thank the Contest Committee, chaired by Wouter Meulemans, for preparing interesting and challenging contest problems. A report on the contest is included in these proceedings.

Many people and organizations contributed to the success of GD 2023. We would like to thank all members of the Program Committee and the external reviewers for carefully reviewing and discussing the submitted papers and posters; this was crucial for putting together a strong and interesting program. Also thanks to all authors who chose GD 2023 as the publication venue for their research.

We are grateful for the support of our "Gold" sponsors Tom Sawyer Software and yWorks, our "Bronze" sponsors Springer and Neo4J, and our "Contributor" Bazuel. Their generous support helped to ensure the continued success of this conference.

Last but not least, the organizing committee did a wonderful job in ensuring a smooth and joyful conference experience both for the scientific and the non-scientific parts. The credit for this remarkable achievement goes wholly to the organizing co-chairs, Emilio Di Giacomo, Fabrizio Montecchiani, and Alessandra Tappini. They in turn would like to express their thanks to the other local organizers and volunteers, including Carla Binucci, Luca Grilli, Giacomo Ortali, and Tommaso Piselli.

The 32nd International Symposium on Graph Drawing and Network Visualization (GD 2024) will take place September 18–20, 2024, in Vienna, Austria. Stefan Felsner and Karsten Klein will co-chair the Program Committee, and Robert Ganian and Martin Nöllenburg will co-chair the Organizing Committee.

October 2023   Michael A. Bekos
                                                               Markus Chimani

Organization

Steering Committee

Patrizio Angelini	John Cabot University, Italy
Michael A. Bekos	University of Ioannina, Greece
Markus Chimani	Osnabrück University, Germany
Giuseppe Di Battista	Roma Tre University, Italy
Emilio Di Giacomo	University of Perugia, Italy
Stefan Felsner	Technische Universität Berlin, Germany
Reinhard von Hanxleden	University of Kiel, Germany
Karsten Klein	University of Konstanz, Germany
Stephen G. Kobourov (Chair)	University of Arizona, USA
Anna Lubiw	University of Waterloo, Canada
Roberto Tamassia	Brown University, USA
Ioannis G. Tollis	ICS-FORTH and University of Crete, Greece
Alexander Wolff	University of Würzburg, Germany

Program Committee

Md. Jawaherul Alam	Amazon, USA
Daniel Archambault	Swansea University, UK
Martin Balko	Charles University in Prague, Czech Republic
Michael A. Bekos (Co-chair)	University of Ioannina, Greece
Steven Chaplick	Maastricht University, The Netherlands
Markus Chimani (Co-chair)	Osnabrück University, Germany
Sabine Cornelsen	University of Konstanz, Germany
Eva Czabarka	University of South Carolina, USA
Emilio Di Giacomo	University of Perugia, Italy
Christian Duncan	Quinnipiac University, USA
Stefan Felsner	TU Berlin, Germany
Fabrizio Frati	Roma Tre University, Italy
Petr Hliněný	Masaryk University in Brno, Czech Republic
Andreas Kerren	Linköping University, Sweden
Fabian Klute	Polytechnic University of Catalonia, Spain
Anna Lubiw	University of Waterloo, Canada
Tamara Mchedlidze	Utrecht University, The Netherlands
Debajyoti Mondal	University of Saskatchewan, Canada
Fabrizio Montecchiani	University of Perugia, Italy
Martin Nöllenburg	Technische Universität Wien, Austria

Yoshio Okamoto	University of Electro-Communications, Japan
Chrysanthi Raftopoulou	National Technical University of Athens, Greece
Lena Schlipf	Universität Tübingen, Germany
Jens M. Schmidt	University of Rostock, Germany
Matthias Stallmann	North Carolina State University, USA
Csaba Toth	California State University Northridge, USA
Torsten Ueckerdt	Karlsruhe Institute of Technology, Germany
Johannes Zink	Universität Würzburg, Germany

Organizing Committee

Carla Binucci	University of Perugia, Italy
Emilio Di Giacomo (Co-chair)	University of Perugia, Italy
Luca Grilli	University of Perugia, Italy
Fabrizio Montecchiani (Co-chair)	University of Perugia, Italy
Giacomo Ortali	University of Perugia, Italy
Tommaso Piselli	University of Perugia, Italy
Alessandra Tappini (Co-chair)	University of Perugia, Italy

Contest Committee

Philipp Kindermann	University of Trier, Germany
Fabian Klute	Polytechnic University of Catalonia, Spain
Tamara Mchedlidze	Utrecht University, The Netherlands
Debajyoti Mondal	University of Saskatchewan, Canada
Wouter Meulemans (Chair)	TU Eindhoven, The Netherlands

External Reviewers

Agarwal, Shivam	Blum, Johannes
Aichholzer, Oswin	Bonichon, Nicolas
Akitaya, Hugo	Brandenburg, Franz
Angelini, Patrizio	Cardinal, Jean
Arleo, Alessio	Čermák, Filip
Aronov, Boris	Chakraborty, Dibyayan
Arseneva, Elena	Cooper, Joshua
Barát, János	D'Elia, Marco
Behrisch, Michael	Da Lozzo, Giordano
Bergold, Helena	Di Bartolomeo, Sara
Bhore, Sujoy	Diatzko, Gregor
Biniaz, Ahmad	Didimo, Walter
Binucci, Carla	Dijk, Thomas C. Van
Blažej, Václav	

Dobler, Alexander
Dumitrescu, Adrian
Dunne, Cody
Dutle, Aaron
Eiben, Eduard
Eppstein, David
Firman, Oksana
Fujiwara, Takanori
Fulek, Radoslav
Förster, Henry
Gethner, Ellen
Goaoc, Xavier
Gonçalves, Daniel
Grilli, Luca
Gronemann, Martin
Grosso, Fabrizio
Gupta, Siddharth
Guspiel, Grzegorz
Hamalainen, Rimma
Hamm, Thekla
Hegemann, Tim
Hoffmann, Michael
Hušek, Radek
Joret, Gwenaël
Jungeblut, Paul
Katheder, Julia
Katsanou, Eleni
Khazaliya, Liana
Kindermann, Philipp
Klawitter, Jonathan
Klemz, Boris
Klesen, Felix
Kobourov, Stephen
Kratochvil, Jan
Kucher, Kostiantyn
Kypridemou, Elektra
Lauff, Robert
Linhares, Claudio
Liu, Kevin
Mann, Ryan
Martínez Sandoval, Leonardo Ignacio
Masařík, Tomáš
McGee, Fintan
Meuwese, Ruben
Mueller, Tobias

Opler, Michal
Ortali, Giacomo
Ortlieb, Christian
Parada, Irene
Patrignani, Maurizio
Piselli, Tommaso
Pokrývka, Filip
Pupyrev, Sergey
Purchase, Helen
Reddy, Meghana M.
Roch, Sandro
Rollin, Jonathan
Rosenke, Christian
Rutter, Ignaz
Samal, Robert
Schaefer, Marcus
Scheibner, Mark
Scheucher, Manfred
Schnider, Patrick
Schröder, Felix
Schulz, André
Shahrokhi, Farhad
Sieper, Marie Diana
Staals, Frank
Steiner, Raphael
Storandt, Sabine
Stumpf, Peter
Szekely, Laszlo
Tappini, Alessandra
Tkadlec, Josef
Toth, Geza
van Goethem, Arthur
van Wageningen, Simon
Vogtenhuber, Birgit
Wallinger, Markus
Wang, Yong
Wang, Zhiyu
Wolff, Alexander
Wood, David R.
Wu, Hsiang-Yun
Wulms, Jules
Xia, Ge
Yip, Chi Hoi
Zeng, Ji

Sponsors

Gold Sponsors

Bronze Sponsors

Contributors

Invited Talks

The CGAL Project – cgal.org

Monique Teillaud

Inria Nancy – Grand Est, LORIA, France
monique.teillaud@inria.fr

Abstract. CGAL, The Computational Geometry Algorithms Library, provides easy access to efficient and reliable geometric algorithms in the form of a C++ library. It is used both in academia and industry, in various areas needing geometric computation, such as geographic information systems, computer aided design, molecular biology, medical imaging, computer graphics, and robotics.

The CGAL Project is a joint effort of several research groups in academia and their industrial partner GeometryFactory. It started in 1996 as a consortium of seven academic sites in Europe and Israel. In 2003, it has become an Open Source project, inviting developers from around the world to join.

This talk is definitely not going to be a CGAL course. Instead I will try to give insight about the history of the project and the multiple ingredients that have made it successful over the years: technical choices, development tools, project organization, license choice, etc, without forgetting the most important: the CGAL people.

On Orthogonal Drawings of Plane and Not So Plane Graphs

Michael Kaufmann

Wilhelm-Schickard-Institut für Informatik,
Universität Tübingen, Tübingen, Germany
michael.kaufmann@uni-tuebingen.de

Abstract. The orthogonal drawing model is one of the basic models in Graph Drawing. Algorithms and variants have developed from the very beginning of the field, over the years and it is still very present in recent works.

In my talk, I will make a personal walk through the last 30 years of orthogonal graph drawing. I will highlight some topics that I like and which you possibly don't know, still containing remarkable techniques. At the end, I will give some open questions and directions from the past that might be remarkable.

Contents – Part II

Contents – Part I

Linear Layouts

Geometric Aspects

Visualization Challenges

Graph Representations

Graph Decompositions

Graph Drawing Contest Report

Topological Aspects

Cops and Robbers on 1-Planar Graphs

Stephane Durocher[1], Shahin Kamali[2], Myroslav Kryven[1], Fengyi Liu[1],
Amirhossein Mashghdoust[1], Avery Miller[1], Pouria Zamani Nezhad[1],
Ikaro Penha Costa[1], and Timothy Zapp[1]

[1] Department Computer Science, University of Manitoba, Winnipeg, MB, Canada
{stephane.durocher,myroslav.kryven,avery.miller,ikaro.costa}@umanitoba.ca,
{liuf3412,mashghda,zamaninp,zappt3}@myumanitoba.ca
[2] Department Electrical Engineering and Computer Science, York University,
Toronto, ON, Canada
kamalis@yorku.ca

Abstract. *Cops and Robbers* is a well-studied pursuit-evasion game in
which a set of cops seeks to catch a robber in a graph G, where cops and
robber move along edges of G. The *cop number* of G is the minimum
number of cops that is sufficient to catch the robber. Every planar graph
has cop number at most three, and there are planar graphs for which
three cops are necessary [Aigner and Fromme, DAM 1984]. We study
the problem for *1-planar graphs*, that is, graphs that can be drawn in
the plane with at most one crossing per edge. In contrast to planar
graphs, we show that some 1-planar graphs have unbounded cop number.
Meanwhile, for maximal 1-planar graphs, we prove that three cops are
always sufficient and sometimes necessary. In addition, we completely
determine the cop number of outer 1-planar graphs.

Keywords: Beyond Planarity · 1-Planar Graphs · Pursuit-Evasion ·
Cops and Robbers

1 Introduction

Pursuit-evasion is a family of problems (also called games) in which one group
seeks to capture members of another group in a given environment. There are
many variants of pursuit-evasion games, depending on the game environment,
information available to each player about the environment and other players,
and restrictions on the freedom or speed at which players can move.

One of the most common and well-studied pursuit-evasion problems is the
game of *Cops and Robbers* on graphs, which was formalized by Quilliot [27] and
Nowakowski and Winkler [26] in the 1980s; see also the recent book by Bonato
and Nowakowski [6]. The game is played in a graph by two players: the robber
player and the cop player. The common assumption that we also adopt here

This research is funded in part by the Natural Sciences and Engineering Research
Council of Canada (NSERC).

is that each player has full information about the graph and the other player's moves. The game consists of rounds (or steps) on a given graph. In the initial round, the cop player selects starting vertices for a set of cops, and then the robber player selects a starting vertex for a robber. In the subsequent rounds, the players alternate turns; during the cops' turn, the cop player may move some of the cops to adjacent vertices. Similarly, the robber player may move the robber to an adjacent vertex during the robber's turn. The cop player wins if the robber and any of the cops are simultaneously on the same vertex; otherwise, when the game continues indefinitely, the robber player wins. If a single cop suffices to catch the robber in a graph G, even when the robber plays adversarially, then G is a *cop-win graph*; otherwise, G is a *robber-win graph*. The minimum number of cops necessary to catch the robber in G, denoted $c(G)$, is called the *cop number* of G, and G is a $c(G)$-*cop win graph*.

Related Work. Various characterizations of c-cop win graphs are known [11]. For example, every cop-win graph has a *domination elimination ordering sequence* [27] (also known as *dismantling ordering*), i.e., an ordering of the vertices such that, if the vertices are removed in this order, each vertex (except the last) is dominated at the time it is removed. This characterization helped identify classes of cop-win graphs, such as chordal graphs (as every elimination ordering of a chordal graph is also a dismantling ordering) and visibility graphs [23].

For some classes of graphs, a cop strategy comes directly from the graph's structural properties. For example, Aigner and Fromme [1] showed that every planar graph has a cop number of at most three by implicitly using the Jordan curve property, and, the fact that a cycle in a *plane graph* (a graph drawn in the plane without edge crossings) partitions the graph into interior and exterior regions. Some classes of planar graphs have cop number two (e.g., series-parallel graphs [20] and outerplanar graphs). On the other hand, planar graphs with cop number three are known, e.g., the *dodecahedron*, which is the skeleton of the platonic solid with faces of degree five. It is conjectured that the dodecahedron is the smallest planar graph with cop number three [1]. Maurer et al. [24] give an example of a *triangulation* (i.e., maximal planar graph) with cop number three.

Aigner and Fromme [1] also asked whether the result concerning planar graphs can be generalized to graphs of higher *orientable genus*, which is the minimum number of handles attached to a sphere so that the graph could be drawn without crossings on the resulting surface. Shortly after, Quilliot [28] gave an upper bound of $2g + 3$ on the cop number of graphs of genus g. Schröder [30] improved this bound to $1.5g + 3$ and conjectured that there is a tighter upper bound of $g + 3$. The current best upper bound of approximately $1.268g$ is due to Erde and Lehner [15]; find details in the survey of Bonato and Mohar [8].

Having established the cop number of planar graphs, a natural next question concerns the cop number of graphs that are *almost planar*. Proximity to being planar for a given graph G can be parameterized by the minimum number of crossings per edge among all drawings of G in the plane, where a planar graph requires zero crossings per edge. In recent years, there has been an increasing interest in the family of graphs that generalize planar graphs, called *beyond-*

planar graphs, that is, graphs that can be drawn with few crossings [19]. A common subclass of beyond-planar graphs is *k-planar graphs*, that is, graphs that can be drawn in the plane with at most k crossings per edge. Various properties of *k-planar graphs* are known: their maximum edge density, relation to other families of beyond-planar graphs, as well as the complexity of many algorithmic problems [12]. The special case of 1-*planar graphs* has been extensively studied, for example, their maximum edge density is $4n - 8$ [31], which is tight, and they can be colored with at most seven colors [29]. Even though this might seem to generalize planar graphs only slightly, in contrast to planar graphs, determining whether a given graph is 1-planar is NP-hard [18,22]. Moreover, their relation to other families of beyond-planar graphs is well understood [4]. In this paper we only consider *simple* drawings, that is, edges can cross at most once and adjacent edges do not cross.

Our Results. The upper bound of Quilliot [28] on the cop number of graphs with bounded genus was one of the first steps in the study of the Cops and Robbers game on beyond-planar graphs. In this work, we extend this direction by studying bounds on the cop number in another prominent family of beyond planar graphs: k-planar graphs. Graphs of bounded genus and k-planar graphs are unrelated; i.e., there are 1-planar graphs with arbitrary large genus, and graphs that have genus one but require arbitrarily many crossings per edge to be drawn in the plane [14]. We show in Sect. 2 that, despite the fact that planar graphs have cop number at most three, 1-planar graphs may have unbounded cop number; see Theorem 1. Constructing 1-planar graphs with large cop number, as described in the proof of Theorem 1, results in sparse 1-planar graphs with many vertices of degree two. With this in mind, we consider dense 1-planar graphs, particularly maximal 1-planar graphs, in Sect. 4. A 1-planar graph is said to be *maximal* if no edge can be added such that the resulting graph remains 1-planar. Maximal 1-planar graphs with n vertices and the maximum number $4n - 8$ of edges [32] are called *optimal 1-planar*. Maximal 1-planar graphs can have fewer than $4n - 8$ edges, e.g., there exist maximal 1-planar graphs that have as few as $2.65n$ edges [9]. In Sect. 3, we construct a *planar quadrangulation* (a plane graph in which each face has exactly four edges) with cop number three. Using this quadrangulation, in Sect. 4.1, we show how to construct a maximal 1-planar graph (which is also an optimal 1-planar graph) for which three cops are necessary to catch a robber. In Sect. 4.2, we show that each maximal 1-planar graph has cop number at most three using the Jordan arc property, as used by Aigner and Fromme [1], and structural properties of maximal 1-planar graphs.

In Sect. 5, we consider *outer k-planar graphs* [2,10], where, in addition to each edge being crossed at most k times, we also restrict the vertices to be in convex position on the outer boundary. The cop number of an outer k-planar graph is bounded by $1.5k + 6.5$ due to its relation to treewidth [20] (if the treewidth of a graph is at most t, then its cop number is at most $t/2 + 1$ [20]). This is an interesting contrast to general k-planar graphs, which can have unbounded cop number. Finally, we examine outer 1-planar graphs, which have cop number at

most two due to the upper bound via their treewidth, and characterize those
that are cop-win.

2   1-Planar Graphs May Have Unbounded Cop Number

In this section we show that there exist 1-planar graphs with unbounded cop
number.

Theorem 1. *For every $c \in \mathbb{N}$, there exists a 1-planar graph with cop number at
least c.*

Proof. For any c, there exists a graph G with cop number c [25, Theorem 2.6].
Choose a k-planar drawing of G for some k. Define a graph G' that is obtained
from G by subdividing each edge $k-1$ times if k is odd, or k times if k is even; see
Fig. 1. Berarducci and Intrigila [5, Theorem 5.6] showed that replacing each edge
with the same odd-length path does not decrease the cop number, so it follows
that $c(G') \geq c(G)$. Also, G' is 1-planar as we can obtain a 1-planar drawing of
G' from the k-planar drawing of G: between every two consecutive crossings on
an edge in G, place one of the added subdivision vertices in G'.   □

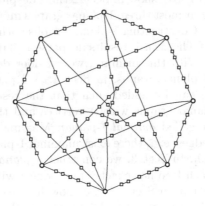

Fig. 1. Subdividing each edge an equal number of times does not reduce the cop number
of the resulting graph.

3   Cop Number of Quadrangulations

In this section, we study the cop number of *quadrangulations*, that is, plane
graphs in which every face has exactly four edges and four vertices. A graph
G can be drawn as a quadrangulation in the plane if and only if G is a maximal planar bipartite graph [21]. Since quadrangulations are planar, they have

cop number at most three [1]. We show that there exists a quadrangulation Q for which three cops are necessary; see Theorem 2. We will use this result in Sect. 4 to prove that there are maximal 1-planar graphs with cop number three. To construct our quadrangulation Q we first construct a triangulation T [24] which is based on the dodecahedron graph D; see Fig. 2. The triangulation T is constructed by adding vertices and edges to each face of D in such a way that each resulting face is a triangle; see Fig. 2b. The original dodecahedron edges are in black and the new triangulation edges are in dark green. We construct our quadrangulation Q by adding one vertex and three edges to each face of this triangulation in such a way that each resulting face is a quadrangle; see Fig. 2b. The new quadrangulation edges are in light green. After quadrangulating the triangulation, we have subdivided the original edges of the dodecahedron once, so the vertices of the dodecahedron that were initially adjacent are now at distance two from each other. Let $V' \subset V(Q)$ be the set of such vertices of Q that subdivide the original edges of the dodecahedron.

Theorem 2. *The quadrangulation Q has cop number three.*

Proof. Because the quadrangulation Q is a planar graph, its cop number is at most three [1]. We will show that Q has cop number at least three. In particular, we show that if there are only two cops, there are vertices in $V(D) \cup V' \subset V(Q)$ that the robber can choose to move to without being caught, using the following strategy. For any vertex v, denote by $N[v]$ the set of vertices within distance at most one from v, and denote by $N^2[v]$ the set of vertices within distance at most two from v. The robber begins on a vertex r of the dodecahedron D, and remains on this vertex until a cop moves to a vertex adjacent to r in Q. For each neighbour d of r in the dodecahedron D, denote by $d' \in V'$ the vertex that subdivides the edge rd. The robber inspects the neighbors of r in the dodecahedron D and chooses such a neighbor d that $N[d'] \cup N^2[d]$ contains no cops, and the robber's next two turns consist of moving to d' then d. Once the robber arrives at the dodecahedron vertex d, it repeats the above strategy.

We now prove by induction that the robber can follow the strategy indefinitely and will never be captured by any of the two cops. Suppose we're at the start of round k, the robber is at a vertex $r \in V(D)$, and it is not yet caught. This means that neither of the two cops is at r. If there are no cops adjacent to r, then the robber stays where it is, so it is not yet caught and it is located at a vertex in $V(D)$ at the start of round $k+1$. Otherwise, suppose there is at least one cop adjacent to r. Let the three neighbors of r in D be d_1, d_2, and d_3 and the corresponding dodecahedron subdivision vertices be d'_1, d'_2, and d'_3; see Fig. 3. For each $i, j \in \{1, 2, 3\}$ such that $i \neq j$, the only vertex shared by $N[d'_i] \cup N^2[d_i]$ and $N[d'_j] \cup N^2[d_j]$ (see the gray regions in Fig. 3) is r which, by the induction assumption, is not occupied by any of the cops. Therefore, each cop can be in at most one vertex in $N[d'_i] \cup N^2[d_i]$, for $i \in \{1, 2, 3\}$. Since there are only two cops, it follows that, for some $j \in \{1, 2, 3\}$, the set of vertices $N[d'_j] \cup N^2[d_j]$ is cop-free. Therefore, the robber can follow the given strategy, i.e., moving to d'_j in round k then d_j in round $k+1$. No cop can capture the robber at d'_j in round k since there were no cops in $N[d'_j]$ at the start of round k, and, no cop can capture the robber at d_j in round $k + 1$ since there were no cops in $N^2[d_j]$ at the start of round k. Thus, the robber is not yet caught and it is located at a vertex in $V(D)$ at the start of round $k + 2$.   □

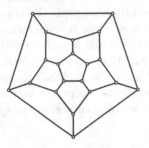

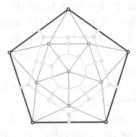

(a) The dodecahedron graph D.

(b) Each face of the dodecahedron graph D in the quadrangulation Q.

Fig. 2. Construction of a quadrangulation Q with cop number three from the dodecahedron graph D (the skeleton of a dodecahedron).

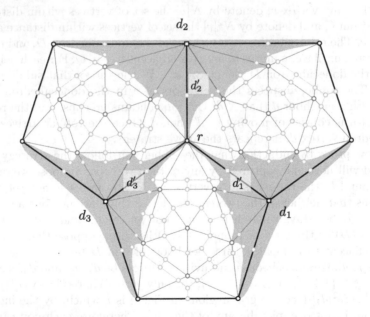

Fig. 3. Three faces of the dodecahedron graph D in quadrangulation Q. The edges of D in black, of T in dark green, and of Q in light green. (Color figure online)

4   Cop Number of Maximal 1-Planar Graphs

4.1   Lower Bound

We prove that there exists a maximal 1-planar graph, which is also optimal 1-planar, with cop number at least three. The structure of optimal 1-planar graphs is well known:

Lemma 1 (Suzuki [33]). *For any embedded graph G, denote by $Q(G)$ the subgraph of G induced by the non-crossing edges. Then: (1) If G is an embedded optimal 1-planar graph, then $Q(G)$ is a plane quadrangulation; and, (2) Let H be an arbitrary simple planar quadrangulation. There exists a simple optimal 1-planar graph G such that $H = Q(G)$ if and only if H is 3-connected.*

Observe that the quadrangulation Q defined in Sect. 3 is 3-connected. Therefore, according to the proof of Lemma 1 in [33], we can construct an optimal 1-planar graph Q' from Q by adding the crossing diagonal edges in each quadrangular face, as illustrated by the brown edges in Fig. 4.

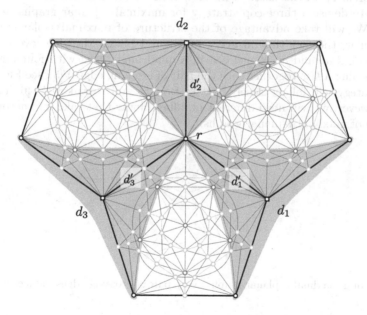

Fig. 4. Q' is formed from Q by adding crossing diagonal edges (brown). (Color figure online)

Theorem 3. *The maximal 1-planar graph Q' has cop number at least three.*

Proof. The proof uses the same robber strategy and induction proof as Theorem 2. The only difference is that the closed neighborhood $N[d'_i]$ of d'_i and the closed distance-two neighborhood $N^2[d_i]$ of d_i, for each $i = 1, 2, 3$, are larger, but it still holds that the only vertex shared by $N[d'_i] \cup N^2[d_i]$ and $N[d'_j] \cup N^2[d_j]$ (see the gray regions in Fig. 4) is r. Therefore, for some $j \in \{1, 2, 3\}$, the set of vertices $N[d'_j] \cup N^2[d_j]$ is cop-free, and the robber can move to vertex $d_j \in V(D)$ via d'_j without being caught. □

4.2   Upper Bound

Aigner and Fromme [1, Theorem 6] proved that planar graphs have cop number at most three. One of the main arguments in their proof is the *Jordan curve property*, i.e., that a Jordan curve partitions the surface into two connected regions: interior and exterior. Bonato and Mohar [8] noted that this approach could also be used for designing a three-cop strategy for other families of graphs that share this property, i.e., graphs with drawings in which there exist special separating cycles (that will play the role of the Jordan curves) that divide the drawing into interior and exterior. Once we introduce crossings, however, it becomes unclear how to identify such separating cycles, and this approach fails for 1-planar graphs (see Theorem 1). In this section, we show that we can still use the Jordan curve property to devise a three-cop strategy for maximal 1-planar graphs; see Theorem 4. We will take advantage of the structure of maximal 1-planar graphs, in particular, the fact that in every maximal 1-planar drawing, every pair of crossing edges ab and cd is enclosed in a quadrangle $acbd$ with four uncrossed edges (forming a *kite* together with the pair of crossing edges) [3]; see Fig. 5. Our proof strategy will follow closely that of Bonato and Nowakowski [6, Theorem 4.25]; however, we will need different building blocks that take into account the structure of maximal 1-planar graphs.

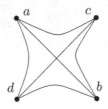

Fig. 5. In a maximal 1-planar graph, every pair of crossing edges induces a kite.

Lemma 2. *Let H be an induced subgraph of a maximal 1-planar graph, let $u, v \in V(H)$ be distinct, and let P be a shortest path between u and v in H. Then no two edges of P cross.*

Proof. Assume for the sake of contradiction that two edges ab and cd of P cross. Then there is a kite $acbd$ and there is a shorter path that takes one of the side edges (ac, cb, bd, or da) of the kite instead of the two crossing edges. □

Lemma 3. *Let H be an induced subgraph of a maximal 1-planar graph such that the robber can only move in H, let $u, v \in V(H)$ be distinct, and let P be a shortest path between u and v in H. Then a single cop on P can, after a finite number of moves: (a) prevent the robber from entering P, that is, the robber will be immediately caught if they move onto P; and, (b) catch the robber immediately after it traverses an edge that crosses an edge of P.*

Proof. The proof of (a) holds for general graphs and can be found in the book of Bonato and Nowakowski [6, Theorem 1.7]. The idea behind the proof is as follows. Because the shortest path P does not have shortcuts, after a finite number of moves, the cop can follow the "projection" of the robber onto the path, and thus, always remain close to the vertex of the path where the robber might enter.

To prove (b), suppose that after sufficiently many moves, one of the cops guards the path P in the sense of (a). Assume the robber is at one of the endpoints of some edge ab, say a, crossing some edge $v_i v_{i+1}$ of the path P. Since H is an induced subgraph of a maximal 1-planar graph, ab together with $v_i v_{i+1}$ form a kite $a v_i b v_{i+1}$. Consider a cop strategy in which P is guarded by one cop in the sense of (a). That is, the cop moves in a way to catch the robber as soon as the robber enters P. So, after a finite number of moves, if the robber is at a, the cop will be at v_i or v_{i+1}, in anticipation of the robber moving to any of these vertices (note that the edges between a and v_i and v_{i+1} exist due to the kite structure). That is, when the robber moves from a to b, the cop will be at either v_i or v_{i+1} and can move to b to catch the robber in the next step.   □

In light of Lemma 3, we say that a cop *guards* the shortest path P between two vertices (referred to also as an *isometric path*) if after a finite number of moves, it can prevent the robber from entering or crossing P. Observe that, according to Lemma 2, no two edges of P cross.

Lemma 4. *Let H be an induced subgraph of a maximal 1-planar graph such that the robber can only move in H, let $u, v \in V(H)$ be distinct, and let P_1 and P_2 be two internally disjoint paths from u to v such that P_1 is isometric in H, P_2 is isometric in $H - (V(P_1) \setminus \{u, v\})$, and no edge of P_1 crosses an edge of P_2. Then $P_1 \cup P_2$ can be guarded by two cops in H.*

Proof. According to Lemma 3, since P_1 is isometric in H, it can be guarded by one cop. Similarly, since P_2 is isometric in $H - (V(P_1) \setminus \{u, v\})$ it can also be guarded by one cop.   □

Theorem 4. *Any maximal 1-planar graph G has cop number at most three.*

Proof. We provide a sketch of the proof here; see the complete proof in the full version of this paper [13].

Given that G is a maximal 1-planar graph, it is 2-connected and thus must contain two internally disjoint shortest paths P_1 and P_2 between two vertices $v, w \in G$, where P_1 is isometric in G, P_2 is isometric in $G \setminus (V(P_1) \setminus \{u, v\})$, and no edge of P_1 crosses an edge of P_2. We let two cops C_1 and C_2 guard the two paths P_1 and P_2, respectively, as described in Lemma 4. Therefore, after a finite number of t steps, the robber must avoid edges that are located on these paths and must avoid crossing them. Define the *robber territory* H as the subgraph induced by the vertices in the interior or exterior of the cycle $P_1 \cup P_2$, whichever contains the robber at step t. Given that the robber territory is non-empty and G is a maximal 1-planar graph, one can consider a third isometric path P_3 between v and w that contains a vertex in the robber territory; this path can share some vertices with P_1 or P_2 (or both). We let the third cop C_3 guard P_3, as in Lemma 3. Therefore, after a finite number of steps, the robber's "safe zone" becomes limited to one of the parts of the partition of the robber's territory created by P_3. We show that two cops can guard this area so that the robber stays confined to it, and the third cop can be used to repeat this process in the smaller subgraph that forms the safe zone of the robber.   □

5   Characterization of Cop-Win Outer 1-Planar Graphs

A graph G is *outer k-planar* if it can be drawn in the plane so that all the vertices are in convex position on the outer boundary and each edge is crossed at most k times. In this section, we provide a complete characterization of the cop number of outer 1-planar graphs.

First, we briefly discuss the cop number of general outer k-planar graphs. Joret et al. [20] showed that the cop number of a graph can be bounded in terms of its treewidth. In particular, if the treewidth of a graph is at most t, then its cop number is at most $t/2 + 1$ [20]. Every outer k-planar graph has treewidth at most $3k + 11$ [34, Proposition 8.5], so, its cop number is at most $1.5k + 6.5$.

Restricting attention to outer 1-planar graphs, Auer et al. [2] showed that every outer 1-planar graph has treewidth at most three, which implies that the cop number of outer 1-planar graphs is at most two. Bonato et al. [7] showed that an outerplanar graph G is cop-win if and only if G is chordal. We generalize this result to outer 1-planar graphs, which completes our characterization of c-cop-win outer 1-planar graphs for all $c \in \{1, 2\}$.

Let $V = (v_1, \ldots, v_n)$ be the cyclic ordering of vertices of an outer 1-planar graph $G = (V, E)$. Let $V[v_i, v_j]$, be the set of vertices $(v_i, v_{i+1}, \ldots, v_j)$, for $i \leq j$, and $V(v_i, v_j)$ be the set of vertices $(v_{i+1}, v_{i+2}, \ldots, v_{j-1})$, for $i < j$.

Proposition 1. *Let $G = (V, E)$ be an outer 1-planar graph. Consider $U \subset V$ such that $G[U]$ forms a cycle. For every pair of vertices $u, w \in U$ such that $V[u, w] \cap U = \{u, w\}$, it holds that if u and w are not adjacent, then there are two edges $e_u, e_w \in E(U)$ incident to u and w, respectively, such that e_u and e_w cross.*

Proof. Consider such a pair $u, w \in V$, and assume that neither of the cycle edges uu_1 nor uu_2 crosses any of the cycle edges ww_1 and ww_2; see Fig. 6. Consider the path $P_1 = (w_1, \ldots, u)$ in $G[U]$ from w_1 to u such that $w \notin P_1$. It follows that $w_2 \notin P_1$, and so there must be an edge $e_1 = w_1' w_1'' \in E(P_1)$ such that $w_1' \in V[w_1, w_2]$ and $w_1'' \in V[w_2, u_1]$ (note that the edge e_1 crosses ww_2). Next, consider the path $P_2 = (w_2, \ldots, u)$ in $G[U]$ from w_2 to u such that $w, w_1 \notin P_2$.

There must be an edge $e_2 = w_2'w_2'' \in E(P_2)$ such that $w_2' \in V[w_1', w_1'']$ and $w_2'' \in V[w_1'', u_1]$ (note that the edge e_2 crosses e_1). Therefore, e_1 is crossed by two edges, ww_2 and $w_2'w_2''$, contradicting the fact that G is outer 1-planar. □

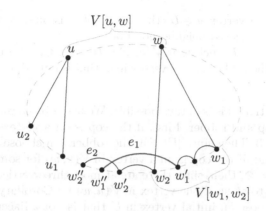

Fig. 6. Illustration in support of Proposition 1.

Corollary 1. *Let $G = (V, E)$ be an outer 1-planar graph. Consider $U \subset V$ such that $G[U]$ forms a cycle of order $k \geq 4$. For every $u, w \in U$ such that $V(u, w) \cap U = \varnothing$, the number of vertices in $V \setminus V(u, w)$ that are adjacent to at least one vertex in $V(u, w)$ is at most three.*

Proof. By Proposition 1, either u and w are adjacent, or, there are two edges e_u and e_w incident to u and w, respectively, that cross. First, suppose that u and w are adjacent. Then, each edge $vv' \in E(G)$ such that $v \in V(u, w)$ and $v' \in V \setminus V(u, w)$ and $v' \notin \{u, w\}$ must cross the edge uw, so there can be at most one such edge vv' since G is outer 1-planar. Next, suppose that u and w are not adjacent. Consequently, for any $v \in V(u, w)$ and any $v' \in V \setminus V(u, w)$ with $v' \notin \{u, w\}$, an edge vv' would have to cross either e_u or e_w. But since G is outer 1-planar and there is a crossing between e_u and e_w, it follows that no such edge vv' exists. So, in both cases above, we proved that there is at most one vertex $v' \in V \setminus V(u, w)$ with $v' \notin \{u, w\}$ such that v' is adjacent to at least one vertex in $V(u, w)$. The fact that u and/or w might be adjacent to at least one vertex in $V(u, w)$ gives the desired upper bound of three. □

Theorem 5. *A connected outer 1-planar graph $G = (V, E)$ is cop-win if and only if it is chordal.*

Proof. It is known that if G is chordal, then it is cop-win (in fact, this is true for any graph [27]). Thus it remains to show that, if G is not chordal, then there is a robber strategy to escape one cop.

Suppose that G is not chordal. Then there is a set $U \subset V$ so that $|U| \geq 4$ and $G[U]$ is a cycle. Let $v_1, \ldots, v_n$ be the cyclic ordering of G. For all $i \leq j$, let $V(v_i, v_j)$ be the set of vertices $(v_{i+1}, v_{i+2}, \ldots, v_{j-1})$. The robber's strategy is to maintain the following invariants at the end of each of its turns, and will only move if one of them is violated at the start of its turn:

1. If the cop is on a vertex $u \in U$, then the robber is on a vertex in $U \setminus N[u]$, where $N[u]$ is the closed neighborhood of u.
2. If, for some $u, w \in U$ such that $V(u, w) \cap U = \varnothing$, the cop is on a vertex in $V(u, w)$, then the robber is on a vertex in U that is not adjacent to any vertex in $V(u, w)$.

Now, we show that this is always possible. We begin by considering the initial position of the cop and robber. First, if the cop is on a vertex $u \in U$, we have that $|N[u] \cap U| = 3$. Then since $|U| \geq 4$, the robber can choose an initial vertex in $U \setminus N[u]$. Second, if the cop is on a vertex in $V(u, w)$ for some $u, w \in U$ such that $V(u, w) \cap U = \varnothing$, then, since there are at most three vertices in $V \setminus V(u, w)$ that are adjacent to at least one vertex in $V(u, w)$ by Corollary 1, and $|U| \geq 4$, the robber can choose an initial vertex in U that is not adjacent to any vertex in $V(u, w)$. Now we must show that the robber can maintain these invariants. Let $r \geq 1$ and suppose that the invariants are true at the end of the robber's turn in the r'th round. This implies that the cop and robber are not adjacent, so the cop cannot catch the robber on its $(r + 1)$'st turn. If neither invariant is violated at the beginning of the robber's $(r + 1)$'st turn, then we are done, so suppose that one of the invariants is violated. Suppose that the first invariant is violated, i.e., the cop is on a vertex $u_1 \in U$ and the robber is on a vertex $u_2 \in U \cap N(u_1)$. Then, since $|U| \geq 4$, there must be a vertex $u \in N(u_2) \setminus N(u_1)$, so the robber can move to u and the first invariant is satisfied at the end of the robber's turn. Next, suppose that the second invariant is violated, i.e., the cop is on a vertex in $V(u, w)$ for some $u, w \in U$ such that $V(u, w) \cap U = \varnothing$, and the robber is on a vertex $v \in U$ that is adjacent to at least one vertex in $V(u, w)$. We will show that there is a vertex $v' \in U \cap N(v)$ that is not adjacent to a vertex in $V(u, w)$. Suppose, for the sake of a contradiction, that there is no such vertex v'. Let $\{v_1, v_2\} = N(v) \cap U$. Then v, v_1, v_2 are all adjacent to vertices in $V(u, w)$ by assumption. Now, since neither of the invariants were violated at the end of the robber's r'th turn, then we know that at the start of the cop's $(r + 1)$'st turn, the cop was not on v_1, v, or v_2, nor was it in $V(u, w)$. But, by assumption, after the cop's $(r + 1)$'st turn, the cop is on a vertex in $V(u, w)$. Thus, at the start of the cop's $(r + 1)$'st turn, it must be the case that the cop was on a vertex in $V \setminus \{v_1, v, v_2\}$ that is adjacent to at least one vertex in $V(u, w)$. However, this implies that there are at least four vertices in $V \setminus V(u, w)$ that are adjacent to at least one vertex in $V(u, w)$, which contradicts Corollary 1. By reaching a contradiction, we have proved that there exists a vertex $v' \in U \cap N(v)$ that is not adjacent to a vertex in $V(u, w)$. Thus, the robber can move to this v' and the second invariant is satisfied at the end of the robber's turn. $\square$

6 Discussion and Directions for Future Research

The cop number of an outer k-planar is at most $3k/2 + 13/2$ due to the fact that its treewidth is bounded by $3k + 11$ and the relation between the cop number and treewidth of a graph [20]. Two questions follow naturally. Is there a non-trivial lower bound expressed as a function of k? Can we reduce the multiplicative factor of $3/2$ in the upper bound?

Determining whether a graph G is k-cop-win is NP-hard, and $W[2]$-hard parameterized by k [16]. Very recently Gahlawat and Zehavi [17] showed that the problem is fixed parameter tractable (FPT) in vertex cover. We would like to restate their open question: whether the problem is FPT in feedback vertex set, treewidth, or treedepth.

References

1. Aigner, M., Fromme, M.: A game of cops and robbers. Discret. Appl. Math. **8**(1), 1–12 (1984)
2. Auer, C., et al.: Outer 1-planar graphs. Algorithmica **74**(4), 1293–1320 (2016)
3. Barát, J., Tóth, G.: Improvements on the density of maximal 1-planar graphs. J. Graph Theory **88**(1), 101–109 (2018)
4. Bekos, M.A., Didimo, W., Liotta, G., Mehrabi, S., Montecchiani, F.: On rac drawings of 1-planar graphs. Theor. Comput. Sci. **689**, 48–57 (2017). https://doi.org/10.1016/j.tcs.2017.05.039. https://www.sciencedirect.com/science/article/pii/S0304397517304851
5. Berarducci, A., Intrigila, B.: On the cop number of a graph. Adv. Appl. Math. **14**, 389–403 (1993)
6. Bonato, A., Nowakowski, R.: The Game of Cops and Robbers on Graphs. Student Mathematical Library. American Mathematical Society (2011)
7. Bonato, A., et al.: Optimizing the trade-off between number of cops and capture time in cops and robbers. J. Comb. **13**(1), 79–203 (2019)
8. Bonato, A., Mohar, B.: Topological directions in cops and robbers. J. Comb. **11**, 47–64 (2020)
9. Brandenburg, F.J., Eppstein, D., Gleißner, A., Goodrich, M.T., Hanauer, K., Reislhuber, J.: On the density of maximal 1-planar graphs. In: Didimo, W., Patrignani, M. (eds.) GD 2012. LNCS, vol. 7704, pp. 327–338. Springer, Heidelberg (2013). https://doi.org/10.1007/978-3-642-36763-2_29
10. Chaplick, S., Kryven, M., Liotta, G., Löffler, A., Wolff, A.: Beyond outerplanarity. In: Frati, F., Ma, K.-L. (eds.) GD 2017. LNCS, vol. 10692, pp. 546–559. Springer, Cham (2018). https://doi.org/10.1007/978-3-319-73915-1_42
11. Clarke, N.E., MacGillivray, G.: Characterizations of k-cop win graphs. Discret. Math. **312**(8), 1421–1425 (2012)
12. Didimo, W., Liotta, G., Montecchiani, F.: A survey on graph drawing beyond planarity. ACM Comput. Surv. **52**(1), 1–37 (2019)
13. Durocher, S., et al.: Cops and robbers on 1-planar graphs (2023). arXiv:2309.01001
14. Eppstein, D.: k-planar graphs and genus. Computer Science Stack Exchange (2015). https://mathoverflow.net/questions/221968/k-planar-graphs-and-genus. Accessed 06 June 2023

15. Erde, J., Lehner, F.: Improved bounds on the cop number of a graph drawn on a surface. In: Nešetřil, J., Perarnau, G., Rué, J., Serra, O. (eds.) Extended Abstracts EuroComb 2021. TM, vol. 14, pp. 111–116. Springer, Cham (2021). https://doi.org/10.1007/978-3-030-83823-2_18

16. Fomin, F.V., Golovach, P.A., Kratochvíl, J., Nisse, N., Suchan, K.: Pursuing a fast robber on a graph. Theor. Comput. Sci. **411**(7), 1167–1181 (2010). https://doi.org/10.1016/j.tcs.2009.12.010. https://www.sciencedirect.com/science/article/pii/S0304397509008342

17. Gahlawat, H., Zehavi, M.: Parameterized analysis of the cops and robber game. In: Leroux, J., Lombardy, S., Peleg, D. (eds.) 48th International Symposium on Mathematical Foundations of Computer Science (MFCS 2023). Leibniz International Proceedings in Informatics (LIPIcs), vol. 272, pp. 49:1–49:17. Schloss Dagstuhl - Leibniz-Zentrum für Informatik, Dagstuhl, Germany (2023). https://doi.org/10.4230/LIPIcs.MFCS.2023.49. https://drops.dagstuhl.de/opus/volltexte/2023/18583

18. Grigoriev, A., Bodlaender, H.L.: Algorithms for graphs embeddable with few crossings per edge. Algorithmica **49**(1), 1–11 (2007)

19. Hong, S.H., Tokuyama, T.: Beyond Planar Graphs. Communications of NII Shonan Meetings (2020)

20. Joret, G., Kamiński, M., Theis, D.O.: The cops and robber game on graphs with forbidden (induced) subgraphs. Contrib. Discret. Math. **5**(2), 40–51 (2010)

21. Kalai, G., Nevo, E., Novik, I.: Bipartite rigidity. Trans. Am. Math. Soc. **368**(8), 5515–5545 (2016). https://www.jstor.org/stable/tranamermathsoci.368.8.5515

22. Korzhik, V.P., Mohar, B.: Minimal obstructions for 1-immersions and hardness of 1-planarity testing. In: Tollis, I.G., Patrignani, M. (eds.) GD 2008. LNCS, vol. 5417, pp. 302–312. Springer, Heidelberg (2009). https://doi.org/10.1007/978-3-642-00219-9_29

23. Lubiw, A., Snoeyink, J., Vosoughpour, H.: Visibility graphs, dismantlability, and the cops and robbers game. Comput. Geom. **66**, 14–27 (2017)

24. Maurer, A., McCauley, J., Valeva, S., Beveridge, A., Islar, V., Saraswat, V.: Cops and robbers on planar graphs. In: Summer 2010 Interdisciplinary Research Experience for Undergraduates (2010)

25. Neufeld, S., Nowakowski, R.: A game of cops and robbers played on products of graphs. Discret. Math. **186**(1), 253–268 (1998)

26. Nowakowski, R.J., Winkler, P.: Vertex-to-vertex pursuit in a graph. Discret. Math. **43**, 235–239 (1983)

27. Quilliot, A.: Jeux et pointes fixes sur les graphes. Ph.D. thesis, Université de Paris VI (1978)

28. Quilliot, A.: A short note about pursuit games played on a graph with a given genus. J. Comb. Theory Ser. B **38**(1), 89–92 (1985)

29. Ringel, G.: Ein sechsfarbenproblem auf der kugel. In: Abhandlungen aus dem Mathematischen Seminar der Universität Hamburg, vol. 29, pp. 107–117. Springer, Heidelberg (1965)

30. Schröder, B.S.W.: The Copnumber of a Graph is Bounded by [3/2 genus (G)] + 3, pp. 243–263. Birkhäuser Boston (2001)

31. Schumacher, V.H.: Zur struktur 1-planarer graphen. Mathematische Nachrichten **125**(1), 291–300 (1986). https://doi.org/10.1002/mana.19861250122. https://onlinelibrary.wiley.com/doi/abs/10.1002/mana.19861250122

32. Schumacher, V.H.: Zur struktur 1-planarer graphen. Math. Nachr. **125**(1), 291–300 (1986)

33. Suzuki, Y.: Re-embeddings of maximum 1-planar graphs. SIAM J. Discret. Math. **24**, 1527–1540 (2010)
34. Wood, D.R., Telle, J.A.: Planar decompositions and the crossing number of graphs with an excluded minor. In: Kaufmann, M., Wagner, D. (eds.) GD 2006. LNCS, vol. 4372, pp. 150–161. Springer, Heidelberg (2007). https://doi.org/10.1007/978-3-540-70904-6_16

Removing Popular Faces in Curve Arrangements

Phoebe de Nooijer[1], Soeren Terziadis[2(✉)] ⓘ, Alexandra Weinberger[3] ⓘ,
Zuzana Masárová[4] ⓘ, Tamara Mchedlidze[1] ⓘ, Maarten Löffler[1] ⓘ,
and Günter Rote[5] ⓘ

[1] Utrecht University, Utrecht, The Netherlands
{t.mtsentlintze,m.loffler}@uu.nl
[2] TU Wien, Vienna, Austria
soeren.nickel@ac.tuwien.ac.at
[3] Graz University of Technology, Graz, Austria
weinberger@ist.tugraz.at
[4] IST Austria, Maria Gugging, Austria
zuzana.masarova@ist.ac.at
[5] Freie Universität Berlin, Berlin, Germany
rote@inf.fu-berlin.de

Abstract. A face in a curve arrangement is called *popular* if it is
bounded by the same curve multiple times. Motivated by the automatic
generation of curved nonogram puzzles, we investigate possibilities to
eliminate the popular faces in an arrangement by inserting a single additional curve. This turns out to be NP-hard; however, it becomes tractable
when the number of popular faces is small: We present a probabilistic
FPT-approach in the number of popular faces.

Keywords: Puzzle generation · Curve arrangements ·
Fixed-parameter tractable (FPT)

1 Introduction

Let $\mathcal{A}$ be a set of curves which lie inside the area bounded by a closed curve,
called the *frame*. All curves in $\mathcal{A}$ are either *closed*, or *open* with endpoints on
the frame. We refer to $\mathcal{A}$ as a *curve arrangement*, see Fig. 1a. We consider only
simple arrangements, where no three curves meet in a point and there are only
finitely many total intersections, which are all crossings (no tangencies).

The arrangement $\mathcal{A}$ can be seen as an embedded multigraph whose vertices
are crossings of curves and whose edges are *curve segments*. $\mathcal{A}$ subdivides the
region bounded by the frame into *faces*. A face is *popular* when it is incident to
multiple curve segments belonging to the same curve in $\mathcal{A}$ (see Figs. 1b–c). We
study the NONOGRAM 1-RESOLUTION (N1R) problem: can one additional curve
ℓ be inserted into $\mathcal{A}$ such that no faces of $\mathcal{A} \cup \{\ell\}$ are popular (see Fig. 1d)?

Authors are sorted by seniority.

M. A. Bekos and M. Chimani (Eds.): GD 2023, LNCS 14466, pp. 18–33, 2023.
https://doi.org/10.1007/978-3-031-49275-4_2

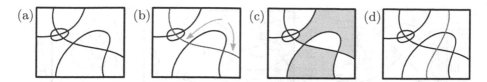

Fig. 1. (a) A curve arrangement in a rectangular frame. (b) The top right face is incident to two disconnected segments of the red curve, making it *popular*. (c) All popular faces are highlighted. (d) After inserting an additional curve, no more popular faces remain. (Color figure online)

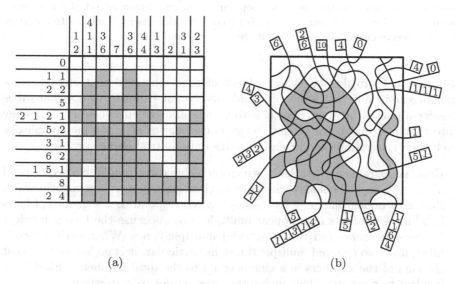

Fig. 2. Two nonogram puzzles in solved state. (a) A classic nonogram. (b) A curved nonogram.

Nonograms. Our question is motivated by the problem of generating *curved nonograms*. Nonograms, also known as *Japanese puzzles*, *paint-by-numbers*, or *griddlers*, are a popular puzzle type where one is given an empty grid and a set of *clues* on which grid cells need to be colored. A clue consists of a sequence of numbers specifying the numbers of consecutive filled cells in a row or column. A solved nonogram typically results in a picture (see Fig. 2a). There is quite some work in the literature on the difficulty of solving nonograms [1,3,6].

Van de Kerkhof et al. [7] introduced *curved* nonograms, in which the puzzle is no longer played on a grid but on an arrangement of curves (see Fig. 2b). In curved nonograms, clues specify numbers of filled faces of the arrangement in the sequence of faces incident to a common curve on one side. Van de Kerkhof et al. focus on heuristics to automatically generate such puzzles from a desired solution picture by extending curve segments to a complete curve arrangement.

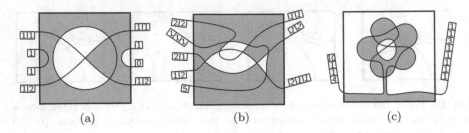

(a)   (b)   (c)

Fig. 3. Three types of curved nonograms of increasing complexity [7], shown with solutions. (a) *Basic* puzzles have no popular faces. (b) *Advanced* puzzles may have popular faces, but no self-intersections. (c) *Expert* puzzles have self-intersecting curves. We can observe closed curves (without clues) in (a) and (c).

Nonogram Complexity. Van de Kerkhof et al. observed that curved nonograms come in different levels of complexity — not in terms of how hard it is to *solve* a puzzle, but how hard it is to understand the rules (see Fig. 3). They state that it would be of interest to generate puzzles of a specific complexity level; their generators can currently do this only by trial and error.

- *Basic* nonograms are puzzles in which each clue corresponds to a sequence of distinct faces. The analogy with clues in classic nonograms is straightforward.
- *Advanced* nonograms may have clues that correspond to a sequence of faces in which some faces may appear multiple times because the face is incident to the same curve (on the *same* side) multiple times. When such a face is filled, it is also counted multiple times; in particular, it is no longer true that the sum of the numbers in a clue is equal to the total number of filled faces incident to the curve. This makes the rules harder to understand.
- *Expert* nonograms may have clues in which a single face is incident to the same curve on *both* sides. They are even more confusing than advanced nonograms. Expert nonograms are only suitable for experienced puzzle freaks.

It is easy to see that arrangements with self-intersecting curves correspond exactly to expert puzzles. The difference between basic and advanced puzzles is more subtle; it is exactly the presence of *popular faces* in the arrangement.

One possibility to generate nonograms of a specific complexity would be to take an existing generator and modify the output. Recently, Brunck et al. [5] have investigated how popular faces in a nonogram might be removed by reconfiguring and/or reconnecting parts of curves at small local areas, which they call switches (e.g. around curve crossings), and they have proved that this problem is NP-hard. As an alternative, one may try to get rid of the popular faces by adding extra curves that cut the popular faces into smaller pieces. In this paper, we explore what we can do by inserting a single new curve into the arrangement. Clearly, inserting curves will not remove self-intersections, so we focus on changing advanced puzzles into basic puzzles; i.e., removing all popular faces.

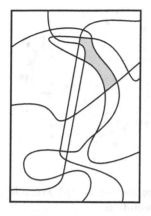 

Fig. 4. Real puzzles (without clues) with all popular faces highlighted.

1.1 Results

After discussing in Sect. 2 how a singular face is resolved, we show in Sect. 3 that deciding whether we can remove all popular faces from a given curve arrangement by inserting a single curve – which we call the N1R problem – is NP-complete. However, often the number of popular faces is small, see Fig. 4. Hence, we are also interested in the problem parametrized by the number of popular faces k. we show in Sect. 4 that the problem can be solved by a randomized algorithm in FPT time.

2 Resolving One popular Face by Adding a Single Curve

As a preparation, we analyze how a single bad face F can be resolved. If F is visited three or more times by some curve, it cannot be resolved with a single additional curve ℓ, and we can immediately abort. Otherwise, there are *popular* edges among the edges of F, which belong to a curve that visits F twice. As a visual aid, we indicate each such pair of edges by connecting them with a red curve (a *curtain*), see Fig. 5a or 8b.

Observation 1. *To ensure that a popular face F becomes unpopular after insertion of a single curve ℓ into the arrangement, it is necessary and sufficient that the curve ℓ has the following properties.*

1. *It visits the face F exactly once;*
2. *It does not enter or exit through a popular edge;*
3. *It separates each pair of popular edges. In other words, ℓ cuts all curtains.* □

The ways how ℓ can traverse a popular face F can be modeled as a graph: We place a vertex on every edge of F except the popular edges. We then connect two such vertices u, v if for every curtain c, the endpoints of c alternate with

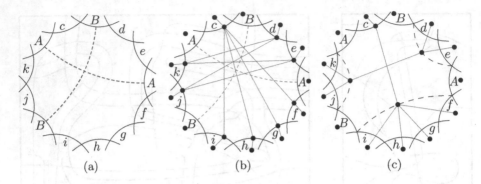

Fig. 5. Resolving a popular face F. (a) Curtains model popular edges. (b) Possible ways how ℓ can pass through F. (c) A more compact representation

the vertices u and v around F, as shown in Fig. 5b. This representation can be condensed as shown in Fig. 5c and explained in the full version [10, Appendix D].

In our arguments, we will often use the dual graph $\mathcal{A}^d$ of a curve arrangement $\mathcal{A}$, where every face of $\mathcal{A}$ is represented by a vertex and edges represent faces which share a common boundary segment (not just a common crossing point). In particular, a curve ℓ traversing $\mathcal{A}$ and crossing a sequence of faces $F_1, \ldots, F_k$ in that order can be expressed as a path $P = (F_1, \ldots, F_k)$ in $\mathcal{A}^d$.

3   N1R is NP-complete

In order to prove NP-hardness, we reduce from *Planar Non-intersecting Eulerian Cycle*. This reduction assumes ℓ to be a closed loop, but it can easily be adapted to work for an open curve ℓ' starting and ending at the frame.

3.1   Non-intersecting Eulerian Cycles

An Eulerian cycle in a graph is a closed walk that contains every edge exactly once. An Eulerian cycle in a graph embedded into the plane (a plane graph) is *non-intersecting* if every pair of consecutive edges $(a, b), (b, c)$ is adjacent in the radial order around b. Intuitively, an Eulerian cycle is non-intersecting if it can be drawn without repeated vertices after replacing each vertex by a small cycle linking the incident edges in circular order (see Figs. 6a and 6b). The Eulerian cycle has to visit all of the original edges, but it does not have to cover the small vertex cycles (see Fig. 6c). The following problem was proved to be NP-complete by Bent and Manber [2, Theorem 1].

Problem 1 (Planar Non-Intersecting Eulerian Cycle PNEC). Given a planar graph embedded into the plane graph G, decide whether G contains a non-intersecting Eulerian cycle.

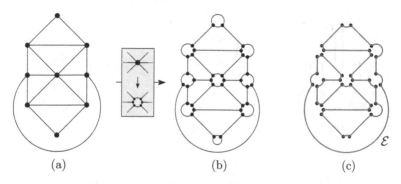

Fig. 6. Vertices of the graph G (a) are replaced by cycles (b). A non-intersecting cycle drawn on the modified graph visiting all original edges (c).

3.2 NP-completeness Reduction

We will present a polynomial-time reduction from PNEC to N1R, i.e., we will create a curve arrangement $\mathcal{A}$ containing popular faces based on a planar input graph G of PNEC, s.t. there exists a curve ℓ for which $\mathcal{A} \cup \ell$ contains no popular faces if and only if G contains a non-intersecting Eulerian cycle. We assume that G is 2-edge-connected and all vertices have even degree, because otherwise, G clearly cannot contain an Eulerian cycle. We also replace every self-loop with a path of length two, without affecting the existence of a planar non-intersecting Eulerian cycle.

The reduction is gadget based. We will represent every vertex $v \in V$ with a vertex gadget $\mathcal{N}(v)$ and every edge $e = (u, v) \in E$ with an edge gadget $\mathcal{L}(e)$ or $\mathcal{L}(u, v)$. Both gadgets are sets of curves starting and ending at the frame, and $\mathcal{A} = \bigcup_{v \in V} \mathcal{N}(v) \cup \bigcup_{e \in E} \mathcal{L}(e)$.

Vertex Gadgets. The vertex gadgets consist of curves in one of three basic shapes shown in Fig. 7a, which we call beakers. We place one beaker per incident edge of v, at the position of v, all rotated, s.t. their *bases* (the lower ends in Fig. 7a) overlap in a specific pattern. The *opening* of each beaker (the upper ends in Fig. 7a) will point outwards. We use three variants of the vertex gadget, depending on the vertex degree.

The vertex gadget for a degree-two vertex is simply made up of two overlaying Type-I beakers (see Fig. 7b). Since ℓ must cross the two curtains c_1 and c_2, it must connect the two points p_1, p_2 by crossing the overlap of the two beakers (a face of degree two, marked in green in Fig. 7b). Since by Observation 1, ℓ can enter any beaker only once, the routing of ℓ as shown in Fig. 7c is forced and corresponds exactly to the traversal of a planar non-intersecting Eulerian cycle through a vertex of degree two.

The vertex gadget $\mathcal{N}(v)$ for a degree-four vertex v consists of four Type-I beakers, one per incident edge, which form the intersection pattern of Fig. 7d. Since ℓ must cross the four curtains, it must enter or exit the gadget at least four

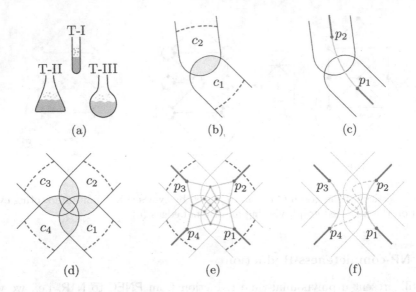

Fig. 7. (a) Basic beaker curve shapes. (b) Degree two gadget (2 Type-I beakers) and (c) its forced resolution. (d) Degree four gadget (4 Type-I beakers). (e) Dual graph of the degree 4 gadget. (f) A possible curve ℓ in light blue; some alternative routings, which connect the same endpoints, in dashed light blue. (Color figure online)

times through the thick blue edges in Fig. 7e and the vertices p_1, p_2, p_3, p_4 of the dual graph $\mathcal{A}^d$. Since ℓ cannot cross itself, there are only two possibilities how ℓ can pass through $\mathcal{N}(v)$: It can connect p_1 with p_2 and p_3 with p_4, as in Fig. 7f, or p_1 with p_4 and p_2 with p_3. Both possibilities can be realized by routings of ℓ, and they correspond precisely to the ways how a non-intersecting Eulerian cycle can pass through the edges incident to v. Note that the exact routing of ℓ can vary inside $\mathcal{N}(v)$ (indicated by the dashed lines in Fig. 7).

The vertex gadget for a vertex v of degree $d \geq 6$ is more complex. We place $d - 1$ Type-II beakers $c_1, \ldots, c_{d-1}$ symmetrically around the location of v (Fig. 8a). Each beaker intersects four adjacent beakers (two on each side), with the exception that $c_{d/2-1}$ and $c_{d/2+1}$ (dark green curves in Fig. 8a) do not intersect. We place an additional Type-III beaker c_d (the light green curve in Fig. 8a) that surrounds all bases of the Type-II beakers and protrudes between c_{d-1} and c_1, such that the intersection pattern of Fig. 8a arises.

All popular faces and curtains in $\mathcal{N}(v)$ are shown in Fig. 8b. The dual of the construction is shown in Fig. 8c. The curtains in the green faces force ℓ to pass from these faces to the adjacent small faces with the blue boundaries. This constrains ℓ to pass through a chain of faces as shown in Fig. 8d. The passages from these faces to other neighboring faces can now be excluded, and the corresponding edges have been removed from the dual graph in Fig. 8d.

The curtains in the openings of the beakers force the outer blue endpoints of ℓ. The endpoint in beaker c_i will be called p_i. Now we analyze which of

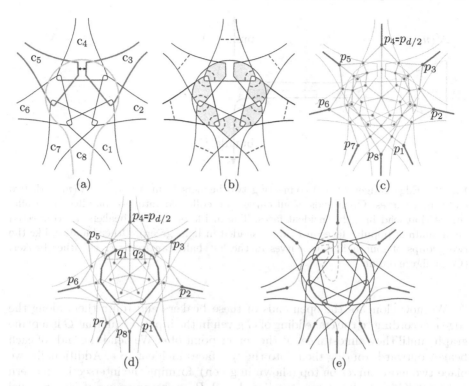

Fig. 8. (a) Vertex gadget $\mathcal{N}(v)$ for a degree-8 vertex v, (b) its curtains and (c) dual graph. (d) Highlighted light green faces in (b) force the dark blue connections in the dual graph and restrict it. (e) One of two symmetric possibilities for the splitting curve ℓ. The dashed lines show different possible routings of ℓ. (Color figure online)

these endpoints can be connected with each other. We see that in most cases, p_i can only be connected to p_{i-1} or p_{i+1} without going through another endpoint. The exception is $p_{d/2-1}$ and $p_{d/2+1}$, which can be connected via the inner loop from q_1 to q_2. However, this connection would cut off $p_{d/2}$ from the remaining points. We conclude that the visits of ℓ to $\mathcal{N}(u)$ must match endpoints p_i that are adjacent in the circular order. There are two matchings, which correspond to the two possibilities how a non-intersecting Eulerian cycle can visit v. Both possibilities can be realized by routings of ℓ; one is shown in Fig. 8e, and the other is symmetric.

We now have placed vertex gadgets for all vertices. They require ℓ to connect to an endpoint in each opening of a beaker. With these openings, we will now construct the edge gadgets.

Edge Gadgets. Let $e = (u, v) \in E$ be an edge in G. Then there are two vertex gadgets $\mathcal{N}(u)$ and $\mathcal{N}(v)$ already placed. In particular, we placed one beaker in the gadgets per incident edge at u or v, i.e. two per edge.

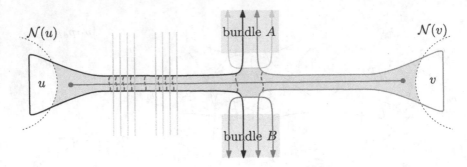

Fig. 9. Edge gadget $\mathcal{L}(u,v)$ connecting two beakers from $\mathcal{N}(u)$ and $\mathcal{N}(v)$ with two additional curves. Open ends of all curves are collected into two bundles of parallel curves that lead into the incident faces. The inside of the two beakers are connected via a chain of popular faces in $\mathcal{L}(u,v)$ (shaded in light green). Other bundles, like the two groups of four light blue curves in the left half, can freely cross either beaker. (Color figure online)

We now elongate the open ends of these beakers and route them along the edge e according to the embedding of G given in the input (recall that G is a plane graph) until they almost meet at the center point of e. We bend the ends of each beaker outward, routing them into the two faces incident to e. Additionally, we place two more curves on top (shown in green), forming the intersection pattern of Fig. 9. This results in two *bundles* A and B, each consisting of four parallel curves. (The light blue curves in the left half are not part of the gadget; they are two bundles that come from other gadgets.)

This connects a popular face in $\mathcal{N}(u)$ to one in $\mathcal{N}(v)$, forming one big popular face in $\mathcal{L}(u,v)$. An arbitrary number of curves may cross the opening of a beaker. The face is then simply divided into a chain of consecutive popular faces. In each of these faces, except the left- and right-most faces, which contain the blue endpoints, ℓ has to leave through two specific edges in order to cut the curtains. This forces ℓ to pass straight through $\mathcal{L}(u,v)$ from $\mathcal{N}(u)$ to $\mathcal{N}(v)$ along the thin dark-blue horizontal axis.

It remains to describe how the open ends of the curves in the bundles are routed to the frame (since all curves other than ℓ have start and end at the frame). This is not difficult because these bundles can cross quite freely without creating popular faces. Each bundle consists of a unique set of curves, except for the two bundles from one edge gadget. A bundle that originates from the edge gadget $\mathcal{L}(e)$ can thus cross any bundle from a different edge gadget without creating popular faces. It can cross a different edge gadget $\mathcal{L}(e')$ by passing over one of its beakers, as shown with the light-blue curves.

Since G does not contain self-loops, we route each bundle along a path in the dual of G to the outer face of G, and then connect it to the frame, see Fig. 10. A popular face might only be created when a bundle crosses the other bundle from the same edge gadget, which can be avoided by routing them in parallel.

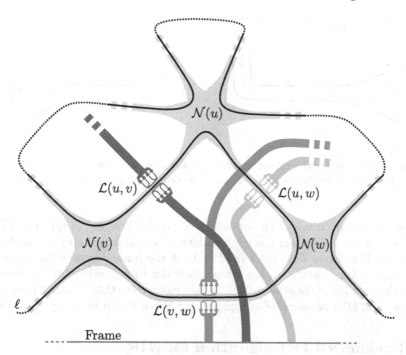

Fig. 10. Schematic representation of three vertex and edge gadgets. The bundles are routed through beakers of other edges ending at the frame (partially shown at the bottom of the figure). A possible routing of ℓ is shown with a black curve.

The curves in a bundle run in parallel. Two bundles originating from an edge gadget $\mathcal{L}(e)$ have different curves as their outside curves. Hence, no popular faces are created between two bundles, and we can make the following statement.·

Observation 2 *All popular faces in $\mathcal{A}$ are contained in vertex gadgets and edge gadgets (the faces with dashed red curtains in Figs. 7b, 7d, 8b, and 9).*

The next theorem follow from the construction and the resulting correspondence between resolving curves and non-crossing Eulerian cycles. The proof can be found in the full version [10, Appendix A].

Theorem 1. *N1R is NP-complete.* □

Adaption to Open Curves. The reduction assumes that ℓ is a closed loop. It can be adapted to work for open curves, i.e., we can create the arrangement $\mathcal{A}$, for which there exists an open curve ℓ' starting and ending at the frame, s.t. $\mathcal{A} \cup \ell'$ does not contain any popular faces, if and only if G contains a planar non-intersecting Eulerian cycle $\mathcal{E}(G)$.

The reduction creates $\mathcal{A}$ in the same fashion as above, except that we do not add the edge gadget for exactly one edge $e_o = (v_o, u_o)$ on the outer face of G.

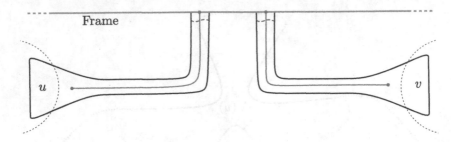

Fig. 11. By routing both ends of an open beaker in parallel to the frame, we force ℓ to start (or end) at the frame between the two connection points of the beaker.

Instead, the open curves of the openings of the beakers in $\mathcal{N}(u_o)$ and $\mathcal{N}(v_o)$, which would normally form the edge gadget $\mathcal{L}(u_o, v_o)$ are simply connected to the frame. This forces ℓ' to start (and end) at the frame between the points at which the beakers connect to the frame, in order to properly split the popular face in the opening of these beakers. It is now easy to see that all other properties still hold and N1R remains NP-complete even when ℓ' can be an open curve.

4   Randomized FPT-algorithm for N1R

In this section, we show that N1R with k popular faces can be solved by a randomized algorithm in $O\left(2^k \text{poly}(n)\right)$ time, placing N1R in the class randomized FPT when parameterized by the number k of popular faces. We model N1R as a problem of finding a simple cycle (i.e., a cycle without repeated vertices) in a modified dual graph G, subject to a constraint that certain edges must be visited.

Problem 2 (Simple Cycle with Edge Set Constraints SNESC). Given an undirected graph $G = (V, E)$ and k subsets $S_1, S_2, \ldots, S_k \subseteq E$ of edges, find a simple cycle, if it exists, that contains exactly one edge from each set S_i.

We start with the dual graph of the given curve arrangement $\mathcal{A}$. We replace the vertex corresponding to the i-th popular face f with a set S_i of edges modeling the ways how an additional curve can cut all curtains of f, as described in Sect. 2 and shown in Fig. 5b. To be specific, we place a vertex on each curve segment s bounding f and connect it to the vertex of the face that is adjacent to f across s. Further we connect two such vertices on curve segments if a curve entering f through one segment and exiting through the other would cut all curtains of f. The latter connecting edges, which run through f, form the set S_i. There is a one-to-one correspondence between the simple cycles containing exactly one edge of every set S_i and the resolving curves for $\mathcal{A}$.

We will describe a randomized algorithm for Problem 2, extending an algorithm of Björklund, Husfeld, and Taslaman [4].

Theorem 2. (a) *The SNESC problem on a graph with n vertices and $m \geq n$ edges can be solved in $O\left(2^k mn^2 \log \frac{2m}{n} \cdot |V(S_1)| \cdot W\right)$ time and $O(2^k n + m)$ space with a randomized Monte-Carlo algorithm, with probability at least $1 - 1/n^W$, for any $W \geq 1$. Here, $V(S_1)$ denotes the set of vertices of the edges in S_1 (note that S_1 can be chosen to be the smallest set among $S_1, \ldots, S_k$). The model of computation is the Word-RAM with words of size $\Theta(k + \log n)$.*

(b) *There is an alternative algorithm (described in the full version [10, Appendix B.3]) needing only polynomial space, namely $O(kn + m)$, at the expense of an additional factor k in the runtime. It uses words of size $\Theta(\log n)$.*

Both algorithms find the cycle with the smallest number of edges if it exists (with high probability).

If $\mathcal{A}$ has n faces, the graph G has $O(n)$ vertices and $m = O(n^2)$ edges. The quadratic blow-up of m results from the construction as shown in Fig. 5b. The number of edges can be reduced to $O(n)$, as shown in Fig. 5c and discussed in the full version [10, Appendix D]. The number k of popular faces is the same as the number k of edge sets S_i.

With the alternative algorithm with polynomial space, since $k \leq n$, we get:

Corollary 1. *The N1R problem with k popular faces in a curve arrangement with n faces can be solved in expected time $O(2^k \mathrm{poly}(n))$ and $O(kn)$ space.*   □

We first give a high-level overview of the algorithm. We start by assigning random weights to the edges from a sufficiently large finite field $\mathbb{F}_q$ of characteristic 2. Such a field exists for every size q that is a power of 2. In a field of characteristic 2, the law $x + x = 0$ holds, and therefore terms cancel when they occur an even number of times. The weight of a *walk* (with vertex and edge repetitions allowed) is obtained by multiplying the edge weights of all visited edges. Our goal is now to compute the sum of weights all closed walks, of given length, that satisfy the edge set constraints. The characteristic-2 property will ensure that the unwanted walks, those which are not simple, cancel, while a simple closed walk makes a nonzero contribution and leads to a nonzero sum with high property. The crucial idea is that, while these sets of closed walks can be very complicated, we can compute the aggregated sum of their weights in polynomial time. We have to anchor these walks at some starting vertex b, and we choose b to be one of the vertices incident to an edge of S_1.

More precisely, for each such vertex b, and for increasing lengths $l = 1, 2, \ldots, n$, the algorithm computes the quantity $\hat{T}_b(l)$, which is the sum of the weights of all closed walks that

- start and end at b,
- have their first edge in S_1,
- use exactly one edge from each set S_i (and use it only once),
- and consist of l edges.

We consider the edge weights as variables and regard $\hat{T}_b(l)$ as a function of these variables. The result is a polynomial where each term is a product of

l variables (possibly with repetition), and hence the polynomial has degree l, unless all terms cancel and it is the zero polynomial. We apply the following lemma, which is a straightforward adaptation of a lemma of Björklund et al. [4].

Lemma 1. *a) Suppose there exists a simple cycle of length l that satisfies the edge set constraints and that goes through an edge of S_1 incident to b. Then the polynomial $\hat{T}_b(l)$ is homogeneous of degree l and is not identically zero.*
b) If there is no such cycle of length $\leq l$, the polynomial $\hat{T}_b(l)$ is identically zero.

The lemma is based on the fact that each term in the polynomial $\hat{T}_b(l)$ represents some closed walk. A term coming from a walk that visits a vertex twice can be matched with another walk, which traverses a loop in the opposite direction and contributes the same term. Since the field has characteristic 2, these terms cancel. A term coming from a simple walk does not cancel. The proof of Lemma 1 is given in the full version [10, Appendix B.5].

In case (a) of Lemma 1, it follows from the Schwartz-Zippel Lemma [12, Corollary 1] (see also [11, Corollary Q1]) that, for randomly chosen weights in $\mathbb{F}_q$, $\hat{T}_b(l)$ is nonzero with probability at least $1 - \text{degree}/|\mathbb{F}_q| = 1 - l/q \geq 1 - n/q$.

Thus, if we choose $q > n^2$, we have a success probability of at least $1 - 1/n$ for finding the shortest cycle when we evaluate the quantities $\hat{T}_b(l)$ for increasing l until they become nonzero. The success probability can be boosted by repeating the experiment with new random weights.

In the unlikely case of a failure, the algorithm may err by not finding a solution although a solution exists, or by finding a solution that is not shortest. The last possibility is not an issue for our original problem, where we just ask about the existence of a cycle, of arbitrary length.

In the following, we will discuss how we can compute the quantities $\hat{T}_b(l)$, and the runtime and space requirement for this calculation. We describe the method for actually recovering the cycle after we have found a nonzero value in the full version [10, Appendix B.4].

4.1   Computing Sums of Path Weights by Dynamic Programming

We cannot compute the desired sums $\hat{T}_b(l)$ directly, but have to do this incrementally via a larger variety of quantities $T_b(R, l, v)$ that are defined as follows:

For $R \subseteq \{1, 2, \ldots, k\}$ with $1 \in R$, $v \in V$, and $l \geq 1$, we define $T_b(R, l, v)$ as the sum of the weights of all walks that

- start at b,
- have their first edge in S_1,
- end at v,
- consist of l edges,
- use exactly one edge from each set S_i with $i \in R$ (and use it only once),
- contain no edge from the sets S_i with $i \notin R$.

The walks that we consider here differ from the walks in $\hat{T}_b(l)$ in two respects: They end at a specified vertex v, and the set R keeps track of the sets S_i that

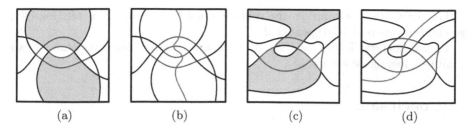

     (a)   (b)   (c)   (d)

Fig. 12. Input (a,c) and resulting output (b,d) generated by the implementation; the green curve resolves the popular faces with the smallest number of crossings. (Color figure online)

were already visited. The quantities $\hat{T}_b(l)$ that we are interested in arise as a special case when we have visited the full range $R = \{1, \ldots, k\}$ of sets S_i and arrive at $v = b$:

$$\hat{T}_b(l) = T_b(\{1, \ldots, k\}, l, b).$$

We compute the values $T_b(R, l, v)$ for increasing values $l = 1, \ldots, n$. The starting values for $l = 1$ are straightforward from the definition.

To compute $T_b(R, l, v)$ for $l \geq 2$, we collect all stored values of the form $T_b(R', l-1, u)$ where (u, v) is an edge of G and R' is derived from R by taking into account the sets S_i to which (u, v) belongs. We multiply these values with the edge weight w_{uv} and sum them up. If (u, v) is in some S_i but $i \notin R$, we don't use this edge. Formally, let $I(u, v) := \{i \mid (u, v) \in S_i\}$ be the index set of the sets S_i to which (u, v) belongs. Then

$$T_b(R, l, v) = \sum_{\substack{(u,v) \in E \\ I(u,v) \subseteq R}} w_{uv} \cdot T_b(R \setminus I(u, v), l-1, u) \tag{1}$$

4.2   Runtime and Space

The finite field additions and multiplications in (1) take constant time, see the full version [10, Appendix B.2]. Similarly, the set operation $R \setminus I(u, v)$ on subsets of $\{1, \ldots, k\}$ and the test $I(u, v) \subseteq R$ can be carried out in constant time, using bit vectors. Thus, for a fixed starting vertex $b \in V(S_1)$ and fixed R, going from $l - 1$ to l by the recursion (1) takes $O(m)$ time in total, because each edge (u, v) appears in at most one of the sums on the right-hand side. The overall runtime is $O(|V(S_1)|2^k mn)$.

As mentioned after Lemma 1, the probability that the algorithm misses the shortest simple path is at most $1/n$. To amplify the probability of correctness, we repeat the computation W times, reducing the failure probability to $1/n^W$.

We consider each starting vertex b separately, and do not need to store entries for lengths $l - 1$ or shorter when proceeding from l to $l + 1$; thus the space requirement is $O(2^k n)$.

For recovering the solution, the runtime must be multiplied by $O(n \log \frac{2m}{n})$, see the full version [10, Appendix B.4].

Figure 12 shows initial results of an implementation of our algorithm on two small test instances; see also [8].

5 Conclusion

In light of our NP-hardness and randomized FPT-algorithm, a natural next step is a deterministic parameterized algorithm. There are $O(n)$ local possibilities of resolving a single popular face, however, this does not immediately lead to an $O(n^k)$ algorithm (which would place N1R in XP), since we might need to branch additionally over all possible connections between these solutions through the dual of $\mathcal{A}$, which can have an unbounded size.

Acknowledgements. This work was initiated at the 16th European Research Week on Geometric Graphs in Strobl in 2019. A.W. is supported by the Austrian Science Fund (FWF): W1230. S.T. has been funded by the Vienna Science and Technology Fund (WWTF) [10.47379/ICT19035]. A preliminary version of this work has been presented at the 38th European Workshop on Computational Geometry (EuroCG 2022) in Perugia [9]. A full version of this paper, which includes appendices but is otherwise identical, is available as a technical report [10].

References

1. Batenburg, K.J., Kosters, W.A.: On the difficulty of nonograms. ICGA J. **35**(4), 195–205 (2012). https://doi.org/10.3233/ICG-2012-35402
2. Bent, S.W., Manber, U.: On non-intersecting Eulerian circuits. Discret. Appl. Math. **18**(1), 87–94 (1987). https://doi.org/10.1016/0166-218X(87)90045-X
3. Berend, D., Pomeranz, D., Rabani, R., Raziel, B.: Nonograms: combinatorial questions and algorithms. Discret. Appl. Math. **169**, 30–42 (2014). https://doi.org/10.1016/j.dam.2014.01.004
4. Björklund, A., Husfeld, T., Taslaman, N.: Shortest cycle through specified elements. In: Proceedings of the Twenty-Third Annual ACM-SIAM Symposium on Discrete Algorithms, SODA 2012, pp. 1747–1753. Society for Industrial and Applied Mathematics, USA (2012). 2095116.2095255
5. Brunck, F., Chang, H.C., Löffler, M., Ophelders, T., Schlipf, L.: Reconfiguring popular faces. Dagstuhl Rep. (Seminar 22062) **12**(2), 24–34 (2022). https://doi.org/10.4230/DagRep.12.2.17
6. Chen, Y., Lin, S.: A fast nonogram solver that won the TAAI 2017 and ICGA 2018 tournaments. ICGA J. **41**(1), 2–14 (2019). https://doi.org/10.3233/ICG-190097
7. van de Kerkhof, M., de Jong, T., Parment, R., Löffler, M., Vaxman, A., van Kreveld, M.J.: Design and automated generation of Japanese picture puzzles. Comput. Graph. Forum **38**(2), 343–353 (2019). https://doi.org/10.1111/cgf.13642
8. de Nooijer, P.: Resolving popular faces in curve arrangements. Master's thesis, Utrecht University (2022). https://studenttheses.uu.nl/handle/20.500.12932/494

9. de Nooijer, P., et al.: Removing popular faces in curve arrangements by inserting one more curve. In: Giacomo, E.D., Montecchiani, F. (eds.) Abstracts of the 38th European Workshop on Computational Geometry (EuroCG 2022), pp. 38:1–38:8, March 2022. https://eurocg2022.unipg.it/booklet/EuroCG2022-Booklet.pdf
10. de Nooijer, P., et al.: Removing popular faces in curve arrangements (2023). arXiv:2202.12175v2 [cs.CG]
11. Rote, G.: The Generalized Combinatorial Lasoń-Alon-Zippel-Schwartz Nullstellensatz Lemma (2023). arXiv:2305.10900 [math.CO]
12. Schwartz, J.T.: Fast probabilistic algorithms for verification of polynomial identities. J. ACM **27**(4), 701–717 (1980). https://doi.org/10.1145/322217.322225

Different Types of Isomorphisms of Drawings of Complete Multipartite Graphs

Oswin Aichholzer[ID], Birgit Vogtenhuber[ID], and Alexandra Weinberger[✉][ID]

Institute of Software Technology, Graz University of Technology, Graz, Austria
{oaich,bvogt,weinberger}@ist.tugraz.at

Abstract. Simple drawings are drawings of graphs in which any two edges intersect at most once (either at a common endpoint or a proper crossing), and no edge intersects itself. We analyze several characteristics of simple drawings of complete multipartite graphs: which pairs of edges cross, in which order they cross, and the cyclic order around vertices and crossings, respectively. We consider all possible combinations of how two drawings can share some characteristics and determine which other characteristics they imply and which they do not imply. Our main results are that for simple drawings of complete multipartite graphs, the orders in which edges cross determine all other considered characteristics. Further, if all partition classes have at least three vertices, then the pairs of edges that cross determine the rotation system and the rotation around the crossings determine the extended rotation system. We also show that most other implications – including the ones that hold for complete graphs – do not hold for complete multipartite graphs. Using this analysis, we establish which types of isomorphisms are meaningful for simple drawings of complete multipartite graphs.

Keywords: Complete multipartite graphs · Isomorphisms · Simple Drawings

1 Introduction

A *simple drawing* of a graph is a drawing in the plane or on the sphere in which vertices are represented as points, edges are non-self-intersecting curves connecting their endpoints and not passing through any other point representing a vertex. Further, every pair of edges intersects at most once (either in a common endpoint or in a proper crossing). The *rotation* of a vertex or crossing in a labeled drawing is the clockwise cyclic order of the endpoints of incident edges around this vertex or crossing. (We remark that for crossings, we always note as endpoints the vertices induced by the respective edge-fragments.) To describe simple drawings, the following are the most commonly used properties.

O. Aichholzer and B. Vogtenhuber partially supported by Austrian Science Fund (FWF) within the collaborative DACH project *Arrangements and Drawings* as FWF project I 3340-N35. O. Aichholzer, B. Vogtenhuber and A. Weinberger partially supported by FWF grant W1230.

We thank the reviewers of EuroCG'21 and GD'23 for their very helpful comments.

- The collection of the rotations of all vertices. This collection is called the *rotation system* [RS];
- The pairs of edges that cross [CE];
- The collection of the rotations of all crossings [CR];
- The collection of the rotations of all vertices and all crossings. This collection is called the *extended rotation system* [ERS];
- The collection of the *crossing orders* of all edges, that is, along each edge, the order in which the edge crosses other edges [CO];

For each such property, we can define a type of isomorphism. Two labeled simple drawings of the same graph are ...

... *RS-isomorphic* if either for each vertex the rotation is the same in both drawings of for each vertex the rotation is inverse between the two drawings.

... *CE-isomorphic* if the same pairs of edges cross. In other literature (e.g. [16]), this property is also called *weak isomorphism.*

... *CR-isomorphic* if either for each crossing the rotation is the same in both drawings or for each crossing the rotation is inverse between the two drawings.

... *ERS-isomorphic* if their extended rotation systems are the same or inverse (where inverse means for each vertex and each crossing the rotation in one drawing is the inverse from the rotation in the other).

... *CO-isomorphic* if for each edge, its crossing order is the same in both drawings.

... *strongly isomorphic* if there exists a homeomorphism of the sphere such that one drawing is mapped to the other.

Unlabeled simple drawings are isomorphic with respect to some type of isomorphism if there exists a labeling such that the labeled drawings are isomorphic with respect to that type. In this paper, when we say that two simple drawings are isomorphic without specifying, the statement holds for both the labeled and the unlabeled case (where often the labeled case follows from the unlabeled case).

Some of the isomorphisms imply other isomorphisms for any graph by definition; see Fig. 1a for a depiction of the trivial implications that hold for all graphs. The extended rotation system combines the information of the rotation system with the information of the rotations around the crossings. Thus, if two drawings are ERS-isomorphic, they are also CR-isomorphic and RS-isomorphic. Since the crossings have to be the same for the rotations around the crossings to be the same (and also for the order of crossings to be the same), each of ERS-isomorphism, CR-isomorphism, and CO-isomorphism implies CE-isomorphism.

Further, Kynčl [17] showed the following useful characterization of strong isomorphism (which we restate here with the above-defined terminology).

Theorem 1. *[17] Two connected, labeled drawings D and D' of the same graph on the sphere are strongly isomorphic if and only if the following properties hold simultaneously: (i) The drawings D and D' are CO-isomorphic. (ii) The drawings D and D' are ERS-isomorphic.*

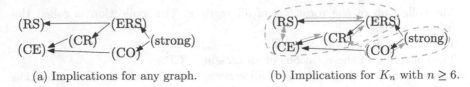

(a) Implications for any graph.   (b) Implications for K_n with $n \geq 6$.

Fig. 1. Implications between isomorphisms. Black arrows hold by definition. Orange arrows hold for K_n with $n \geq 6$. Dashed curves group the isomorphism classes by equivalences (for $K_n, n \geq 6$). (Color figure online)

Complete Graphs. For drawings of complete graphs, more implications and also equivalences are known; see Fig. 1b. Concretely, RS-isomorphism, CE-isomorphism, and ERS-isomorphism (and thus also CR-isomorphism) are all equivalent to each other [11,12,16]. Further, as CO-isomorphism implies CE-isomorphism (and thus ERS-isomorphism), by Theorem 1, CO-isomorphism implies strong isomorphism. For $n \leq 5$, the order of crossings along the edges can be derived from the pairs of crossing edges [5,11,12,22]. For $n \geq 6$ this is no longer the case; but any two simple drawings that are CE-isomorphic can be transformed into each other by a sequence of local operations called triangle flips (a.k.a. Reidemeister moves of type III) [5,11,12,22]. In conclusion, there are only two relevant classes of isomorphisms: One contains RS-, CE-, CR-, and ERS-isomorphism, as well as any combination of them; the other contains strong and CO-isomorphism (plus any combination of the others). We remark that K_n coincides with the complete multipartite graph on n vertices that has n partition classes, each of them containing exactly one vertex.

Complete Multipartite Graphs. The main goal of this paper is to classify which types of isomorphism are relevant for drawings of complete multipartite graphs. Many of the implications that hold for the complete graph do not hold for non-complete graphs, including complete multipartite graphs. For example, it is known that for simple drawings of complete multipartite graphs, RS-isomorphism does not imply CE-isomorphism; see Fig. 7. On the positive side, it still holds that two simple drawings of a complete multipartite graph that are ERS-isomorphic can be transformed into each other by a sequence of triangle flips [2]. However, to the best of our knowledge, except this latter property and the implications that follow directly from the definitions, no implications between the different properties of simple drawings of (general) complete multipartite graphs have been known.

In this paper, we give a complete characterization which implications do or do not hold for drawings of complete multipartite graphs, depending on the cardinalities of their partition classes; cf. Fig. 2. To obtain this characterization, we mostly focus on drawings of complete multipartite graphs in which every partition class consists of at least three vertices, and drawings of $K_{1,n}$ and $K_{2,n}$.

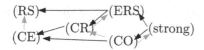

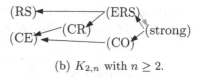

(a) Complete multipartite graphs with at least three vertices in each class.

(b) $K_{2,n}$ with $n \geq 2$.

Fig. 2. Implications that hold for simple drawings of complete multipartite graphs. The implications which don't follow by definition are drawn orange. (Color figure online)

Further Related Work. The majority of previous work on simple drawings is focused on the special case of complete graphs. Most of the work on more general complete multipartite graphs is focused on how to draw these graphs with as few crossings as possible. In 1954, Zarankiewicz [23] gave a (straight-line) drawing construction for complete bipartite graphs that is still conjectured to reach the minimum number of crossings over all simple drawings of such graphs. In 1971, Harborth [13] extended this result and gave a simple drawing of complete multipartite graphs that is conjectured to reach the minimum number of crossings (and can be drawn straight-line if there are at most three partition classes). For further work on the crossing number of complete multipartite graphs see also [6,9,10,15,19,21] and references therein.

Beyond the crossing number problem, Cardinal and Felsner [8] studied rotation systems of complete bipartite graphs and their realization as special simple drawings. Further, there has been work on plane subdrawings [3,18], and on triangle flips [2] in simple drawings of complete multipartite graphs, as well as on the enumeration of simple drawings of $K_{m,n}$ with $m \leq 3, n \leq 3$ [14].

1.1 Obtained Results

An overview of our results is given in Fig. 3a, where we use the following symbols: In areas marked with $\emptyset$, there are no two labeled or unlabeled simple drawings of complete multipartite graphs such that the two drawings share exactly the intersecting properties (and do not share any other properties). In the areas marked with $\exists^L_{=2}$, no labeled or unlabeled simple drawings of complete multipartite graphs with at least three vertices in each partition class such that the drawings share exactly the intersecting properties, but there are labeled simple drawings of $K_{2,n}$ that share exactly the intersecting properties; and in areas marked with $\exists_{=2}$, there are labeled and unlabeled simple drawings of $K_{2,n}$ that share exactly the intersecting properties. In areas marked with $\exists_{\geq x}$ for $x \in \{1,2,3\}$, there are labeled and unlabeled simple drawings sharing exactly the intersecting properties of complete multipartite graphs in which all partition classes have at least x vertices, but no such drawings if the smallest partition class has less than x vertices. Especially, for $x = 1$, there are such simple drawings independent of the sizes of the partition class sharing exactly the intersecting properties. For comparison, Fig. 3b shows the analogous diagram for K_n. In this figure, the label $\exists_{\geq 6}$ means that there exist labeled and unlabeled drawings of

K_n with $n \geq 6$, that share exactly the intersecting properties, while $\exists$ means that there exist such drawings of K_n for any n.

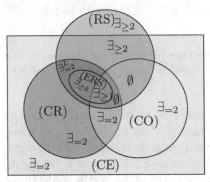

(a) Complete multipartite graphs.

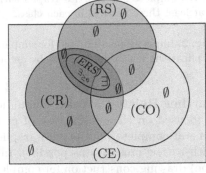

(b) Complete graphs.

Fig. 3. Classification of all possible combinations of different properties. The used notation is explained in the text above.

We list our results by first stating implications between (combinations of) isomorphisms that hold, and ending with a list of isomorphisms that do not imply each other. We consider all combinations except for implications that follow from our statements by definition. All statements that do not explicitly specify whether the considered drawings are labeled or unlabeled hold for both settings. Together, this gives a complete characterization which implications hold.

Theorem 2. *Let G be a complete multipartite graph. Then any two unlabeled simple drawings of G that are RS-isomorphic and CO-isomorphic are strongly isomorphic. If G has at least five vertices, then also any two labeled simple drawings of G that are RS-isomorphic and CO-isomorphic are strongly isomorphic.*

Theorem 3. *Let G be a complete multipartite graph in which each partition class has at least three vertices. Then any two simple drawings of G that are CE-isomorphic are also RS-isomorphic.*

Theorem 4. *Let G be a complete multipartite graph in which each partition class has at least three vertices. Then any two simple drawings of G that are CR-isomorphic are ERS-isomorphic.*

Corollary 5. *Let G be a complete multipartite graph in which each partition class has at least three vertices. Then any two simple drawings of G that are CO-isomorphic are strongly isomorphic.*

Simple drawings of $K_{2,n}$ behave differently in many ways. While Theorem 2 still holds, the other three statements do not. This different behavior of simple drawings of $K_{2,n}$ is in most parts due to the fact that for all n vertices of the larger partition class, the rotation of the vertex does not contain any information because it is a cyclic order of two elements. On the other hand, ERS-isomorphism does now imply strong isomorphism. This follows from the before-mentioned result on triangle flips [2] but can also be shown directly; see [4, Appendix C]. Figure 4 compares the implications for complete multipartite graphs in which all partition classes contain at least three vertices with the ones for $K_{2,n}$. The combination marked with $\exists^L$ in Fig. 4b exists at least for labeled drawings of $K_{2,n}$.

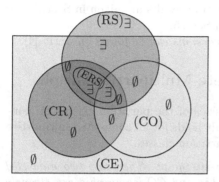

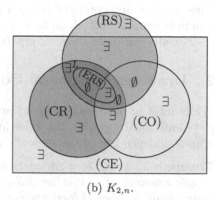

(a) Complete multipartite graphs with at least three vertices in each class.

(b) $K_{2,n}$.

Fig. 4. Classification of all possible combinations of different properties.

Simple drawings of $K_{1,n}$ again behave differently since all such drawings are plane. Thus, for $K_{1,n}$, RS-isomorphism implies strong isomorphism for labeled drawings, and all unlabeled simple drawings are strongly isomorphic.

We now turn to types of isomorphisms and combinations of types that do not imply others.

(1) There are simple drawings of $K_{m,n}$ that are RS-isomorphic but not CE-isomorphic; see Figs. 6 and 7.
(2) For $K_{m,n}$ with $m \geq 2$ and $n \geq 3$, there are simple drawings which are CE-isomorphic and RS-isomorphic but neither CR-isomorphic nor CO-isomorphic; see Figs. 8 and 9.
(3) For $K_{m,n}$ with $m \geq 3$ and $n \geq 3$, there are simple drawings which are ERS-isomorphic but not CO-isomorphic; see Fig. 10.
(4) For $K_{2,n}$, there are simple drawings which are CE-isomorphic but neither CO-isomorphic nor CR-isomorphic nor RS-isomorphic; see Fig. 11.
(5) For $K_{2,n}$, there are simple drawings which are CR-isomorphic but neither CO-isomorphic nor RS-isomorphic; see Fig. 12.
(6) For $K_{2,n}$, there are labeled simple drawings which are CR-isomorphic and RS-isomorphic but not ERS-isomorphic or CO-isomorphic; see Fig. 13.

(7) For $K_{2,n}$, there are simple drawings which are CO-isomorphic but not CR-isomorphic or RS-isomorphic; see Fig. 14.

(8) For $K_{2,n}$, there are simple drawings which are CE-isomorphic, CR-isomorphic, and CO-isomorphic but not RS-isomorphic; see Fig. 15.

(9) For $K_{1,n}$, there are labeled simple drawings which are CO-isomorphic, but not ERS-isomorphic (by relabeling the vertices in the larger partition class).

Outline. We prove Theorem 2 in Sect. 2, Sect. 3 is devoted to proving Theorems 3 and 4, and details of the non-implication results are given in Sect. 4. We conclude with an outlook on future work in Sect. 5.

For all omitted or sketched proofs, full versions can be found on arxiv [4].

2 Drawings of General Complete Multipartite Graphs

While most of our results depend on the sizes of the partition classes of complete multipartite graphs, the following result holds for all complete multipartite graphs, independent of the sizes of their partition classes.

Theorem 2. *Let G be a complete multipartite graph. Then any two unlabeled simple drawings of G that are RS-isomorphic and CO-isomorphic are strongly isomorphic. If G has at least five vertices, then also any two labeled simple drawings of G that are RS-isomorphic and CO-isomorphic are strongly isomorphic.*

We first state Theorem 2 for special complete bipartite graphs and small complete multipartite graphs in Lemma 1 and then use this result for proving the full theorem. We sketch the proofs of Lemma 1 and Theorem 2; for full proofs see [4, Appendix B].

Lemma 1. *Theorem 2 holds for $K_{1,n}$ and for complete multipartite graphs with at most four vertices (in the graph) and for $K_{2,3}$.*

Proof (sketch). For simple drawings of $K_{1,n}$ and $K_{1,1,1} = K_3$, the proof is trivial since the drawings are plane. For each complete multipartite graph on four vertices, there is, up to strong isomorphism and relabeling, only one unique drawing with a crossing and one unique drawing without a crossing. Hence, all unlabeled simple drawings of such graphs which are crossing CE-isomorphic are also strongly isomorphic and thus also all such drawings that are CO-isomorphic. There are, up to strong isomorphism and relabeling, only six drawings of $K_{2,3}$ [14]. They are depicted in Fig. 5. The only two non strongly isomorphic unlabeled drawings that are CE-isomorphic are the ones in Fig. 5e and Fig. 5f. They are not CO-isomorphic and we show that there is no labeling such that the drawings are CO-isomorphic. For each labeled drawing, there are twelve possible labelings, but we show that there is at most one relabeling of the original drawing such that the relabeled drawing is CO-isomorphic to the originally labeled drawing. For this pair of drawings we then show that they are ERS-isomorphic. Since ERS-isomorphism together with CO-isomorphism implies strong isomorphism by Theorem 1, this concludes the proof of Lemma 1.   □

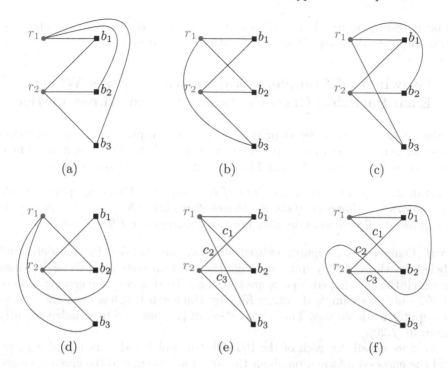

Fig. 5. The (up to relabeling or strong isomorphism) only 6 drawings of $K_{2,3}$, sorted by non-decreasing number of crossings. (Color figure online)

Proof (sketch of Theorem 2). Let D and D' be two simple drawings of a complete multipartite graph G and let L_D and $L_{D'}$ be labelings of D and D', respectively, such that the thus labeled drawings are CE-isomorphic, RS-isomorphic, and CO-isomorphic. Consider a crossing in D and the subdrawing H induced by the four vertices involved by that crossing. Since G is complete multipartite and thus H is a drawing of a complete multipartite subgraph of G, there is a subdrawing $H_{2,2}$ of H that is a simple drawing of $K_{2,2}$ and there is a subdrawing $H_{2,3}$ of D that contains $H_{2,2}$ and is a simple drawing of $K_{2,3}$. (For $H_{2,2}$ the endvertices of the two crossing edges get split into the bipartition classes such that each edge still has two vertices of different bipartition classes; and for $H_{2,3}$ any arbitrary fifth vertex can be used. See [4, Appendix B] for a rigorous proof.)

We then compare $H_{2,3}$ in D to its corresponding subdrawing $H'_{2,3}$ of D', that is, the drawing induced by the same vertices and edges. Since D and D' are CO-isomorphic and RS-isomorphic with respect to the labeling $L(D)$ and $L(D')$ also $H_{2,3}$ and $H'_{2,3}$ are CO-isomorphic and RS-isomorphic and thus, by Lemma 1, are strongly isomorphic. In particular, the ERS is the same or inverse. Thus, if the rotation system is the same in both drawings, so is the rotation of the crossing and analogously if the rotation system is inverse, so is the rotation of the crossing. Since this holds for all crossings of the drawings, D and D' have to be ERS-isomorphic. Hence, D and D' are also ERS-isomorphic. Thus, by Theorem 1, the drawings D and D' are strongly isomorphic. $\square$

3   Drawings of Complete Multipartite Graphs Where Each Partition Class Contains at Least Three Vertices

In this section, we consider complete multipartite graphs where each partition class contains at least three vertices. We will prove both Theorem 3 and Theorem 4 via subdrawings of $K_{3,3}$ and in particular use the following result.

Lemma 2. *Any two labeled drawings of $K_{3,3}$ that are CE-isomorphic are also RS-isomorphic. Moreover, if the two labeled drawings of $K_{3,3}$ are CR-isomorphic in addition to RS-isomorphic, then the two drawings are ERS-isomorphic.*

Proof. Lemma 2 is computer-assisted and has been verified by considering all labeled drawings of $K_{3,3}$ and comparing the rotation systems of those drawings for which the crossing edge pairs are the same. To this end, the enumeration of all 102 unlabeled simple drawings $K_{3,3}$ by Harborth [14] has been encoded in a computer readable way. The enumeration of [14] has also been independently shown in [7, 20].

In more detail, for each of the 102 different unlabeled drawings of $K_{3,3}$ we read the encoded information about the rotation system and the crossing edges, including their rotation, from a given text file. For each drawing we then generate all possible labelings (72 per drawing, 3! different labelings for each color class, times the exchange of red and blue). This results in a set S of a total of 7344 labeled drawings. For all 26 963 496 pairs of drawings from S we check if the two labeled drawings are CE-isomorphic by simply comparing their list of crossed edges. Next, for the 10332 CE-isomorphic pairs we check if both drawings have the same or inverse rotation systems (which is straightforward as we have labeled drawings). It turns out that this is in fact the case for all CE-isomorphic pairs. So in total this implies that regardless which unlabeled two drawings of $K_{3,3}$ we take, and regardless how we label them, if they have the same crossing edge pairs, then they have the same or inverse rotation system, which implies the first part of Lemma 2. Finally, for all CE-isomorphic pairs, we test if they are CR-isomorphic. For the resulting 4680 CR-isomorphic pairs, we verify that either the rotation system and the crossing rotations are both the same, or both are inverse. In other words, they are ERS-isomorphic. The total running time of this routine is less than two seconds on a standard computer.

The program code, the input data, and a short description how to use and verify the correctness of the program are available online at [1]. $\square$

Using Lemma 2, we first show Theorem 3 for bipartite graphs.

Lemma 3. *Theorem 3 holds for $K_{m,n}$ with $n \geq m \geq 3$.*

Proof. For simple drawings of $K_{m,n}$ with $m \geq 3, n \geq 3$ we will show how to determine the rotation system from the information on crossing edge-pairs. Let the vertex set be split in a red set $\{r_1, \ldots, r_m\}$ and a blue set $\{b_1, \ldots, b_n\}$. We will determine the rotation system from the information of small subsets in three steps. In the first step, we find the rotation system of a subdrawing that is a $K_{3,3}$ and includes vertices r_1 and b_1 by using Lemma 2. This rotation system will determine the global orientation of the rotation system. In the second step, we determine the rotations around vertices r_1 and b_1. We find them by using Lemma 2 to sort incident edges in the rotation. In the third step, we determine the rotations around the remaining vertices.

Step 1: We first consider the subdrawing induced by $\{r_1, r_2, r_3, b_1, b_2, b_3\}$ and find the rotation system of this subdrawing in the following way: It is a simple drawing of $K_{3,3}$, so we know from Lemma 2 that there are only two possible rotation systems, which are inverse to each other. There are only two possibilities for the rotation around r_1: either $\{b_1, b_2, b_3\}$ or $\{b_1, b_3, b_2\}$. We choose the rotation such that r_1 has rotation $\{b_1, b_2, b_3\}$. (This will be the only choice for our proof, thus determining the rotation system. Choosing the inverse rotation would give the inverse rotation system.)

Step 2: We now find the rotation around vertex r_1: We look at subdrawings induced by $\{r_1, r_2, r_3, b_1, b_i, b_j\}$. These are simple drawings of $K_{3,3}$ so by Lemma 2 there are only two possible rotation systems. We have already fixed the rotation around b_1, thus the rotation system of these subgraphs is fixed. This way we learn in which order b_i and b_j are in the rotation around r_1 (using b_1 as reference point). Doing this for different pairs $\{b_i, b_j\}$ we have a way to sort the vertices $\{b_1, b_2, ..., b_n\}$ around r_1, that is, to determine the rotation around r_1.

We find the rotation around vertex b_1 analogously by considering the subdrawings induced by $\{r_1, r_i, r_j, b_1, b_2, b_3\}$ and proceeding as described for r_1.

Step 3: Finally, we find the rotation around the remaining vertices in the following way: To find the rotation around vertex r_i for $i > 2$, we look at subdrawings induced by $\{r_1, r_2, r_i, b_1, b_j, b_k\}$. These are simple drawings of $K_{3,3}$ and so by Lemma 2 the rotation system is determined up to inversion. As we already know the rotation around b_1, we can obtain the rotation system of this subgraph. As in Step 2, we sort the vertices in the rotation around r_i, thus determining the rotation around r_i. We repeat this process for all remaining vertices r_i for $i > 2$ and for r_2 we do the same process with $\{r_1, r_2, r_3, b_1, b_j, b_k\}$. Analogously, we find the rotation around all vertices b_i. $\square$

This result for complete bipartite graphs implies the statement for complete multipartite graphs.

Theorem 3. *Let G be a complete multipartite graph in which each partition class has at least three vertices. Then any two simple drawings of G that are CE-isomorphic are also RS-isomorphic.*

Proof. Let G be a complete multipartite graph in which each partition class has at least three vertices and let D and D' be two CE-isomorphic drawings of G. Let A be one of the partition classes. Then the subgraph G_A of G consisting of all vertices of G and exactly the edges with one endpoint in A is a complete bipartite graph where both bipartition classes contain at least three vertices. Thus, by Lemma 3, the rotation in the drawing of G_A that is a subdrawing of D and the drawing of G_A that is a subdrawing of D' are either the same or inverse. Assume without loss of generality that they are the same (the case in which they are inverse is analogous). Since all edges incident to A are in G_A, this determines all rotations of vertices in A (to be the same in D and D'). Let B be a partition class of G that is different from A. Analogously to before, the two drawings of subgraph G_B of G consisting of all vertices of G and exactly the edges with one endpoint in B have the same or inverse rotation systems by Lemma 3. G_B and G_A both contain the complete bipartite graph $G_{A,B}$ induced by the partition classes A and B. Since we assumed that the drawings of G_A have the same rotation system, the subdrawing induced by $G_{A,B}$ must have the same rotation system also in G_B. Thus, the drawing of G_B that is a subdrawing of D and the one that is a subdrawing of D' have the same rotation system. This determines the rotations of all vertices in B (to be the same rotation in D and D'). Continuing analogously for each partition class, the rotations of all vertices are determined and D and D' must have the same rotation system. $\square$

Theorem 4. *Let G be a complete multipartite graph in which each partition class has at least three vertices. Then any two simple drawings of G that are CR-isomorphic are ERS-isomorphic.*

Proof. Let G be a complete multipartite graph in which each partition class has at least three vertices, let D and D' be two drawings with labelings such that the two labeled drawings are CR-isomorphic drawings of G. Then the labeled drawings are also CE-isomorphic and by Theorem 3 RS-isomorphic.

Assume, without loss of generality, that the rotation system in both drawings is the same (the case if they are inverse follows analogously, only with the drawings mirrored). If the rotations around all crossings are also the same in both drawings, then both drawings are ERS-isomorphic. So assume all crossings are inverse. Then also in every subdrawing of D that is a $K_{3,3}$, all crossings are inverse while all rotations are the same, which contradicts Lemma 2. $\square$

Corollary 5 is a near-direct consequence of Theorems 2 and 3. Any two CO-isomorphic drawings are trivially CE-isomorphic. Thus, by Theorem 3, they are RS-isomorphic and consequently, by Theorem 2 be strongly isomorphic.

4 Examples of Simple Drawings Showing Specific Properties

In this section, we look at relations where no implications hold. In particular, we depict the examples mentioned in the introduction. That is, for any combination of the characteristics RS, CE, CR, CO, and ERS for which there are simple drawings of complete multipartite graphs sharing exactly those, but no others, we give an example of such simple drawings. The depicted drawings are labeled. However, except for Fig. 13, all statements also hold when seeing the drawings as unlabeled, that is, for every relabeling such that the isomorphisms between the relabeled drawings is a super-set of the isomorphism between the originally labeled drawings, the isomorphisms are actually the same; see [4, Appendix D] for proofs of the unlabeled cases. The proofs for the labeled cases can be derived from the depicted drawings via their rotation and crossing properties. We list the proof details only for Fig. 6, other proofs are similar; for details on the remaining proofs see [4, Appendix D].

In each of the drawings, we highlight edges that behave differently by drawing them bold and orange. Further, where relevant, we indicate that crossings have the same rotation in the compared drawings by drawing a solid, green arc around them and indicate that crossings have inverse rotations between the compared drawings by drawing an orange dashed-dotted arc around them. Finally, the vertices that have a dash in their label and are incident to dashed lines indicate how to extend the drawing by making arbitrary copies of that vertex, e.g. the $K_{2,3}$ drawn solid in Fig. 6 can be extended via b' and copies of it to $K_{2,n}$ for $n \geq 4$.

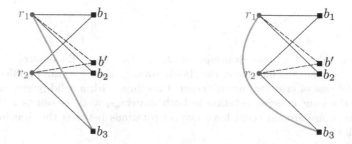

Fig. 6. Simple drawings of $K_{2,3}$ that are RS-isomorphic, but not CE-isomorphic. In particular, in both drawings, the rotations of both r_1 and r_2 are b_1, b_2, b_3, and since the blue vertices have only degree two there is only one possible rotation for each of them. However, in the left drawing there exist crossings between the (bold, orange) edge $r_1 b_3$ and edges $r_2 b_1$ and $r_2 b_2$, while in the right drawing no edge crosses $r_1 b_3$. The drawings can be extended via b' (and the dashed edges incident to it) and copies of it to $K_{2,n}$ for $n \geq 4$. (Color figure online)

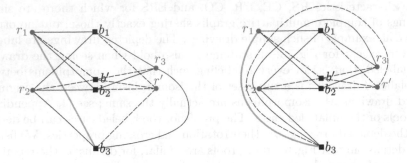

Fig. 7. Simple drawings of $K_{3,3}$ that are RS-isomorphic, but not CE-isomorphic. The drawings can be extended via the vertices b', r' and copies of them to $K_{m,n}$ for $m \geq 4, n \geq 4$. (Color figure online)

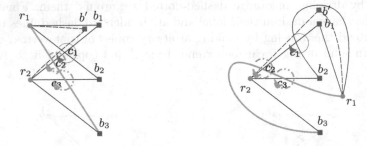

Fig. 8. Two labeled simple drawings of $K_{2,3}$ that are CE-isomorphic and RS-isomorphic, but the crossings along the (bold, orange) edge $r_1 b_3$ are in different order and the rotations of crossings are different. Crossings with a solid, green arc around them have the same crossing rotation in both drawings, while crossings with a dash-dotted, orange arc around them have inverse rotations between the drawings.(Color figure online)

Fig. 9. Two labeled simple drawings of $K_{3,3}$ that are CE-isomorphic and RS-isomorphic, but the crossings along the (bold, orange) edge $r_1 b_3$ an edge ($r_1 b_3$ drawn in bold, orange) are in different order and the rotation around crossings is different. (Color figure online)

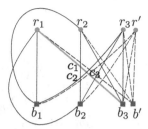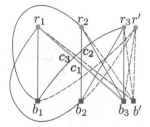

Fig. 10. Two simple drawings of $K_{3,3}$, which are ERS-isomorphic, but the crossings along the (bold, orange) edge $r_1 r_3$ is different. (Color figure online)

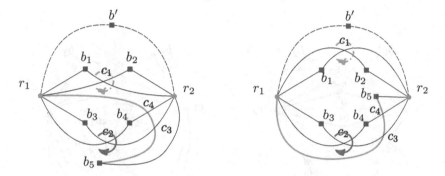

Fig. 11. Two simple drawings of $K_{2,5}$, which are CE-isomorphic, but neither RS-isomorphic nor CO-isomorphic nor CR-isomorphic. (Color figure online)

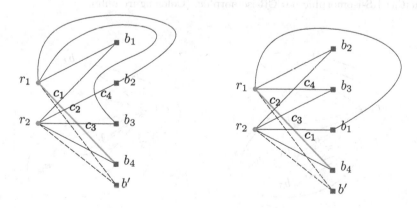

Fig. 12. Two simple drawings of $K_{2,4}$, which are CR-isomorphic, but neither RS-isomorphic nor CO-isomorphic. (Color figure online)

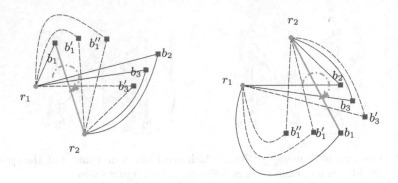

Fig. 13. Two labeled simple drawings of $K_{2,3}$, which are CR-isomorphic and RS-isomorphic, but not ERS-isomorphic. (Color figure online)

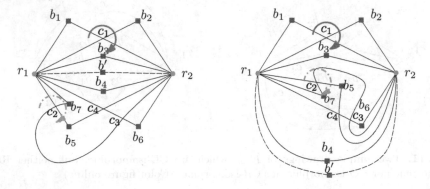

Fig. 14. Two simple drawings of $K_{2,7}$, which are CE-isomorphic and CO-isomorphic, but neither RS-isomorphic nor CR-isomorphic. (Color figure online)

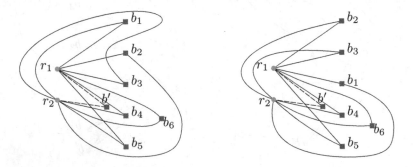

Fig. 15. Two simple drawings of $K_{2,6}$, which are CO-isomorphic and CR-isomorphic, but not RS-isomorphic. (Color figure online)

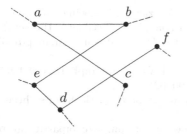 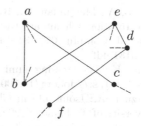

Fig. 16. Two simple drawings of a connected graph on six vertices, which are CE-isomorphic, RS-isomorphic, and CO-isomophic, but not CR-isomorphic.

5 Conclusion

From our results on implication between different types of isomorphism it follows that for drawings of complete multipartite graphs in which each partition class contains at least three vertices, there are four relevant classes of isomorphism: One class contains strong and CO-isomorphism; one contains CR- and ERS-isomorphism; one contains CE-isomorphism; and the last one contains RS-isomorphism. We remark that there are graphs for which less implications hold; see the labeled drawings in Fig. 16 (this is extendable to unlabeled drawings by vertices along the dashed lines).

In addition to relevantly improving the understanding of drawings of complete multipartite graphs, we believe that our characterization can serve as a basis for further studying these graphs and drawings and obtaining new results on them.

It might also be helpful for research on simple drawings of other graphs. In this context, the question arises which other relevant graph classes admit or do not admit which kind of isomorphism implications.

References

1. https://figshare.com/s/97449727daca12d6bbf6
2. Aichholzer, O., Chiu, M.K., Hoang, H.P., Hoffmann, M., Kynčl, J., Maus, Y., Vogtenhuber, B., Weinberger, A.: Drawings of complete multipartite graphs up to triangle flips. In: 39th International Symposium on Computational Geometry, LIPIcs. Leibniz International Proceedings in Informatics, vol. 258, pp. 6:1–6:16. Schloss Dagstuhl. Leibniz-Zent. Inform. Wadern (2023). https://doi.org/10.4230/lipics.socg.2023.6
3. Aichholzer, O., García, A., Parada, I., Vogtenhuber, B., Weinberger, A.: Shooting stars in simple drawings of $K_{m,n}$. In: Angelini, P., von Hanxleden, R. (eds.) GD 2022. LNCS, vol. 13764, pp. 49–57. Springer, Cham (2023). https://doi.org/10.1007/978-3-031-22203-0_5
4. Aichholzer, O., Vogtenhuber, B., Weinberger, A.: Different types of isomorphisms of drawings of complete multipartite graphs (2023). http://arxiv.org/abs/2308.10735v1

5. Arroyo, A., McQuillan, D., Richter, R.B., Salazar, G.: Drawings of K_n with the same rotation scheme are the same up to triangle-flips (Gioan's theorem). Australas. J. Combin. **67**, 131–144 (2017). https://ajc.maths.uq.edu.au/pdf/67/ajc-v67_p131.pdf

6. Asano, K.: The crossing number of $K_{1,3,n}$ and $K_{2,3,n}$. J. Graph Theory **10**(1), 1–8 (1986). https://doi.org/10.1002/jgt.3190100102

7. Brötzner, A.: Isomorphism Classes of Drawings of $K_{3,3}$. Bachelor's thesis, Graz University of Technology (2021)

8. Cardinal, J., Felsner, S.: Topological drawings of complete bipartite graphs. J. Comput. Geom. **9**(1), 213–246 (2018). https://doi.org/10.20382/jocg.v9i1a7

9. Fabila-Monroy, R., Paul, R., Viafara-Chanchi, J., Weinberger, A.: On the rectilinear crossing number of complete balanced multipartite graphs and layered graphs. In: Abstracts of XX Encuentros de Geometría Computacional (EGC 2023), pp. 33–36 (2023). https://egc23.web.uah.es/wp-content/uploads/2023/07/EGC2023_Booklet.pdf#page=45

10. Gethner, E., Hogben, L., Lidický, B., Pfender, F., Ruiz, A., Young, M.: On crossing numbers of complete tripartite and balanced complete multipartite graphs. J. Graph Theory **4**(84), 552–565 (2017). https://onlinelibrary.wiley.com/doi/full/10.1002/jgt.22041

11. Gioan, E.: Complete graph drawings up to triangle mutations. In: Kratsch, D. (ed.) WG 2005. LNCS, vol. 3787, pp. 139–150. Springer, Heidelberg (2005). https://doi.org/10.1007/11604686_13

12. Gioan, E.: Complete graph drawings up to triangle mutations. Discrete Comput. Geom. **67**, 985–1022 (2022). https://doi.org/10.1007/s00454-021-00339-8

13. Harborth, H.: Über die Kreuzungszahl vollständiger, n-geteilter Graphen. Math. Nachr. **48**, 179–188 (1971). https://doi.org/10.1002/mana.19710480113

14. Harborth, H.: Parity of numbers of crossings for complete n-partite graphs. Mathematica Slovaca **26**(2), 77–95 (1976). http://eudml.org/doc/33976

15. Ho, P.T.: The crossing number of $K_{2,4,n}$. Ars Combin. **109**, 527–537 (2013)

16. Kynčl, J.: Simple realizability of complete abstract topological graphs in P. Discrete Comput. Geom. **45**(3), 383–399 (2011). https://doi.org/10.1007/s00454-010-9320-x

17. Kynčl, J.: Improved enumeration of simple topological graphs. Discrete Comput. Geom. **50**(3), 727–770 (2013). https://doi.org/10.1007/s00454-013-9535-8

18. Mengersen, I.: Kreuzungsfreie Kanten in vollständigen n-geteilten Graphen. Ph.D. thesis, Technische Universität Braunschweig (1975)

19. Ouyang, Z., Wang, J., Huang, Y.: Two recursive inequalities for crossing numbers of graphs. Front. Math. China **12**(3), 703–709 (2017). https://doi.org/10.1007/s11464-016-0618-8

20. Prinoth, K.: Computing exhaustive lists of complete bipartite simple drawings. Bachelor's thesis, Graz University of Technology (2021)

21. Schaefer, M.: Crossing numbers of graphs. Discrete Mathematics and its Applications (Boca Raton), CRC Press, Boca Raton, FL (2018). https://doi.org/10.1201/9781315152394

22. Schaefer, M.: Taking a detour; or, Gioan's theorem, and pseudolinear drawings of complete graphs. Discrete Comput. Geom. **66**, 12–31 (2021). https://doi.org/10.1007/s00454-021-00296-2

23. Zarankiewicz, K.: On a problem of P. Turan concerning graphs. Fund. Math. **41**, 137–145 (1954). https://doi.org/10.4064/fm-41-1-137-145

Parameterized Complexity for Drawings

On the Parameterized Complexity
of Bend-Minimum Orthogonal Planarity

Emilio Di Giacomo[ID], Walter Didimo[ID], Giuseppe Liotta[ID],
Fabrizio Montecchiani[ID], and Giacomo Ortali$^{(\boxtimes)}$[ID]

Dipartimento di Ingegneria, Università degli Studi di Perugia, Perugia, Italy
{emilio.digiacomo,walter.didimo,giuseppe.liotta,fabrizio.montecchiani,
giacomo.ortali}@unipg.it

Abstract. Computing planar orthogonal drawings with the minimum number of bends is one of the most relevant topics in Graph Drawing. The problem is known to be NP-hard, even when we want to test the existence of a rectilinear planar drawing, i.e., an orthogonal drawing without bends (Garg and Tamassia, 2001). From the parameterized complexity perspective, the problem is fixed-parameter tractable when parameterized by the sum of three parameters: the number of bends, the number of vertices of degree at most two, and the treewidth of the input graph (Di Giacomo et al., 2022). We improve this last result by showing that the problem remains fixed-parameter tractable when parameterized only by the number of vertices of degree at most two plus the number of bends. As a consequence, rectilinear planarity testing lies in FPT parameterized by the number of vertices of degree at most two.

Keywords: Orthogonal drawings · bend-minimization · parameterized complexity

1 Introduction

Orthogonal drawings of planar graphs represent a foundational graph drawing paradigm [4,5,20] and an active field of research [6,9,12,14]. Given a planar 4-graph G, i.e., a planar graph of vertex-degree at most 4, a *planar orthogonal drawing* Γ of G is a planar drawing that maps the vertices of G to distinct points of the plane and each edge of G to a chain of horizontal and vertical segments connecting the two endpoints of the edge. A *bend* of Γ is the meeting point of two consecutive segments along the same edge. Given a graph G and $b \in \mathbb{N}$, the b-ORTHOGONAL PLANARITY problem asks for the existence of an orthogonal

Research partially supported by: (i) University of Perugia, Ricerca Base 2021, Proj. "AIDMIX - Artificial Intelligence for Decision Making: Methods for Interpretability and eXplainability"; (ii) MUR PRIN Proj. 2022TS4Y3N - "EXPAND: scalable algorithms for EXPloratory Analyses of heterogeneous and dynamic Networked Data"; (iii) MUR PRIN Proj. 2022ME9Z78 - "NextGRAAL: Next-generation algorithms for constrained GRAph visuALization".

M. A. Bekos and M. Chimani (Eds.): GD 2023, LNCS 14466, pp. 53–65, 2023.
https://doi.org/10.1007/978-3-031-49275-4_4

drawing of G with at most b bends in total. While every n-vertex planar 4-graph admits an orthogonal drawing with $O(n)$ bends [1], b-ORTHOGONAL PLANARITY is well-known to be NP-complete already when $b = 0$ [17]. In this case, the b-ORTHOGONAL PLANARITY problem is also known in the literature as RECTLINEAR PLANARITY. It is even NP-hard to approximate the minimum number of bends in an orthogonal drawing with an $O(n^{1-\varepsilon})$ error for any $\varepsilon > 0$ [17].

From a parameterized complexity perspective, the hardness of RECTLINEAR PLANARITY rules out the possibility of finding a tractable solution for b-ORTHOGONAL PLANARITY parameterized solely by the number of bends. On the positive side, a recent result of Di Giacomo et al. [6] shows that b-ORTHOGONAL PLANARITY lies in the XP class when parameterized by the treewidth tw of the input graph. In fact, the result in [6] is based on a fixed-parameter tractable algorithm in the parameter $b + k + \text{tw}$, where k is the number of vertices of degree at most two. A natural question is thus whether a subset of these three parameters suffices to establish tractability, as asked in [6]. We also remark that, in a recent Dagstuhl seminar on parameterized complexity in graph drawing [15, pag. 94], understanding whether RECTLINEAR PLANARITY is in FPT parameterized by only one of k and tw was identified as a prominent challenge.

Contribution. We address the questions posed in [6,15] and prove that b-ORTHOGONAL PLANARITY is fixed-parameter tractable parameterized by $b + k$, hence dropping the dependency on tw. Consequently, RECTLINEAR PLANARITY is in FPT when parameterized by k alone.

From a technical point of view, the algorithm by Di Giacomo et al. [6] exploits treewidth to apply dynamic programming. The main crux of that algorithm lies on two main ingredients. First, a fixed treewidth (and fixed vertex-degree) bounds the number of interesting embeddings in which different faces have vertices in the current bag. Second, a classic result by Tamassia [21] states that the existence of an orthogonal drawing is equivalent to the existence of a combinatorial representation describing the angles at the vertices and the bends along the edges. In this respect, a fixed number of bends and of vertices of degree at most two bounds the possible combinatorial descriptions of a face, and hence makes the definition of small records feasible for the sake of dynamic programming. To avoid the dependency on treewidth, in our work we exploit additional insight into the structure of the problem. First, we adopt SPQ*R-trees to decompose the input graph and to keep track of its possible embeddings. While it was already known how to deal with S- and P-nodes efficiently by exploiting dynamic programming on SPQ*R-trees [5,10], the main obstacle is represented by R-nodes, i.e., the rigid components of the graph. To overcome this obstacle, we first prove that the number of children of an R-node depends on the number of degree-2 vertices plus the number of bends. Then, we carefully combine flow-network based techniques [21] for planar embedded graphs with the notion of orthogonal spirality [5], which measures how much a component of an orthogonal drawing is rolled-up. A full version of the paper with all the proofs can be found in [7].

The most natural question that remains open is to settle the complexity of b-ORTHOGONAL PLANARITY parameterized by k alone, which we conjecture to be at least W[1]-hard due to the fact that bends and low-degree vertices have interchangeable behaviours in terms of spirality.

2 Preliminaries

We only consider planar 4-graphs, i.e., graphs with vertex-degree at most 4. A *plane graph* is a planar graph with a fixed planar embedding, i.e., prescribed clockwise orderings of the edges around the vertices and a given external face.

Orthogonal Representations. A *planar orthogonal representation H* of a planar graph G describes the shape of a class of orthogonal drawings in terms of left/right bends along the edges and angles at the vertices. A drawing Γ of H (i.e., a planar orthogonal drawing of G that preserves H) can be computed in linear time [21]. If H has no bends, it is a *rectilinear representation*. Since we only deal with planar drawings, we just use the term rectilinear (orthogonal) representation in place of planar rectilinear (orthogonal) representation.

Let G be a plane graph with n vertices. In [21], it is proved that a bend-minimum orthogonal representation of G can be constructed in $O(n^2 \log n)$ time by finding a min-cost flow on a suitable flow-network constructed from the planar embedding of G. The feasible flows on this network correspond to the possible orthogonal representations of the graph, and the total cost of the flow equals the number of bends of the corresponding representation. Fixing the values of the flow along some arcs of the network, one can force desired vertex-angles or left/right bends along the edges in the corresponding orthogonal representation. The time complexity in [21] has been subsequently improved to $O(n^{\frac{7}{4}}\sqrt{\log n})$ [16] time, and later to $O(n^{\frac{3}{2}})$ [3] if there is no constraints on the orthogonal representation and one can use a flow-network whose arcs all have infinite capacities.

Let H' be an orthogonal representation of a (not necessarily connected) subgraph G' of G. An orthogonal representation H of G is H'-*constrained* if its restriction to G' equals H'. By the considerations above, the following holds.

Lemma 1. *Let G be an n-vertex plane 4-graph and let H' be an orthogonal representation of a subgraph G' of G. There exists an $O(n^{\frac{7}{4}}\sqrt{\log n})$-time algorithm that computes an H'-constrained orthogonal representation of G with the minimum number of bends, if it exists, or that rejects the instance otherwise.*

SPQR-Trees. Let G be a biconnected planar graph. The *SPQR-tree T* of G, introduced in [4], represents the decomposition of G into its triconnected components [19]. Each triconnected component corresponds to a non-leaf node ν of T; the triconnected component itself is called the *skeleton* of ν and is denoted as skel(ν). Node ν can be: (*i*) an *R-node*, if skel(ν) is a triconnected graph; (*ii*) an *S-node*, if skel(ν) is a simple cycle of length at least three; (*iii*) a *P-node*, if skel(ν) is a bundle of at least three parallel edges. A degree-1 node of T is a *Q-node* and represents a single edge of G. A *real edge* (resp. *virtual edge*) in skel(ν) corresponds to a Q-node (resp., to an S-, P-, or R-node) adjacent to ν in T. Neither two S- nor two P-nodes are adjacent in T. The SPQR-tree of a biconnected graph can be computed in linear time [4,18].

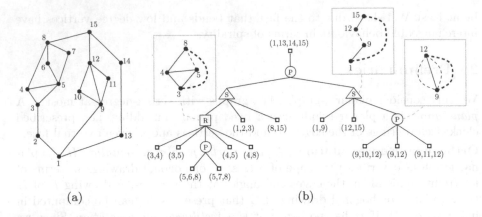

      (a)   (b)

Fig. 1. (a) A biconnected planar graph G. (b) The SPQ*R-tree of G rooted at the Q*-node of the chain (1,13,14,15). The skeletons of three nodes are shown; the virtual edges are dashed and the reference edge is thicker.

Let e be a designated edge of G, called the *reference edge* of G, let ρ be the Q-node of T corresponding to e, and let T be rooted at ρ. For any P-, S-, or R-node ν of T, skel(ν) has a virtual edge, called *reference edge* of skel(ν) and of ν, associated with a virtual edge in the skeleton of its parent. The reference edge of the root child of T is the edge corresponding to ρ. The rooted tree T describes all planar embeddings of G with its reference edge on the external face; they are obtained by combining the different planar embeddings of the skeletons of P- and R-nodes with their reference edges on the external face. Namely, for a P- or R-node ν, denote by skel$^-$(ν) the skeleton of ν without its reference edge. If ν is a P-node, the embeddings of skel(ν) are the different permutations of the edges of skel$^-$(ν); If ν is an R-node, skel(ν) has two possible embeddings, obtained by flipping skel$^-$(ν) at its poles. For every node $\nu \neq \rho$, the *pertinent graph* G_ν of ν is the subgraph of G whose edges correspond to the Q-nodes in the subtree of T rooted at ν. We also say that G_ν is a *component of* G. The pertinent graph G_ρ of the root ρ coincides with the reference edge of G. If H is an orthogonal representation of G, its restriction H_ν to G_ν is an *orthogonal component* of H.

3   The Biconnected Case

In this section we assume that G is a biconnected graph with k degree-2 vertices. Similar to previous works [5,9–11,13], our approach exploits a variant of SPQR-tree, called *SPQ*R-tree*, and the notion of *spirality* for orthogonal components. We recall below these two concepts; see Fig. 1 for an illustration.

SPQ*R-trees. Assume that G is not a simple cycle (otherwise computing a bend-minimum orthogonal representation of G is trivial). In an SPQ*R-tree of G we do not distinguish between Q-nodes and series of edges, and represent both of them with a type of node called *Q*-node*. More precisely, each degree-1 node of T is a Q*-node that corresponds to a maximal chain of edges of G (possibly a single edge) starting and ending at vertices of degree larger than two and passing through a sequence of degree-2 vertices only (possibly none). If ν is an S-, a P-, or an R-node, and if μ is a Q*-node child of ν, the edge of skel(ν) corresponding to μ is virtual if μ corresponds to a chain of at least two edges, else it is a real edge. If T is rooted at a certain Q*-node ρ, the chain corresponding to ρ is the *reference chain of* G. The definitions of pertinent graphs (or components) for the nodes of a rooted SPQ*R-tree extend naturally from those of a rooted SPQR-tree. In particular G_ρ is the pertinent graph of ρ, i.e., the reference chain of G. Moreover, as in [2,5,8–11,13], we assume to work with a *normalized* SPQ*R-tree, in which every S-node has exactly two children. Note that every SPQ*R-tree can be easily transformed into a normalized SPQ*R-tree by recursively splitting an S-node with more than two children into multiple S-nodes with two children. Observe that in this case an S-node may have an S-node as a child. If G has n vertices, a normalized SPQ*R-tree of G still has $O(n)$ nodes and can be easily computed in $O(n)$ time from an SPQ*R-tree of G.

Orthogonal Spirality. Let H be an orthogonal representation of G. If P^{uv} is a path from a vertex u to a vertex v in H, the *turn number of* P^{uv}, denoted by $n(P^{uv})$, is the number of right turns minus the number of left turns along P^{uv}, while moving from u to v. Note that a turn can occur at either a bend or a vertex. Let T be a rooted normalized SPQ*R-tree of G and let ν be a node of T. Let H_ν be the restriction of H to G_ν, and denote by $\{u, v\}$ the poles of ν, conventionally ordered according to an st-numbering of G, where s and t are the poles of the root of T. For each $w \in \{u, v\}$, let $\mathrm{indeg}_\nu(w)$ and $\mathrm{outdeg}_\nu(w)$ be the degree of w inside and outside H_ν, respectively. Define two, possibly coincident, *alias vertices* of w, denoted by w' and w'', as follows: (*i*) if $\mathrm{indeg}_\nu(w) = 1$, then $w' = w'' = w$; (*ii*) if $\mathrm{indeg}_\nu(w) = \mathrm{outdeg}_\nu(w) = 2$, then w' and w'' are dummy vertices, each splitting one of the two distinct edge segments incident to w outside H_ν; (*iii*) if $\mathrm{indeg}_\nu(w) > 1$ and $\mathrm{outdeg}_\nu(w) = 1$, then $w' = w''$ is a dummy vertex that splits the edge segment incident to w outside H_ν.

Let A^w be the set of distinct alias vertices of a pole w. Let P^{uv} be any simple path from u to v in H_ν, and let u' and v' be alias vertices of u and v, respectively. The path $S^{u'v'}$ obtained by concatenating (u', u), P^{uv}, and (v, v') is a *spine* of H_ν. The *spirality* $\sigma(H_\nu)$ of H_ν, introduced in [5], is either an integer or a semi-integer number, defined based on the following cases: (*i*) if $A^u = \{u'\}$ and $A^v = \{v'\}$ then $\sigma(H_\nu) = n(S^{u'v'})$; (*ii*) if $A^u = \{u'\}$ and $A^v = \{v', v''\}$ then $\sigma(H_\nu) = \frac{1}{2}\big(n(S^{u'v'}) + n(S^{u'v''})\big)$; (*iii*) if $A^u = \{u', u''\}$ and $A^v = \{v'\}$ then $\sigma(H_\nu) = \frac{1}{2}\big(n(S^{u'v'}) + n(S^{u''v'})\big)$; (*iv*) if $A^u = \{u', u''\}$ and $A^v = \{v', v''\}$, assume w.l.o.g. that (u, u') is the edge before (u, u'') counterclockwise around u and that (v, v') is the edge before (v, v'') clockwise around v; then $\sigma(H_\nu) = \frac{1}{2}\big(n(S^{u'v'}) + n(S^{u''v''})\big)$.

Figure 2 shows the graph G of Fig. 1(a), two P-components G_{μ_1} and G_{μ_2} highlighted, and an orthogonal representation H of G. The orthogonal components H_{μ_1} and H_{μ_2} in H have respectively spiralities $\sigma_{\mu_1} = -1$, see Case (iv), and $\sigma_{\mu_2} = 0$, see Case (i). Figure 3(a) shows an orthogonal S-component with spirality $\frac{3}{2}$, see Case (ii).

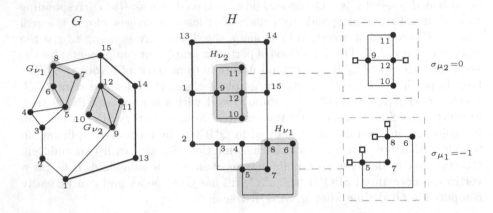

Fig. 2. Illustration of the concept of spirality.

Note that, the spirality of H_ν does not vary if we choose a different path P^{uv}. Also, it is proved in [5] that a component H_ν of H can always be substituted by any other representation H'_ν of G_ν with the same spirality, getting a new valid orthogonal representation with the same set of bends on the edges of H that are not in H_ν (see [5] and also Theorem 1 in [10]). For brevity, in the remainder of the paper we often denote by σ_ν the spirality of a component H_ν.

Testing Algorithm. Let T be a rooted normalized SPQ*R-tree of the input graph G, and let b be a non-negative integer. An orthogonal representation of G (resp. of a component G_ν of G) with at most b bends is a *b-orthogonal representation* of G (resp. of G_ν). Theorem 1 will prove that testing if G admits a b-orthogonal representation is FPT parametrized by $k+b$. We start with a key property.

Lemma 2. *Let ν be a node of T and H be a b-orthogonal representation of G. The spirality σ_ν of the restriction H_ν of H to G_ν belongs to $[-k-b-2, k+b+2]$.*

Proof. We consider the case $\sigma_\nu > 0$ and prove that $\sigma_\nu \leq k+b+2$. When σ_ν is negative, the proof that $-k-b-2 \leq \sigma_\nu$ is symmetric. If $0 \leq \sigma_\nu \leq 2$, the lemma is trivially true. Assume now that $\sigma_\nu > 2$. Denote by u, v the two poles of ν. The boundary of H_ν can be split into two paths from u to v, which we call the left path P^l and the right path P^r of H_ν. Namely, P^l (resp. P^r) is the path from u to v while walking clockwise (resp. counterclockwise) on the external boundary of H_ν. Note that, these paths may share some edges. By the definition of spirality, we have that $\sigma_\nu \leq n(P^l) + 2$, where $n(P^l)$ is the turn number of P^l. Since we are assuming $\sigma_\nu > 2$, this implies $n(P^l) > 0$. Also, each right turn of P^l is a $270°$ angle in the external face of H_ν, hence it is either a degree-2 vertex or a bend.

Hence, $n(P^l) \leq k+b$ and, consequently, $\sigma_\nu \leq k+b+2$. See for example Fig. 3(a), which depicts an orthogonal representation H_ν of G_ν with spirality $\sigma_\nu = \frac{3}{2}$ and $n(P^l) = 3$. Path P^l has one degree-2 vertex and three bends.   □

Spirality Sets. For each node ν of T, we define a function X_ν that maps each target spirality value σ_ν for a representation of G_ν to a pair (b_ν, H_ν) such that b_ν is the minimum number of bends over all orthogonal representations of G_ν with spirality σ_ν, and H_ν is a b_ν-orthogonal representation of G_ν with spirality σ_ν. We say that b_ν is the *cost* of σ_ν for ν and that H_ν is the *representative* for ν given σ_ν. If the minimum number of bends of an orthogonal representation of G_ν having spirality σ_ν is higher than a target maximum number of bends b, we write $X_\nu(\sigma_\nu) = (\infty, \emptyset)$. The set $\Sigma_\nu = \{(\sigma_\nu, X_\nu(\sigma_\nu)) \mid X_\nu(\sigma_\nu) \neq (\infty, \emptyset)\}$ is called the *spirality set of ν*.

If for some node ν the spirality set Σ_ν is empty, the instance can be safely rejected. Also, by Lemma 2, for each node ν of T, we can assume $|\sigma_\nu| \leq k+b+2$. Hence, we have $|\Sigma_\nu| = O(k+b)$. In [5] it is shown that Σ_ν can be computed in time: (i) $O(|\Sigma_\nu|)$ if ν is a Q*-node; (ii) $O(|\Sigma_\nu|)$ if ν is a P-node and $O(|\Sigma_\nu|^2)$ if ν is an S-node, and if we know the spirality sets of the children of ν. For each pair $(\sigma_\nu, X_\nu(\sigma_\nu))$ in Σ_ν, the representative H_ν for ν is encoded in constant time by suitably linking the representatives for the children of ν and adding a constant-size information that specifies the values of the angles at the poles of ν. By Lemma 2 we can restate the results in [5] as follows.

Lemma 3. *If ν is a Q*-node of T then Σ_ν can be computed in $O(k+b)$ time.*

Lemma 4. *Let ν be a P-node of T with children $\mu_1, ..., \mu_h$ ($h \in \{2,3\}$). Given Σ_{μ_i} for each $i \in \{1, ..., h\}$, there is an $O(k+b)$-time algorithm that computes Σ_ν.*

Lemma 5. *Let ν be an S-node of T with children μ_1 and μ_2. Given Σ_{μ_1} and Σ_{μ_2}, there is an $O((k+b)^2)$-time algorithm that computes Σ_ν.*

Let ρ be the root of T and let H be an orthogonal representation of G where G_ρ (i.e., the reference chain) is on the external face. Without loss of generality, we assume that G_ρ is to the right of H, i.e., for each simple cycle containing the end-vertices u and v of G_ρ, visiting this cycle clockwise we encounter u, v, and G_ρ in this order. In any orthogonal representation H of G, the number of right turns minus the number of left turns encountered traversing clockwise any cycle of G in H equals four [4]. Hence, the following lemma immediately holds.

Lemma 6. *Let H be an orthogonal representation of G, let T be the $SPQ*R$-tree rooted at ρ, and let ν be the root child of T. Denoted by $n(H_\rho^{uv})$ the turn number of the chain G_ρ in H, moving from u to v, we have $\sigma(H_\nu) - n(H_\rho^{uv}) = 4$.*

Figure 3b illustrates Lemma 6, where $n(H_\rho^{uv}) = -5$ and $\sigma(H_\nu) = -1$. The next lemma gives the time complexity needed to compute Σ_ν when ν is an R-node.

Lemma 7. *Let ν be an R-node of T and let $\mu_1, \ldots, \mu_d$ be its children that are not Q*-nodes. Given Σ_{μ_i} for each $i \in \{1, \ldots, d\}$, there exists an algorithm that computes Σ_ν in $O(2^{p \cdot \log p}) \cdot n^{O(1)}$ time, with $p = k+b$.*

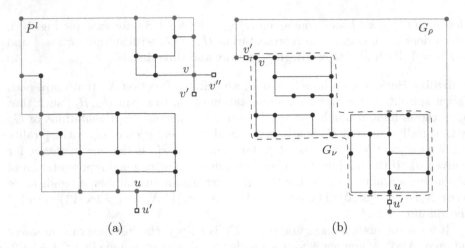

Fig. 3. (a) Illustration for the proof of Lemma 2. (b) Illustration for Lemma 6.

Proof (Sketch). We prove the lemma in the case where G_ν has two alias vertices. The other cases can be proved with similar arguments, see [7]. For every target value of spirality $\sigma_\nu \in [-k - b - 2, k + b + 2]$, we compute $X_\nu(\sigma_\nu)$ as follows. For each of the two possible embeddings of skel(ν), consider the component G_ν enhanced with a dummy edge e, connecting the poles u and v of ν, in such a way that e is on the external face of G_ν and to the right of G_ν. Let G' be the resulting graph. A *valid tuple* is a tuple $(\sigma_1, ..., \sigma_d)$ such that $\sum_{i=1}^{d} b_i \leq b$, where $X_{\mu_i}(\sigma_i) = (b_i, H_i)$ for every $i \in \{1, ..., d\}$. For each valid tuple, we perform the following procedure. Let H_e be an orthogonal representation of e such that $n(H_e^{uv}) = \sigma_\nu - 4$, where H_e^{uv} is the turn number of H_e going from u to v. This means that H_e^{uv} has $\sigma_\nu - 4$ bends, each turning to the left if $\sigma_\nu - 4 < 0$, or each turning to the right if $\sigma_\nu - 4 > 0$. Also, let $J = e \cup G_{\mu_1} \cup \cdots \cup G_{\mu_d}$ and let H_J be an orthogonal representation of the graph J such that: (i) the restriction of H_J to G_{μ_i} coincides with H_{μ_i} for each $i \in \{1, ..., d\}$; and (ii) the restriction of H_J to e coincides with H_e. We compute an H_J-constrained orthogonal representation H' of G' with the minimum number of bends in polynomial time by Lemma 1, if it exists. Consider an SPQ*R-tree T' of G' rooted at the Q*-component ξ representing e. Note that ν is the root child of T'. Hence, by Lemma 6, if H' exists then its restriction H_ν to G_ν has spirality σ_ν. By Lemma 1, H_ν is a bend-minimum orthogonal representation of G_ν having spirality σ_ν and the spirality of the restriction of H_ν to G_{μ_i} is σ_i, for each $i \in \{1, ..., d\}$. Denote by b_ν the number of bends of H_ν. If $b_\nu > b$ for each of the two planar embeddings of skel(ν), then $X_\nu(\sigma_\nu) = (\infty, \emptyset)$ and we do not insert $(\sigma_\nu, X_\nu(\sigma_\nu))$ in the spirality set Σ_ν. Else, for the two embeddings of skel(ν) we retain the representation H_ν of minimum cost b_ν, set $X_\nu(\sigma_\nu) = (b_\nu, H_\nu)$, and insert $(\sigma_\nu, X_\nu(\sigma_\nu))$ in Σ_ν.

Correctness. The correctness of the procedure above follows by these facts: (i) to construct a bend-minimum representation H_ν with spirality σ_ν, we consider all possible combinations of values of spiralities for $G_{\mu_1}, \ldots, G_{\mu_d}$; (ii) thanks to the interchangeability of orthogonal components with the same spirality, for each such combination we aim to construct H_ν so that it contains a minimum-bend representation of each G_{μ_i} with its target value of spirality; (iii) if $b_\nu \leq b$, the spirality determined by H_ν for each child μ of ν that corresponds to a Q*-component is surely in the spirality set of μ; (iv) By Lemma 2 it suffices to test the existence of an H_ν for each target value of spirality $\sigma_\nu \in [-k-b-2, k+b+2]$.

Time-complexity. By Lemma 1, for each tuple the time required for the computation is $O(n^{\frac{7}{4}}\sqrt{\log n})$. By Lemma 2, we consider $O(d^{k+b})$ spirality values. Moreover, we can assume that $d \leq k + b$, otherwise the instance can be rejected. More precisely, since G_{μ_i} ($i \in \{1, \ldots, d\}$) contains at least one cycle, any orthogonal representation of G_{μ_i} requires four 270° (i.e., reflex) angles on the external face, and at most two of these angles can occur at the poles of G_{μ_i}. Since each reflex angle that does not occur at a pole of G_{μ_i} requires either a degree-2 vertex or a bend, we necessarily have $d \leq k + b$ if H_ν exists. Hence, there are $O(d^{k+b}) = O((k+b)^{k+b}) = O(2^{(k+b)\cdot\log(k+b)})$ valid tuples.   □

We can now prove the main result of this section.

Theorem 1. *Let G be a biconnected graph with k degree-2 vertices, let b be an integer, and let $p = k + b$. There exists an $O(2^{p \cdot \log p}) \cdot n^{O(1)}$-time algorithm that tests whether G admits a b-orthogonal representation, and that computes one with the minimum number of bends in the positive case.*

Proof. If G is a simple cycle the test is trivial. Otherwise, let T be the SPQ*R-tree of G. For each Q*-node ρ of T, root T at ρ. Recall that this rooted tree describes all planar embeddings of G with the reference chain G_ρ on the external face. We perform a post-order visit of T. We first compute Σ_ν for each leaf ν of T, that is, for each Q*-node distinct from ρ. We can do this in $O(n(k+b))$ time by Lemma 3. During this visit of T, for every internal non-root node ν of T the algorithm computes Σ_ν by using the spirality sets of the children of ν, exploiting Lemmas 4, 5 and 7, depending on whether ν is a P-, an S-, or an R-node, respectively. If the spirality set Σ_ν of ν is empty, then G does not have a b-orthogonal representation with the given reference edge G_ρ on the external face. In this case the algorithm stops visiting T rooted at ρ, and starts visiting T rooted at another Q*-node. Suppose that the algorithm reaches the root child ν of T, computes the spirality set Σ_ν, and this spirality set is not empty. Denote by u and v the poles of ν. Then, the algorithm considers each pair $(\sigma_\nu, X_\nu(\sigma_\nu) = (b_\nu, H_\nu))$ in the spirality set of ν. Lemma 6 implies that G admits an orthogonal representation H with G_ρ on the external face whose restriction to G_ν in H has spirality σ_ν if and only if the restriction to G_ρ in H has an orthogonal representation H_ρ such that $n(H_\rho^{uv}) = \sigma_\nu - 4$. Let n_ρ be the number of vertices of G_ρ different from u and v. Since G_ρ is a chain of edges, we have that such H_ρ has $b_\rho = 0$ bends if $n_\rho \geq |\sigma_\nu - 4|$; otherwise it has $b_\rho = |\sigma_\nu - 4| - n_\rho$ bends and we just check if $b_\nu + b_\rho \leq b$. By the reasoning above, G admits a b-orthogonal representation H if and only if such a test is positive for some σ_ν. Also, if this is true, we can construct H by attaching H_ν to H_ρ, and since for each node ν we have used the representation H_ν with minimum number of bends among those with the same spirality, H has the minimum number of bends over all b-orthogonal representations of G.

By Lemmas 4, 5 and 7 for each node we spend $O(2^{p \cdot \log p}) \cdot n^{O(1)}$ time, and the condition of Lemma 6 at the root level can be tested in $O(n)$ time. Since T has $O(n)$ nodes and G has $O(n)$ rooted SPQ*R-trees, the statement follows. □

4   The General Case

We assume first that G is connected but not biconnected; we then consider non-connected graphs in the proof of Theorem 2, the proof of which is in [7]. A biconnected component of G is also called a *block*. There are two main difficulties in extending the result of Theorem 1 when G is not biconnected. In terms of drawability, some angle constraints may be required at the cutvertices of the input graph G. Namely, one cannot simply test the representability of each single block independently, as it might be impossible to merge the orthogonal representations of the different blocks into an orthogonal representation of G without additional angle constraints at the cutvertices. In terms of vertex-degree, G might also have degree-1 vertices and, more importantly, a vertex of degree three or four in G can be a degree-2 vertex in a block; hence, the sum of the degree-2 vertices in all blocks can be larger than the number of degree-2 vertices in G. To manage angle constraints at the cutvertices, we adopt a variant of a strategy used in a previous work on series-parallel graphs [9], and extend it to also consider rigid components. About the number of degree-2 vertices, similarly to [6], we extend the parameter k to include both vertices of degree one and two, and we observe that each block of G must have $O(k + b)$ degree-2 vertices if G admits a b-orthogonal representation.

Block-Cutvertex Tree. Let G be a connected graph having k vertices of degree at most two, and let b be a non-negative integer. To deal with the possible planar embeddings of G, we use the *BC-tree (block-cutvertex tree)* of G, and combine it with the SPQ*R-trees of its blocks. The BC-tree T of G describes the decomposition of G in terms of its blocks and cutvertices; refer to Fig. 4. Each node of T represents either a block or a cutvertex of G. A *block-node* (resp. a *cutvertex-node*) of T represents a block (resp. a cutvertex) of G. There is an edge between two nodes of T if one of them represents a cutvertex and the other represents a block that contains the cutvertex. If $B_1, \ldots, B_q$ are the blocks of G ($q \geq 2$), we denote by $\beta(B_i)$ the block-node of T corresponding to B_i ($1 \leq i \leq q$) and by T_{B_i} the tree T rooted at $\beta(B_i)$. For a cutvertex c of G, we denote by $\chi(c)$ the node of T that corresponds to c. Each T_{B_i} describes a class of planar embeddings of G such that, for each non-root node $\beta(B_j)$ ($1 \leq j \leq q$) with parent node $\chi(c)$ and grandparent node $\beta(B_k)$, the cutvertex

c and B_k lie on the external face of B_j. An *orthogonal representation of G with respect to $\mathcal{T}_{B_i}$* is an orthogonal representation whose planar embedding belongs to the class described by $\mathcal{T}_{B_i}$. Therefore, to test whether G admits a b-orthogonal representation we have to test if G admits a b-orthogonal representation with respect to $\mathcal{T}_{B_i}$ for some $i \in \{1, \ldots, q\}$.

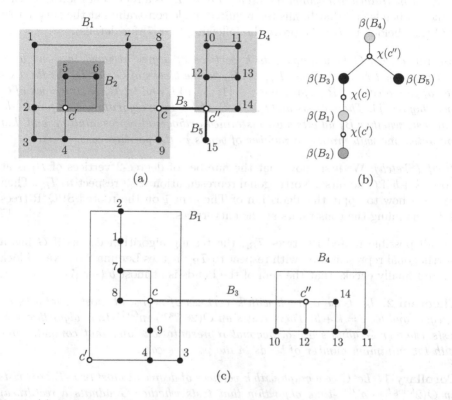

(a)   (b)

(c)

Fig. 4. (a) An orthogonal representation H of a 1-connected graph and its blocks. (b) The BC-tree $\mathcal{T}_{B_4}$; the embedding of H is in the class described by $\mathcal{T}_{B_4}$. (c) Orthogonal representations of B_1 and B_4 that cannot be glued planarly.

Testing Algorithm. For any fixed $i \in \{1, \ldots, q\}$, testing whether G has a b-orthogonal representation with respect to $\mathcal{T}_{B_i}$, requires to compute a bend-minimum representation for each block B_j subject to angle constraints at the cutvertices of B_j, so that we can merge the representation of B_j with those of its adjacent blocks. For example, consider the orthogonal representation H in Fig. 4(a), whose embedding is in the class described by the rooted BC-tree $\mathcal{T}_{B_4}$ of Fig. 4(b). H has blocks $B_1, \ldots B_5$ (where B_3 and B_5 are trivial blocks) and cutvertices c, c', and c''. Note that the representation of B_1 must have c on its external face with a flat angle, so to accommodate the representation of B_3. Also, c' must have a reflex angle in the representation of B_1, so to accommodate the representation of B_2. Conversely, the two orthogonal representations in Fig. 4(c) have embeddings that do not adhere to any rooted version of $\mathcal{T}$, and they cannot be merged, because neither c nor c'' lie on the external face of their components. A detailed description of all types of angle constraints that may be needed at the cutvertices of a block is given in [7].

A *b-constrained orthogonal representation* of B_j is a representation H_{B_j} with at most b bends and that fulfills the required angle constraints at the cutvertices of G that belong to B_j. We prove the following (see [7] for details).

Lemma 8. *Let G be a graph with k vertices of degree at most two and let b be a non-negative integer. Let $B_1, \ldots, B_q$ be the blocks of G, let $\mathcal{T}_{B_i}$ be the BC-tree of G rooted at $\beta(B_i)$ (for some $i \in \{1, \ldots, q\}$), and let B_j be any block of G (possibly $i = j$). There exists an $O(2^{p \cdot \log p}) \cdot n^{O(1)}$-time algorithm, with $p = k + b$, that tests whether B_j admits a b-constrained orthogonal representation, and that computes one with minimum number of bends in the positive case.*

Proof (Sketch). We first show that the number of degree-2 vertices of B_j is at most $2k + b$ if G admits a b-orthogonal representation with respect to $\mathcal{T}_{B_i}$. Then we show how to apply the algorithm of Theorem 1 on the rooted SPQ*R-trees of B_j, handling the constraints at the cutvertices. □

For all possible rooted BC-trees $\mathcal{T}_{B_i}$, the testing algorithm checks if G has a b-orthogonal representation with respect to $\mathcal{T}_{B_i}$: it uses Lemma 8 on every block B_j and finally checks that the sum of the bends is at most b (see [7]).

Theorem 2. *Let G be a graph with k vertices of degree at most 2, let b be an integer, and let $p = k + b$. There exists an $O(2^{p \cdot \log p}) \cdot n^{O(1)}$-time algorithm that tests whether G admits a b-orthogonal representation, and that computes one with the minimum number of bends in the positive case.*

Corollary 1. *Let G be a graph with k vertices of degree at most two. There exists an $O(2^{k \cdot \log k}) \cdot n^{O(1)}$-time algorithm that tests whether G admits a rectilinear representation, and that computes one in the positive case.*

References

1. Biedl, T.C., Kant, G.: A better heuristic for orthogonal graph drawings. Comput. Geom. **9**(3), 159–180 (1998). https://doi.org/10.1016/S0925-7721(97)00026-6
2. Chaplick, S., Di Giacomo, E., Frati, F., Ganian, R., Raftopoulou, C.N., Simonov, K.: Parameterized algorithms for upward planarity. In: Goaoc, X., Kerber, M. (eds.) 38th International Symposium on Computational Geometry, SoCG 2022, Berlin, Germany, 7–10 June 2022. LIPIcs, vol. 224, pp. 26:1–26:16. Schloss Dagstuhl - Leibniz-Zentrum für Informatik (2022). https://doi.org/10.4230/LIPIcs.SoCG.2022.26
3. Cornelsen, S., Karrenbauer, A.: Accelerated bend minimization. J. Graph Algorithms Appl. **16**(3), 635–650 (2012). https://doi.org/10.7155/jgaa.00265

4. Di Battista, G., Eades, P., Tamassia, R., Tollis, I.G.: Graph Drawing: Algorithms for the Visualization of Graphs. Prentice-Hall (1999)
5. Di Battista, G., Liotta, G., Vargiu, F.: Spirality and optimal orthogonal drawings. SIAM J. Comput. **27**(6), 1764–1811 (1998)
6. Di Giacomo, E., Liotta, G., Montecchiani, F.: Orthogonal planarity testing of bounded treewidth graphs. J. Comput. Syst. Sci. **125**, 129–148 (2022)
7. Didimo, W., Giacomo, E.D., Liotta, G., Montecchiani, F., Ortali, G.: On the parameterized complexity of bend-minimum orthogonal planarity. CoRR abs/2308.13665 (2023). https://doi.org/10.48550/arXiv.2308.13665
8. Didimo, W., Giordano, F., Liotta, G.: Upward spirality and upward planarity testing. SIAM J. Discret. Math. **23**(4), 1842–1899 (2009)
9. Didimo, W., Kaufmann, M., Liotta, G., Ortali, G.: Rectilinear planarity of partial 2-trees. In: Angelini, P., von Hanxleden, R. (eds.) Graph Drawing and Network Visualization - 30th International Symposium, GD 2022. LNCS, vol. 13764, pp. 157–172. Springer, Cham (2022). https://doi.org/10.1007/978-3-031-22203-0_12
10. Didimo, W., Kaufmann, M., Liotta, G., Ortali, G.: Computing bend-minimum orthogonal drawings of plane series-parallel graphs in linear time. Algorithmica (2023). https://doi.org/10.1007/s00453-023-01110-6
11. Didimo, W., Liotta, G.: Computing orthogonal drawings in a variable embedding setting. In: Chwa, K.-Y., Ibarra, O.H. (eds.) ISAAC 1998. LNCS, vol. 1533, pp. 80–89. Springer, Heidelberg (1998). https://doi.org/10.1007/3-540-49381-6_10
12. Didimo, W., Liotta, G., Ortali, G., Patrignani, M.: Optimal orthogonal drawings of planar 3-graphs in linear time. In: Chawla, S. (ed.) SODA 2020, pp. 806–825. SIAM (2020). https://doi.org/10.1137/1.9781611975994.49
13. Didimo, W., Liotta, G., Patrignani, M.: HV-planarity: algorithms and complexity. J. Comput. Syst. Sci. **99**, 72–90 (2019). https://doi.org/10.1016/j.jcss.2018.08.003
14. Frati, F.: Planar rectilinear drawings of outerplanar graphs in linear time. Comput. Geom. **103**, 101854 (2022)
15. Ganian, R., Montecchiani, F., Nöllenburg, M., Zehavi, M.: Parameterized complexity in graph drawing (Dagstuhl Seminar 21293). Dagstuhl Rep. **11**(6), 82–123 (2021). https://doi.org/10.4230/DagRep.11.6.82
16. Garg, A., Tamassia, R.: A new minimum cost flow algorithm with applications to graph drawing. In: North, S. (ed.) GD 1996. LNCS, vol. 1190, pp. 201–216. Springer, Heidelberg (1997). https://doi.org/10.1007/3-540-62495-3_49
17. Garg, A., Tamassia, R.: On the computational complexity of upward and rectilinear planarity testing. SIAM J. Comput. **31**(2), 601–625 (2001). https://doi.org/10.1137/S0097539794277123
18. Gutwenger, C., Mutzel, P.: A linear time implementation of SPQR-trees. In: Marks, J. (ed.) GD 2000. LNCS, vol. 1984, pp. 77–90. Springer, Heidelberg (2001). https://doi.org/10.1007/3-540-44541-2_8
19. Hopcroft, J.E., Tarjan, R.E.: Dividing a graph into triconnected components. SIAM J. Comput. **2**(3), 135–158 (1973). https://doi.org/10.1137/0202012
20. Rahman, M.S., Nakano, S., Nishizeki, T.: A linear algorithm for bend-optimal orthogonal drawings of triconnected cubic plane graphs. J. Graph Algorithms Appl. **3**(4), 31–62 (1999). http://www.cs.brown.edu/publications/jgaa/accepted/99/SaidurNakanoNishizeki99.3.4.pdf
21. Tamassia, R.: On embedding a graph in the grid with the minimum number of bends. SIAM J. Comput. **16**(3), 421–444 (1987). https://doi.org/10.1137/0216030

Fixed-Parameter Algorithms
for Computing RAC Drawings of Graphs

Cornelius Brand[ID], Robert Ganian[✉][ID], Sebastian Röder, and Florian Schager

Algorithms and Complexity Group, TU Wien (Vienna University of Technology),
Vienna, Austria
{cbrand,rganian}@ac.tuwien.ac.at, sebastian.roeder@student.tuwien.ac.at,
florian.schager@tuwien.ac.at

Abstract. In a right-angle crossing (RAC) drawing of a graph, each edge is represented as a polyline and edge crossings must occur at an angle of exactly $90°$, where the number of bends on such polylines is typically restricted in some way. While structural and topological properties of RAC drawings have been the focus of extensive research, little was known about the boundaries of tractability for computing such drawings. In this paper, we initiate the study of RAC drawings from the viewpoint of parameterized complexity. In particular, we establish that computing a RAC drawing of an input graph G with at most b bends (or determining that none exists) is fixed-parameter tractable parameterized by either the feedback edge number of G, or b plus the vertex cover number of G.

Keywords: RAC drawings · fixed-parameter tractability · vertex cover number · feedback edge number

1 Introduction

Today we have access to a wealth of approaches and tools that can be used to draw planar graphs, including, e.g., Fáry's Theorem [29] which guarantees the existence of a planar straight-line drawing for every planar graph and the classical algorithm of Fraysseix, Pach and Pollack [28] that allows us to obtain straight-line planar drawings on an integer grid of quadratic size. However, much less is known about the kinds of drawings that can be achieved for non-planar graphs. The study of combinatorial and algorithmic aspects of such drawings lies at the heart of a research direction informally referred to as "beyond planarity" (see, e.g., the relevant survey and book chapter [18,21]).

An obvious goal when attempting to visualize non-planar graphs would be to obtain a drawing which minimizes the total number of crossings. This question is widely studied within the context of the crossing number of graphs, and while obtaining such a drawing is NP-hard [33] it is known to be fixed-parameter tractable when parameterized by the total number of crossings required thanks to a seminal result of Grohe [34]. However, research over the past twenty years has shown that drawings which minimize the total number of crossings are not necessarily optimal in terms of human readability. Indeed, the topological and geometric properties of such drawings may have a significantly larger impact than the total number of crossings, as was observed, e.g., by the initial informal experiment of Mutzel [41] and the pioneering set of user experiments carried out by the graph drawing research lab at the University of Sydney [36–38]. The latter works demonstrated that "large-angle drawings" (where edge crossings have larger angles) are significantly easier to read than drawings where crossings occur at acute angles.

Motivated by these findings, in 2011 Didimo, Eades, and Liotta investigated graph drawings where edge crossings are only permitted at 90° angles [20] (see Fig. 1 for an illustration). Today, these *right-angle crossing* (or *RAC*) drawings are among the best known and most widely studied beyond-planar drawing styles [18,21], with the bulk of the research to date focusing on understanding necessary and sufficient conditions for the existence of such drawings as well as the space they require [1–4,8,16,17,27]. A prominent theme in the context of RAC drawings concerns the number of times edges are allowed to be bent: it has been shown that every graph admits a RAC drawing if each edge can be bent 3 times [20], and past works have considered straight-line RAC drawings as well as RAC drawings where the number of bends per edge is limited to 1 or 2.

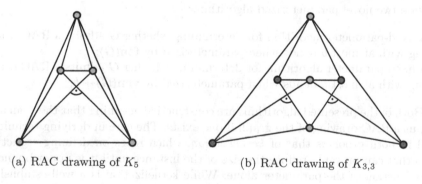

(a) RAC drawing of K_5 (b) RAC drawing of $K_{3,3}$

Fig. 1. Examples of RAC drawings.

And yet, in spite of the considerable body of work concentrating on combinatorial and topological properties of such drawings, so far almost nothing is known about the complexity of computing a RAC drawing of a given graph. Indeed, while the problem of determining whether a graph admits a straight-line RAC drawing is NP-hard [4] and was recently shown to be ∃R-complete [44], there is a surprising lack of known algorithms that can compute such drawings for special classes of graphs or, more generally, parameterized algorithms that exploit quantifiable properties of the input graph to guarantee the tractability of computing RAC drawings (either without or with limited bends). This gap in our understanding starkly contrasts the situation for so-called 1-planar drawings—another prominent beyond-planar drawing style for which a number of fixed-parameter algorithms are known [6, 24, 25]—as well as recent advances mapping the boundaries of tractability for other graph drawing problems [9, 10, 35].

Contribution.   We initiate an investigation of the parameterized complexity of determining whether a graph G admits a RAC drawing. Given the well-motivated focus of previous works on limiting the amount of bends in such drawings, an obvious first choice for a parameterization would be to consider an upper bound b on the total number of bends permitted in the drawing. However, on its own such a parameter cannot suffice to achieve fixed-parameter tractability in view of the NP-hardness of the problem for $b = 0$, i.e., for straight-line RAC drawings.

Hence, we turn towards identifying structural parameters of G that guarantee fixed-parameter RAC drawing algorithms. While established decompositional parameters such as treewidth [43] and clique-width [14] represent natural choices of parameterizations for purely combinatorial problems, the applicability of these parameters in solving graph drawing problems is complicated by the inherent difficulty of performing dynamic programming when the task is to obtain a drawing of the graph. This is why the parameters often used in this setting are non-decompositional, with the most notable examples being the *vertex cover number* **vcn** (i.e., the size of a minimum vertex cover) and the *feedback edge number* **fen** (i.e., the edge deletion distance to acyclicity); further details are available in the overview of related work below. As our main contributions, we provide two novel parameterized algorithms:

1. a fixed-parameter algorithm for determining whether G admits a RAC drawing with at most b bends when parameterized by **fen**(G);
2. a fixed-parameter algorithm for determining whether G admits a RAC drawing with at most b bends when parameterized by **vcn**$(G) + b$;

Both of the presented algorithms are constructive, meaning that they can also output a RAC drawing of the graph if one exists. The core underlying technique used in both proofs is that of *kernelization*, which relies on defining reduction rules that can provably reduce the size of the instance until it is upper-bounded by a function of the parameter alone. While kernelization is a well-established and generic technique, its use here requires non-trivial insights into the structural properties of optimal solutions in order to carefully identify parts of the graph which can be simplified without impacting the final outcome.

We prove that both algorithms in fact hold for the more general case where each edge is marked with an upper bound on the number of bends it can support, allowing us to capture the previously studied 1- and 2-bend RAC drawings. Moreover, we show that the latter algorithm can be lifted to establish fixed-parameter tractability when parameterized by b plus the *neighborhood diversity* (i.e., the number of maximal modules) of G [30, 39, 40]. In the concluding remarks, we also discuss possible extensions towards more general parameterizations and apparent obstacles on the way to such results.

Related Work.   Didimo, Eades and Liotta initiated the study of RAC drawings by analyzing the interplay between the number of bends per edge and the total number of edges [20]. Follow-up works also considered extensions and variants of the initial concept, such as upward RAC drawings [3], 2-layer RAC drawings [16,17] and 1-planar RAC drawings [8]. More recent works investigated the existence of RAC drawings for bounded-degree graphs [2], and RAC drawings with at most one bend per edge [1]. It is known that every graph admits a RAC drawing with at most three bends per edge [20], and that determining whether a graph admits a RAC drawing with zero bends per edge is NP-hard [4].

The vertex cover number has been used as a structural graph parameter to tackle a range of difficult problems in graph drawing as well as other areas. Fixed-parameter algorithms for drawing problems based on the vertex cover number are known for, e.g., computing the obstacle number of a graph [5], computing the stack and queue numbers of graphs [9,10], computing the crossing number of a graph [35] and 1-planarity testing [6]. Similarly, the feedback edge number (sometimes called the *cyclomatic number*) has been used to tackle problems which are not known to be tractable w.r.t. treewidth, including 1-planarity testing [6] and the EDGE DISJOINT PATHS problem [32] (see also Table 1 in [31]).

These two parameterizations are incomparable: there are problems which remain NP-hard on graphs of constant vertex cover number while being FPT when parameterized by the feedback edge number (such as EDGE DISJOINT PATHS [26,32]), and vice-versa. That being said, the existence of a fixed-parameter algorithm parameterized by the feedback edge number is open for a number of graph drawing problems that are known to be FPT w.r.t. the vertex cover number; examples include computing the aforementioned stack, queue and obstacle numbers.

2   Preliminaries

We assume familiarity with standard concepts in graph theory [22], and with basic parameterized complexity theory [15,23]. All graphs considered in this manuscript are assumed to be simple and undirected.

The *feedback edge number* of a graph G, denoted by $\mathbf{fen}(G)$, is the size of a minimum edge set F such that $G - F$ is acyclic; it is well-known that such a set F (and hence also the feedback edge number) can be computed in linear time. The *vertex cover number* of G, denoted $\mathbf{vcn}(G)$, is the size of a minimum vertex cover of G, i.e., of a minimum set X such that $G - X$ is edgeless. Such a minimum set X can be computed in time $\mathcal{O}(1.2738^{|X|} + |X| \cdot |V(G)|)$ [13], and a vertex cover of size at most $2|X|$ can be computed in linear time by a trivial approximation algorithm. The third structural parameter considered here is the *neighborhood diversity* $\mathbf{nd}(G)$ of G, which is the minimum size of a partition $\mathcal{P}$ of $V(G)$ such that for each pair a, b in the same part of $\mathcal{P}$, it holds that $N(a) \setminus \{b\} = N(b) \setminus \{a\}$ where $N(a)$ and $N(b)$ are the open neighborhoods of a and b, respectively. It is known that each part in such a partition $\mathcal{P}$ must be a clique or an independent set, and such a minimum partition can be computed in polynomial time [40].

RAC Drawings.   Given a graph $G = (V, E)$ on n vertices with m edges, a *drawing* of G is a mapping δ that takes vertices V to points in the Euclidean plane $\mathbb{R}^2$, and assigns to every edge $e = uv \in E$ the image of a simple plane curve $[0,1] \to \mathbb{R}^2$ connecting the points $\delta(u), \delta(v)$ corresponding to u and v. We require that δ is injective on V, and furthermore that for all vertices v and edges e not incident to v, the point $\delta(v)$ is not contained in $\text{int}(\delta(e))$, where $\text{int}(\delta(e))$ is the image of $(0,1)$ under δ.

A *polyline drawing* of G is a drawing such that for each edge $e \in E$, $\delta(e)$ can be written as a union $\delta(e) = \lambda_1^e \cup \cdots \cup \lambda_t^e$ of closed straight-line segments $\lambda_1^e, \ldots, \lambda_t^e$

such that (1) for each $1 \le i \le t-1$, the segments λ_i^e and λ_{i+1}^e intersect in precisely one of their shared end-points and moreover close an angle different than $180°$, and (2) every other pair of segments is disjoint. The shared intersection points between consecutive segments are called the *bends* of e in the drawing δ.

For two edges e and f, their set of *crossings* in the drawing δ is the set $\text{int}(\delta(e)) \cap \text{int}(\delta(f))$. We will assume without loss of generality that any drawing δ of G has a finite number of crossings.

The central type of drawing studied in this paper are those that allow only *right-angle crossings* between edge drawings (so-called *RAC drawings*): We say that the edges $e, f \in E$ have a *right-angle crossing* in a polyline drawing δ of G if the crossing lies in the relative interiors of the respective line segments defining $\delta(e)$ and $\delta(f)$, and most crucially, the intersecting line segments of $\delta(e)$ and $\delta(f)$ are orthogonal to each other (i.e., they meet at a right angle). Let δ be a polyline-drawing of a graph, $\beta : E \mapsto \{0,1,2,3\}$ a mapping, and $b \in \mathbb{N}$ a number. If every crossing of δ is a right-angle crossing, the number of bends counted over *all* edges is at most b, and every edge itself has at most $\beta(e)$ bends, δ is called a *b-bend β-restricted RAC drawing* of G. We note that

- 0-bend RAC drawings are straight-line RAC drawings (for any choice of β),
- m and $2m$-bend drawings with $\beta(e) = 1$ or $\beta(e) = 2$ for each edge e gives the usual notion of 1-bend and 2-bend RAC drawings, respectively, and
- similarly, $3m$-bend drawings with $\beta(e) = 3$ for each edge e gives rise to the notion of 3-bend RAC drawings, which exist for every graph [20].

Based on the above, we can now formally define our problem of interest:

BEND-RESTRICTED RAC DRAWING (BRAC)
Input: A graph G, an integer $b \ge 0$, and an edge-labelling $\beta : E \mapsto \{0,1,2,3\}$.
Question: Does G admit a b-bend β-restricted RAC drawing?

It has been shown that b-BEND β-RESTRICTED RAC DRAWING is $\exists\mathbb{R}$-complete [11,44] even when restricted to the case where $b = 0$. Without loss of generality, we will assume that the input graph G is connected. We remark that while BRAC is defined as a decision problem, every algorithm provided in this paper is constructive and can output a drawing as a witness for a yes-instance.

3   An Explicit Algorithm for **BRAC**

As already pointed out above, our results for fixed-parameter tractability come as kernels. While there is a generic formal equivalence between the existence of a kernel and a decidable problem being fixed-parameter tractable, this doesn't by itself yield explicit bounds on the running time of the algorithm that results from this generic strategy. In order to derive concrete upper bounds on the running time of our algorithms, we provide an algorithm that solves *b-bend β-restricted RAC drawing* with a specific running time bound. We do so via a combination of branching and an encoding in the existential theory of the reals.

Theorem 1. *An instance* (G, b, β) *of* **BRAC** *can be solved in time* $m^{\mathcal{O}(m^2)}$, *where m is the number of edges of* G.

Proof (Sketch). We begin with a branching step in which we exhaustively consider all possible allocations of the bends to edges. In each branch, we alter the graph G by subdividing each edge precisely the number of times it is assumed to be bent in that branch. At this point, it remains to decide whether this new graph G' admits a straight-line RAC drawing. To do this, one can construct a sentence in the existential theory of the reals that is true if and only if G' admits such a drawing [11]. To conclude the proof, we use the known fact that an existential sentence over the reals in N variables over M polynomials of maximal degree D can be decided in time $(M \cdot D)^{\mathcal{O}(N)}$ (see, e.g., [7, Theorem 13.13]). $\square$

4   A Fixed-Parameter Algorithm via fen(*G*)

We begin our investigation by establishing a kernel for BEND-RESTRICTED RAC DRAWING when parameterized by the feedback edge number. Our kernel is based on the exhaustive application of two reduction rules.

Let us assume we are given an instance (G, b, β) of BRAC and that we have already computed a minimum feedback edge set F of G in linear time. The first reduction rule is trivial: we simply observe that vertices of degree one can always be safely removed since they never hinder the existence of a RAC drawing.

Observation 2. *Let* $v \in V(G)$ *be a vertex with degree one.* $G - \{v\}$ *admits a b-bend β-restricted RAC drawing if and only if G does as well.*

Iteratively applying the reduction rule provided by Observation 2 results in a graph of the form $G' = (V', E' \cup F)$, where $T := (V', E')$ is a tree with at most $2 \cdot \mathbf{fen}(G)$ leaves and where each leaf of T is incident to at least one edge in F. We mark a vertex in T as *special* if it is an endpoint of an edge in F or if it has degree at least 3 in T (see Fig. 2 for an illustration). The total number of special vertices can be upper-bounded by $4 \cdot \mathbf{fen}(G)$.

In order to define the crucial second reduction rule, we will partition the edges of T into edge-disjoint paths such that each special vertex can only appear as an endpoint in such paths.

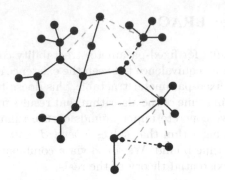 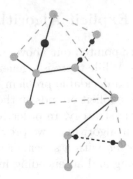

(a) Before removing degree-one vertices.   (b) After removing degree-one vertices.

Fig. 2. Reduction rule one: Degree-one vertices can be pruned. Orange lines represent feedback edges, dashed black lines represent long paths and special vertices are marked in turquoise. (Color figure online)

Definition 3. *We define the* path partition *of T in G' as the unique partition $P_1 \dot{\cup} \cdots \dot{\cup} P_\ell = E'$ such that all P_i are pairwise edge-disjoint paths in T whose endpoints are both special vertices, but with no special vertices in their interior. We call ℓ the size of the path partition.*

An illustration is provided in Fig. 3. Given the established bound on the number of special vertices, the size of the path partition is bounded by $4 \cdot \mathbf{fen}(G)$.

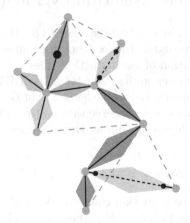

Fig. 3. Path partition of T with feedback edges in orange. (Color figure online)

At this point, let us assume that we have a path partition $P_1 \dot{\cup} P_2 \dot{\cup} \cdots \dot{\cup} P_\ell$ of T in G', where we index the paths in increasing order of length. Our next task is to divide these paths into short and long paths by identifying whether there exists a large gap in the lengths of these paths.

Definition 4. *Define $p_i := |P_i|$ for $i = 1, \ldots, \ell$, and moreover define $P_0 := F$ and $p_0 := |F|$. Let i_0 be the minimal $i = 1, \ldots, \ell$ such that $p_i > 9\ell \cdot p_{i-1}$, if one such i exists, otherwise we set $i_0 := \ell$. We call all paths P_i with $1 \leq i \leq i_0$ short and all other paths long. Then we define the subgraph G_{short} as the edge-induced subgraph of $\bigcup_{i=0}^{i_0} P_i$ of G' (i.e., G_{short} arises by removing all long paths from G').*

Our aim is now to argue that if δ_{short} is a RAC drawing of G_{short}, then we can always extend δ_{short} to a RAC drawing of G'. Without loss of generality we assume that all vertices in $V(G')$ have already been drawn in δ_{short}. First we create an intermediate drawing δ' of G', which will in general not be a RAC drawing. We define δ' as an extension of δ_{short}, where each long path P with endpoints s and t is represented as a simple straight-line segment from $\delta_{\text{short}}(s)$ to $\delta_{\text{short}}(t)$ with all interior vertices distributed arbitrarily along that line segment. Doing this will in general violate the RAC property of δ', hence in the next step we need to alter this straight-line segment in order to ensure that the drawing of P crosses only at right angles. For this we observe that any vertex on P can be moved to effectively act as a bend in a polyline drawing of P. We show that these "additional bends" can be used to turn all crossings into right-angle crossings.

Lemma 5. *Let P be a long path with endpoints s and t and consider its straight-line representation L in δ'. Assume L intersects k straight-line segments in δ'. Then, there exists a polyline segment $L^\star$ from $\delta'(s)$ to $\delta'(t)$ with at most $3k$ bends that intersects precisely the line segments intersected by L, where each such segment is crossed precisely once and at a right angle.*

Proof (Sketch). The proof considers several ways in which L can intersect a line segment in δ' (such as touching an endpoint of the segment, or crossing through its interior, or containing the line segment). In each of these cases, we show that it is possible to add a small number of bends in the immediate vicinity of the respective segment to create a right-angle crossing with the interior of the segment. An intuitive illustration of the main case is provided in Fig. 4.   □

Lemma 6. *Each long path intersects at most $3\ell \cdot p_{i_0}$ straight-line edge segments in δ'.*

Theorem 7. *b-BEND β-RESTRICTED RAC DRAWING admits a kernel of size at most $(36 \cdot \mathbf{fen}(G))^{4 \cdot \mathbf{fen}(G)}$. The kernel can be constructed in linear time.*

Proof (Sketch). Consider an input (G, b, β) with a feedback edge set F of G. We begin by exhaustively removing all vertices of degree one, as per Observation 2. Next, we construct a path partition $\mathcal{P} = (P_1, \ldots, P_\ell)$ of the tree $T = (V', E')$ in G' of size at most $4 \cdot \mathbf{fen}(G)$. Then we split the paths in $\mathcal{P}$ into short and long paths, and we construct the subgraph G_{short} of G' by removing all long paths from G'. To conclude the proof, it suffices to establish that G_{short} is a kernel, and in particular that (G, b, β) is a yes-instance of BRAC if and only if $(G_{\text{short}}, b, \beta')$ is as well (where β' is a restriction of β to the edges in G_{short}). This can be shown by applying Lemmas 5 and 6.   □

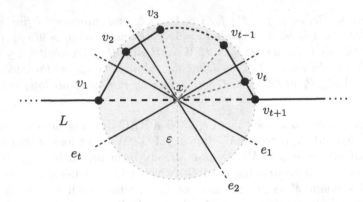

Fig. 4. An illustration of how bends can be used to deal with the situation where multiple straight-line segments in δ' intersect at a single point.

Using Theorem 7, the runtime guarantee given by Theorem 1 and the fact that a feedback edge set of size $\mathbf{fen}(G)$ can be computed in linear time, we obtain:

Corollary 8. *b-BEND β-RESTRICTED RAC DRAWING   is   fixed-parameter tractable parameterized by* $\mathbf{fen}(G)$, *and in particular can be solved in time* $2^{\mathbf{fen}(G)^{\mathcal{O}(\mathbf{fen}(G))}} + \mathcal{O}(|V(G)|)$.

5   Fixed-Parameter Tractability via vcn(G)

As in Sect. 4, the core tool used to establish fixed-parameter tractability for this parameterization is a kernelization procedure, although the ideas and reduction rules used here are very different. Let us assume we are given an instance (G, b, β) of BRAC; as our first step, we compute a vertex cover C of size $k \leq 2 \cdot \mathbf{vcn}(G)$ using the standard approximation algorithm.

We now partition the vertices of our instance G outside of the vertex cover C into *types*, as follows. Two vertices in $G \setminus C$ are of the same type if they have the same set of neighbors in C; observe that the property of "being in the same type" is an equivalence relation, and when convenient we also use the term *type* to refer to the equivalence classes of this relation. To avoid any confusion, we explicitly remark that two vertices may have the same type even when their incident edges are assigned different values by β. The number of types is upper-bounded by 2^k.

We distinguish types by the number of neighbors in C; an illustration is provided in Fig. 5. Let a *member* of a type T be defined as a vertex in T as well as its incident edges. By an exhaustive application of the first reduction rule introduced in Sect. 4 (cf. Observation 2), we may assume that there is no type with less than 2 neighbors in C.

Turning to types with at least 3 neighbors in C, we provide a bound on the size of each such type in a yes-instance of BRAC.

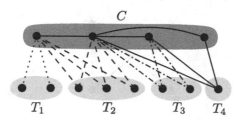

Fig. 5. A graph split into its vertex cover C (in turquoise) and its different types $T_1, \ldots, T_4$ (in orange). (Color figure online)

Lemma 9. *If (G, b, β) is a yes-instance of* BRAC, *then each type T with $i \geq 3$ neighbors in C has at most* $\max(2, 7 - i) + b$ *members.*

Proof. Didimo, Eades and Liotta showed that no complete bipartite graph $K_{c,d}$ with $c + d > 7$ and $\min(c, d) > 2$ admits a straight-line RAC drawing [19]. Hence, if vertices in T have 3 neighbors in C then a b-bend β-restricted RAC drawing of G can contain at most 4 members of T without bends; otherwise, the drawing of 5 members of T and their 3 neighbors in C would contradict the first sentence. Similarly, if vertices in T have at least 5 neighbors in C then a b-bend β-restricted RAC drawing of G cannot contain 3 members of T without bends. $\square$

Lemma 9 implies that we can immediately reject instances containing types with more than 3 neighbors whose cardinality is greater than $4 + b$ (or, for the purposes of kernelization, one may replace these with trivial no-instances). Hence, it now remains to deal with types with precisely two neighbors in C.

Lemma 10. *Consider a b-bend β-restricted RAC drawing δ of G, and let T be a type containing vertices with precisely two neighbors in C. Let T' be the subset of T containing all members of T which do not have bends in δ. Then T' contains at most four members involved in crossings with other members of T' in δ.*

Proof (Sketch). Let u and v be the neighbors of T' in the vertex cover C. Let us consider the vertices lying on one side of the half plane induced by the line $\overleftrightarrow{\delta(u)\delta(v)}$ going through u and v. According to Thales's theorem, every right-angle crossing formed by two edges originating in u and v respectively, has to lie on the semicircle with diameter $\overline{\delta(u)\delta(v)}$. Suppose the edges (u, x) and (v, w) cross at a right angle. Then there cannot be another edge incident to u which crosses the semicircle to the right of the first crossing (see Fig. 6). $\square$

Next, we use the above statement to obtain a bound on the total number of crossings that such a type T can be involved in. We do so by showing that the members of T which themselves do not have bends are only involved in a bounded number of crossings.

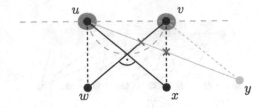

Fig. 6. Illustration for Lemma 10.

Lemma 11. *Consider a b-bend β-restricted RAC drawing δ of G, and let T be a type containing vertices with precisely two neighbors in C. Then at most $3k + 6 + b$ members of T can be involved in a crossing in δ.*

Proof (Sketch). Let T' be the subset of T containing all members of T which do not have bends in $δ$, and let $γ = |T| - |T'|$. Further, let T_0 be the set of members of T' which are pairwise crossing free, but which all cross at least some other edge in $δ$. T_0 forms a layering structure in $δ$, as depicted in Fig. 7. Moreover, if T_0 contains two members that are incident to the same inner face in this layering structure and whose edges are drawn in parallel in $δ$, we remove one of these members from T_0; observe that this may only reduce the size of T_0 by one. Let $α$ be the number of members that remain in T_0 at this point. To complete the proof, it suffices to use the fact that no straight-line segment may cross more than one edge in this layering structure in order to upper-bound $α$ by $3k + 1 + b - γ$, and apply Lemma 10 to bound the number of members in T' involved in crossings.   □

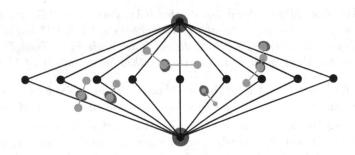

Fig. 7. Illustration for the proof of Lemma 11 with vertices in the vertex cover marked in turquoise.

In particular, Lemma 11 implies that in a *b*-bend *β*-restricted RAC drawing, every sufficiently large type T with precisely 2 neighbors in C must contain a member that is not involved in any crossings. The next lemma highlights why this is useful in the context of our kernelization.

Lemma 12. *Let T be a type with two neighbors in C and assume that G admits a b-bend β-restricted RAC drawing δ. If there is a member in T whose edges are drawn without crossings in δ, then the graph obtained from G by adding a vertex w' to T admits a b-bend β-restricted RAC drawing as well.*

At this point, we have all the ingredients for the main result of this section:

Theorem 13. b-BEND β-RESTRICTED RAC DRAWING *admits a kernel of size $\mathcal{O}(b \cdot 2^k)$, where k is the size of a provided vertex cover of the input graph.*

Proof. Consider an input (G, b, β) and let C be the provided vertex cover of G. We apply the simple reduction rule of deleting vertices of degree 1 from G, resulting in an instance where each type has either 2 or at least 3 neighbors in C. For each type of the latter kind, we check if it contains more members than $\max(3, 7 - i) + b$; if yes, we reject (or, equivalently, replace the instance with a trivial constant-size no-instance), and this is correct by Lemma 9. Moreover, for each type T with precisely 2 neighbors in C containing more than $3k + 6 + b + 1$ many members, we delete members from T until its size is precisely $3k+6+b+1$— the correctness of this step follows from Lemma 11 and 12.

In the resulting graph, each of the at most 2^k many types with at least 3 neighbors in C has size at most $b + 4$, while each of the at most k^2 types with precisely 2 neighbors has size at most $3k + 6 + b + 1$. The kernel bound follows. □

From Theorem 13, the runtime bound given by Theorem 1 and the fact that a vertex cover of size at most $2 \cdot \mathbf{vcn}(G)$ can be obtained in linear time, we obtain:

Corollary 14. b-BEND β-RESTRICTED RAC DRAWING *is fixed-parameter tractable parameterized by $b + \mathbf{vcn}(G)$, and in particular can be solved in time $2^{2^{\mathcal{O}(\mathbf{vcn}(G)+\log b)}} + \mathcal{O}(|V(G)|)$.*

As our final contribution, we show that the above result also implies fixed-parameter tractability with respect to *neighborhood diversity*. This is made possible by the following lemma.

Lemma 15. *Let G be a b-bend RAC drawable graph with a neighborhood diversity $\mathbf{nd}(G)$. Then $\mathbf{vcn}(G) \leq 5 \cdot \mathbf{nd}(G) + b$.*

6   Concluding Remarks

We have established the fixed-parameter tractability of b-BEND β-RESTRICTED RAC DRAWING when parameterized by the feedback edge number $\mathbf{fen}(G)$, or by the vertex cover number $\mathbf{vcn}(G)$ plus an upper bound b on the total number of bends. We have also shown that the latter result implies the fixed-parameter tractability of the problem w.r.t. the neighborhood diversity $\mathbf{nd}(G)$ plus b.

A next step in the computational study of RAC Drawings would be to consider whether the problem is fixed-parameter tractable w.r.t. $\mathbf{vcn}(G)$ alone. Interestingly, a reduction rule for degree-2 vertices without a bound on b is the main obstacle towards obtaining such a fixed-parameter algorithm, and dealing with this case seems to be required if one wishes to generalize the result towards fixed-parameter tractability w.r.t. *treedepth* [42] plus b. A different question one may ask is whether the fixed-parameter algorithm w.r.t. $\mathbf{fen}(G)$ can be generalized towards the recently introduced parameter *slim tree-cut width* [31], which can be equivalently seen as a local version of the feedback edge number [12]. A natural long-term goal within this research direction is then to obtain an understanding of the complexity of BRAC w.r.t. treewidth [43]. Last but not least, it would be interesting to see whether our fixed-parameter tractability results can be strengthened by obtaining polynomial kernels for the same parameterizations.

Acknowledgments. The authors graciously accept support from the WWTF (Project ICT22-029) and the FWF (Project Y1329) science funds.

References

1. Angelini, P., Bekos, M.A., Förster, H., Kaufmann, M.: On RAC drawings of graphs with one bend per edge. Theor. Comput. Sci. **828–829**, 42–54 (2020). https://doi.org/10.1016/j.tcs.2020.04.018
2. Angelini, P., Bekos, M.A., Katheder, J., Kaufmann, M., Pfister, M.: RAC drawings of graphs with low degree. In: Szeider, S., Ganian, R., Silva, A. (eds.) 47th International Symposium on Mathematical Foundations of Computer Science, MFCS 2022, August 22–26, 2022, Vienna, Austria. LIPIcs, vol. 241, pp. 11:1–11:15. Schloss Dagstuhl - Leibniz-Zentrum für Informatik (2022). https://doi.org/10.4230/LIPIcs.MFCS.2022.11
3. Angelini, P., et al.: On the perspectives opened by right angle crossing drawings. J. Graph Algorithms Appl. **15**(1), 53–78 (2011). https://doi.org/10.7155/jgaa.00217
4. Argyriou, E.N., Bekos, M.A., Symvonis, A.: The straight-line RAC drawing problem is np-hard. J. Graph Algorithms Appl. **16**(2), 569–597 (2012). https://doi.org/10.7155/jgaa.00274
5. Balko, M., et al.: Bounding and computing obstacle numbers of graphs. In: Chechik, S., Navarro, G., Rotenberg, E., Herman, G. (eds.) 30th Annual European Symposium on Algorithms, ESA 2022, September 5–9, 2022, Berlin/Potsdam, Germany. LIPIcs, vol. 244, pp. 11:1–11:13. Schloss Dagstuhl - Leibniz-Zentrum für Informatik (2022). https://doi.org/10.4230/LIPIcs.ESA.2022.11
6. Bannister, M.J., Cabello, S., Eppstein, D.: Parameterized complexity of 1-planarity. J. Graph Algorithms Appl. **22**(1), 23–49 (2018). https://doi.org/10.7155/jgaa.00457
7. Basu, S., Pollack, R., Roy, M.F.: Algorithms in Real Algebraic geometry, Algorithms and Computation in Mathematics, vol. 10. Springer, Cham (2006). https://doi.org/10.1007/3-540-33099-2, http://link.springer.com/10.1007/3-540-33099-2
8. Bekos, M.A., Didimo, W., Liotta, G., Mehrabi, S., Montecchiani, F.: On RAC drawings of 1-planar graphs. Theor. Comput. Sci. **689**, 48–57 (2017). https://doi.org/10.1016/j.tcs.2017.05.039

9. Bhore, S., Ganian, R., Montecchiani, F., Nöllenburg, M.: Parameterized algorithms for book embedding problems. J. Graph Algorithms Appl. **24**(4), 603–620 (2020). https://doi.org/10.7155/jgaa.00526

10. Bhore, S., Ganian, R., Montecchiani, F., Nöllenburg, M.: Parameterized algorithms for queue layouts. J. Graph Algorithms Appl. **26**(3), 335–352 (2022). https://doi.org/10.7155/jgaa.00597

11. Bieker, N.: Complexity of graph drawing problems in relation to the existential theory of the reals. Ph.D. thesis, Bachelor's thesis, Karlsruhe Institute of Technology (August 2020) (2020)

12. Brand, C., Ceylan, E., Ganian, R., Hatschka, C., Korchemna, V.: Edge-cut width: An algorithmically driven analogue of treewidth based on edge cuts. In: Bekos, M.A., Kaufmann, M. (eds.) Graph-Theoretic Concepts in Computer Science - 48th International Workshop, WG 2022, Tübingen, Germany, June 22–24, 2022, Revised Selected Papers. LNCS, vol. 13453, pp. 98–113. Springer, Cham (2022). https://doi.org/10.1007/978-3-031-15914-5_8

13. Chen, J., Kanj, I.A., Xia, G.: Improved upper bounds for vertex cover. Theor. Comput. Sci. **411**(40–42), 3736–3756 (2010). https://doi.org/10.1016/j.tcs.2010.06.026

14. Courcelle, B., Makowsky, J.A., Rotics, U.: Linear time solvable optimization problems on graphs of bounded clique-width. Theory Comput. Syst. **33**(2), 125–150 (2000). https://doi.org/10.1007/s002249910009

15. Cygan, M., et al.: Parameterized Algorithms. 1st edn. Springer Publishing Company, Inc., Berlin (2015). https://doi.org/10.1007/978-3-319-21275-3

16. Di Giacomo, E., Didimo, W., Eades, P., Liotta, G.: 2-layer right angle crossing drawings. Algorithmica **68**(4), 954–997 (2014). https://doi.org/10.1007/s00453-012-9706-7

17. Di Giacomo, E., Didimo, W., Grilli, L., Liotta, G., Romeo, S.A.: Heuristics for the maximum 2-layer RAC subgraph problem. Comput. J. **58**(5), 1085–1098 (2015). https://doi.org/10.1093/comjnl/bxu017

18. Didimo, W.: Right angle crossing drawings of graphs. In: Hong, S.-H., Tokuyama, T. (eds.) Beyond Planar Graphs, pp. 149–169. Springer, Singapore (2020). https://doi.org/10.1007/978-981-15-6533-5_9

19. Didimo, W., Eades, P., Liotta, G.: A characterization of complete bipartite RAC graphs. Inf. Process. Lett. **110**(16), 687–691 (2010). https://doi.org/10.1016/j.ipl.2010.05.023

20. Didimo, W., Eades, P., Liotta, G.: Drawing graphs with right angle crossings. Theoret. Comput. Sci. **412**(39), 5156–5166 (2011). https://doi.org/10.1016/j.tcs.2011.05.025

21. Didimo, W., Liotta, G., Montecchiani, F.: A survey on graph drawing beyond planarity. ACM Comput. Surv. **52**(1), 4:1-4:37 (2019). https://doi.org/10.1145/3301281

22. Diestel, R.: Graph Theory. 5th Edn., Graduate Texts in Mathematics, vol. 173. Springer, Cham (2017). https://doi.org/10.1007/978-3-662-53622-3

23. Downey, R.G., Fellows, M.R.: Fundamentals of Parameterized Complexity. Texts in Computer Science, Springer, London (2013). https://doi.org/10.1007/978-1-4471-5559-1

24. Eiben, E., Ganian, R., Hamm, T., Klute, F., Nöllenburg, M.: Extending nearly complete 1-planar drawings in polynomial time. In: Esparza, J., Král', D. (eds.) 45th International Symposium on Mathematical Foundations of Computer Science, MFCS 2020, August 24–28, 2020, Prague, Czech Republic. LIPIcs, vol. 170, pp.

31:1–31:16. Schloss Dagstuhl - Leibniz-Zentrum für Informatik (2020). https://doi.org/10.4230/LIPIcs.MFCS.2020.31

25. Eiben, E., Ganian, R., Hamm, T., Klute, F., Nöllenburg, M.: Extending partial 1-planar drawings. In: Czumaj, A., Dawar, A., Merelli, E. (eds.) 47th International Colloquium on Automata, Languages, and Programming, ICALP 2020, July 8–11, 2020, Saarbrücken, Germany (Virtual Conference). LIPIcs, vol. 168, pp. 43:1–43:19. Schloss Dagstuhl - Leibniz-Zentrum für Informatik (2020). https://doi.org/10.4230/LIPIcs.ICALP.2020.43

26. Fleszar, K., Mnich, M., Spoerhase, J.: New algorithms for maximum disjoint paths based on tree-likeness. Math. Program. **171**(1–2), 433–461 (2018). https://doi.org/10.1007/s10107-017-1199-3

27. Förster, H., Kaufmann, M.: On compact RAC drawings. In: Grandoni, F., Herman, G., Sanders, P. (eds.) 28th Annual European Symposium on Algorithms, ESA 2020, September 7–9, 2020, Pisa, Italy (Virtual Conference). LIPIcs, vol. 173, pp. 53:1–53:21. Schloss Dagstuhl - Leibniz-Zentrum für Informatik (2020). https://doi.org/10.4230/LIPIcs.ESA.2020.53

28. de Fraysseix, H., Pach, J., Pollack, R.: Small sets supporting fáry embeddings of planar graphs. In: Simon, J. (ed.) Proceedings of the 20th Annual ACM Symposium on Theory of Computing, May 2–4, 1988, Chicago, Illinois, USA, pp. 426–433. ACM (1988). https://doi.org/10.1145/62212.62254

29. Fáry, I.: On straight lines representation of planar graphs. Acta Sci. Math. (Szeged) **11**, 229–233 (1948)

30. Ganian, R.: Using neighborhood diversity to solve hard problems. CoRR abs/1201.3091 (2012). https://doi.org/10.48550/arXiv.1201.3091

31. Ganian, R., Korchemna, V.: Slim tree-cut width. In: Dell, H., Nederlof, J. (eds.) 17th International Symposium on Parameterized and Exact Computation, IPEC 2022, September 7–9, 2022, Potsdam, Germany. LIPIcs, vol. 249, pp. 15:1–15:18. Schloss Dagstuhl - Leibniz-Zentrum für Informatik (2022). https://doi.org/10.4230/LIPIcs.IPEC.2022.15

32. Ganian, R., Ordyniak, S.: The power of cut-based parameters for computing edge-disjoint paths. Algorithmica **83**(2), 726–752 (2021). https://doi.org/10.1007/s00453-020-00772-w

33. Garey, M.R., Johnson, D.S.: Crossing number is np-complete. SIAM J. Algebraic Discret. Methods **4**(3), 312–316 (1983). https://doi.org/10.1137/0604033

34. Grohe, M.: Computing crossing numbers in quadratic time. J. Comput. Syst. Sci. **68**(2), 285–302 (2004). https://doi.org/10.1016/j.jcss.2003.07.008

35. Hlinený, P., Sankaran, A.: Exact crossing number parameterized by vertex cover. In: Archambault, D., Tóth, C.D. (eds.) Graph Drawing and Network Visualization - 27th International Symposium, GD 2019, Prague, Czech Republic, September 17–20, 2019, Proceedings. LNCS, vol. 11904, pp. 307–319. Springer (2019). https://doi.org/10.1007/978-3-030-35802-0_24

36. Huang, W.: Using eye tracking to investigate graph layout effects. In: Hong, S., Ma, K. (eds.) APVIS 2007, 6th International Asia-Pacific Symposium on Visualization 2007, Sydney, Australia, 5–7 February 2007, pp. 97–100. IEEE Computer Society (2007). https://doi.org/10.1109/APVIS.2007.329282

37. Huang, W., Eades, P., Hong, S.: Larger crossing angles make graphs easier to read. J. Vis. Lang. Comput. **25**(4), 452–465 (2014). https://doi.org/10.1016/j.jvlc.2014.03.001

38. Huang, W., Hong, S., Eades, P.: Effects of crossing angles. In: IEEE VGTC Pacific Visualization Symposium 2008, PacificVis 2008, Kyoto, Japan, March 5–7, 2008, pp. 41–46. IEEE Computer Society (2008). https://doi.org/10.1109/PACIFICVIS. 2008.4475457

39. Knop, D., Koutecký, M., Masařík, T., Toufar, T.: Simplified algorithmic metatheorems beyond MSO: treewidth and neighborhood diversity. Log. Methods Comput. Sci. **15**(4), 1–32 (2019). https://doi.org/10.23638/LMCS-15(4:12)2019

40. Lampis, M.: Algorithmic meta-theorems for restrictions of treewidth. In: de Berg, M., Meyer, U. (eds.) Algorithms - ESA 2010, 18th Annual European Symposium, Liverpool, UK, September 6–8, 2010. Proceedings, Part I. LNCS, vol. 6346, pp. 549–560. Springer, Cham (2010). https://doi.org/10.1007/978-3-642-15775-2_47

41. Mutzel, P.: An alternative method to crossing minimization on hierarchical graphs. SIAM J. Optim. **11**(4), 1065–1080 (2001). https://doi.org/10.1137/S1052623498334013

42. Nešetřil, J., Ossona de Mendez, P.: Sparsity: Graphs, Structures, and Algorithms, Algorithms and Combinatorics, vol. 28. Springer, Heidelberg (2015). https://doi.org/10.1007/978-3-642-27875-4

43. Robertson, N., Seymour, P.D.: Graph minors. III. Planar Tree-width. J. Comb. Theory, Ser. B. **36**(1), 49–64 (1984). https://doi.org/10.1016/0095-8956(84)90013-3

44. Schaefer, M.: RAC-drawability is ∃ℝ-complete. In: Graph Drawing and Network Visualization: 29th International Symposium, GD 2021, Tübingen, Germany, September 14–17, 2021, Revised Selected Papers, pp. 72–86. Springer-Verlag, Heidelberg (2021). https://doi.org/10.1007/978-3-030-92931-2_5

Parameterized Complexity
of Simultaneous Planarity

Simon D. Fink⬡, Matthias Pfretzschner^(✉)⬡, and Ignaz Rutter⬡

Fakultät für Informatik und Mathematik, Universität Passau, Passau, Germany
{finksim,pfretzschner,rutter}@fim.uni-passau.de

Abstract. Given k input graphs $G^{①}, \ldots, G^{⑥}$, where each pair $G^{①}, G^{②}$ with $i \neq j$ shares the same graph G, the problem SIMULTANEOUS EMBEDDING WITH FIXED EDGES (SEFE) asks whether there exists a planar drawing for each input graph such that all drawings coincide on G. While SEFE is still open for the case of two input graphs, the problem is NP-complete for $k \geq 3$ [18].

In this work, we explore the parameterized complexity of SEFE. We show that SEFE is FPT with respect to k plus the vertex cover number or the feedback edge set number of the union graph $G^{\cup} = G^{①} \cup \cdots \cup G^{⑥}$. Regarding the shared graph G, we show that SEFE is NP-complete, even if G is a tree with maximum degree 4. Together with a known NP-hardness reduction [1], this allows us to conclude that several parameters of G, including the maximum degree, the maximum number of degree-1 neighbors, the vertex cover number, and the number of cutvertices are intractable. We also settle the tractability of all pairs of these parameters. We give FPT algorithms for the vertex cover number plus either of the first two parameters and for the number of cutvertices plus the maximum degree, whereas we prove all remaining combinations to be intractable.

Keywords: Simultaneous Planarity · SEFE · Parameterized Complexity

1 Introduction

Let $G^{①} = (V, E^{①}), \ldots, G^{⑥} = (V, E^{⑥})$ denote k graphs on the same vertex set, where each pair $(G^{①}, G^{②})$ has a common *shared graph* $G^{①} \cap G^{②} = (V, E^{①} \cap E^{②})$. The problem SIMULTANEOUS EMBEDDING WITH FIXED EDGES (SEFE) asks whether there exist planar drawings $\Gamma^{①}, \ldots, \Gamma^{⑥}$ of $G^{①}, \ldots, G^{⑥}$, respectively, such that each pair $\Gamma^{①}, \Gamma^{②}$ induces the same drawing on the shared graph $G^{①} \cap G^{②}$ [10]. We refer to such a k-tuple of drawings as a *simultaneous drawing*. Unless stated otherwise, we assume that k is part of the input and not necessarily a constant. In this work, we focus on the restricted *sunflower case* of SEFE,

Funded by the Deutsche Forschungsgemeinschaft (German Research Foundation, DFG) under grant RU-1903/3-1.

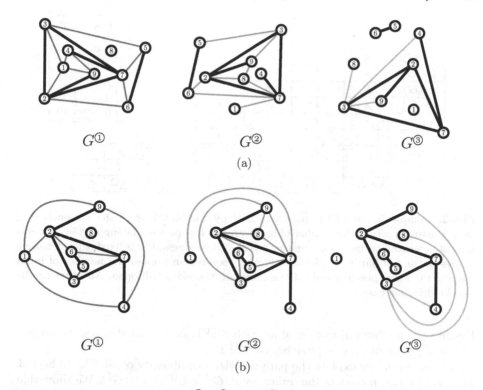

Fig. 1. (a) Three planar graphs $G^{\textcircled{1}}$, $G^{\textcircled{2}}$, and $G^{\textcircled{3}}$ with the shared graph G highlighted in black. (b) Planar drawings of the same three graphs, but the subgraph G is drawn the same way in all three drawings. (Color figure online)

where every pair of input graphs has the same shared graph G. For more than two input graphs, SEFE has been proven to be NP-complete [13], even in the sunflower case [18]. For two input graphs, however, SEFE remains open.

One of the main applications for the SEFE problem is dynamic graph drawing. Given a graph that changes over time, a visualization of k individual snapshots of the graph should aesthetically display the changes between successive snapshots. To this end, it is helpful to draw unchanged parts of the graph consistently. Figure 1 gives an example illustrating this for $k = 3$ snapshots. Using the same layout for the shared graph in Fig. 1b notably simplifies recognizing similarities and differences when compared to the varying layouts in Fig. 1a.

In recent years, SEFE received much attention and many algorithms solving restricted cases have been developed [6,17]. Most notably, this includes the cases where every connected component of G is biconnected [5] or has a fixed embedding [7], and the case where G has maximum degree 3 [5]. Very recently, Fulek and Tóth [12] solved SEFE for $k = 2$ input graphs if the shared graph is connected and Bläsius et al. [4] later improved the running time from $O(n^{16})$ to quadratic using a reduction to the problem SYNCHRONIZED PLANARITY. For $k \geq 3$ input graphs, however, the same restricted case remains NP-complete [1].

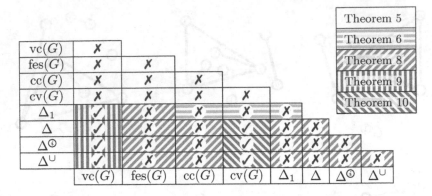

Fig. 2. Complexity of SEFE parameterized by combinations of the parameters of Sect. 4, assuming that the number of input graphs k is part of the input. Check marks indicate tractability and crosses indicate para-NP-hardness. If k is fixed, then Theorem 8 does not imply hardness, as the reduction requires an unbounded number of input graphs; the corresponding cells of the table are therefore still open. All other results also hold in this case.

Despite the plethora of work dealing with SEFE in restricted cases, we are not aware of parameterized approaches to SEFE.

In this work, we explore the parameterized complexity of SEFE. In Sect. 3, we consider parameters of the *union graph* $G^{\cup} = G^{①} \cup \cdots \cup G^{⑧}$. We show that SEFE is FPT with respect to k plus the vertex cover number or the feedback edge set number of $G^{\cup}$. Additionally, we prove intractability for SEFE parameterized by the twin cover number of $G^{\cup}$. In Sect. 4, we turn to parameters of the shared graph G. We consider as parameters the vertex cover number $\mathrm{vc}(G)$, the feedback edge set number $\mathrm{fes}(G)$, the number of connected components $\mathrm{cc}(G)$, the number of cutvertices $\mathrm{cv}(G)$, the maximum number of degree-1 neighbors Δ_1 in G, the maximum degree Δ of G, the maximum degree $\Delta^{①}$ among all input graphs, and the maximum degree $\Delta^{\cup}$ of the union graph. For the latter four parameters, note that $\Delta_1 \leq \Delta \leq \Delta^{①} \leq \Delta^{\cup}$. Figure 2 gives an overview over these parameters together with their individual and pairwise tractabilities. In Sect. 4.1, we show that SEFE is NP-complete even if the shared graph is a tree with maximum degree 4. This allows us to conclude that most parameters and their combinations are intractable. The only exceptions are due to the FPT-algorithms from Sect. 4.2 for $\mathrm{vc}(G) + \Delta_1$ and from Sect. 4.3 for $\mathrm{cv}(G) + \Delta$. Due to space constraints, proofs for statements marked with a star are deferred to the full version [11].

2   Preliminaries

For $k \in \mathbb{N}$, we define $[k] := \{1, \ldots, k\}$. Let $G = (V, E)$ be a simple graph. The *open neighborhood* $N_G(v)$ of a vertex $v \in V$ denotes the set of vertices adjacent to v in G. The *closed neighborhood* $N_G[v] := N_G(v) \cup \{v\}$ additionally contains v.

If the graph is clear from context, we simply write $N(v)$ and $N[v]$, respectively. An *induced subgraph* $H = (V', E')$ of G contains all edges $E' \subseteq E$ with both endpoints in a given set $V' \subseteq V$. The graph G is *connected*, if, for every pair $u, v \in V$, there exists a path between u and v in G. A *separating k-set* is a set $S \subseteq V$ with $|S| = k$ such that the graph $G - S$ obtained by removing S is disconnected. The *split components* of a separating k-set S are the maximal subgraphs of G that are not disconnected by removing S. A separating 1-set is also called a *cutvertex*, a separating 2-set is a *separating pair*. We say that G is *biconnected* if it contains no cutvertex and *triconnected* if it contains no separating pair. A maximal induced subgraph of G that is (bi-)connected is called a *(bi-)connected component* of G. A biconnected component is also called a *block*. A *wheel* is a graph that consists of a cycle and an additional vertex adjacent to every vertex of the cycle.

Parameterized Complexity. A parameterized problem $L \subseteq \Sigma^* \times \mathbb{N}$ is *fixed-parameter tractable (FPT)*, if L can be solved in time $f(k) \cdot n^{O(1)}$, where f is some computable function and k is the parameter. The problem L is *para-NP-hard* with respect to k if L is NP-hard even for constant values of k. For a graph $G = (V, E)$, a *vertex cover* is a vertex set $C \subseteq V$ such that every edge $e \in E$ is incident to a vertex in C. The *vertex cover number* $\mathrm{vc}(G)$ is the size of a minimum vertex cover of G. The following lemma states that the vertex cover number of a planar graph gives an upper bound for the number of its high-degree vertices.

Lemma 1 (*). *Let $G = (V, E)$ be a planar graph and let $N_3 \subseteq V$ denote the set of vertices of degree at least 3 in G. Then $|N_3| \leq 3 \cdot \mathrm{vc}(G)$.*

An edge $\{u, v\} \in E$ is a *twin edge* if $N[u] = N[v]$, that is, u and v have the same neighborhood. A set $C \subseteq V$ is a *twin cover* of G if every edge $e \in E$ is a twin edge or incident to a vertex of C. The *twin cover number* of G is the size of a minimum twin cover of G. A *feedback edge set* of G is an edge set $F \subseteq E$ such that $G - F$ is acyclic. The *feedback edge set number* $\mathrm{fes}(G)$ is the minimum size of a feedback edge set of G.

SEFE. The graph $G^\cup = (V, \bigcup_{i \in [k]} E^{\circledcirc})$ consisting of the edges of all input graphs is called the *union graph*. For brevity, we describe instances of SEFE using the union graph by marking every edge of $G^\cup$ with the input graphs it is contained in. Every edge of $G^\cup$ is either contained in exactly one input graph $G^{\circledcirc}$, or in all of them. In the former case, we say that the edge is *$\textcircled{i}$-exclusive*, in the latter case it is *shared*. The connected components of the shared graph are also called *shared components*.

Jünger and Schulz [14] showed that an instance of SEFE admits a simultaneous embedding if and only if there exist embeddings of the input graphs that are *compatible*, i.e., they satisfy the following two requirements: (1) The cyclic ordering of the edges around every vertex of G must be identical in all embeddings. (2) For every pair C and C' of connected components in G, the face of C that C' is embedded in must be the same in all embeddings. We call the former property *consistent edge orderings* and the latter property *consistent relative positions* [17]. Note that any SEFE instance with a non-planar input graph is a no-instance, we thus assume all input graphs to be planar. Furthermore, Bläsius et al. [5] showed that one can assume that the union graph is biconnected.

3   Parameters of the Union Graph

In this section, we study the parameters vertex cover number, feedback edge set number, and twin cover number of the union graph $G^\cup = G^① \cup \cdots \cup G^Ⓚ$. We give an FPT algorithm for each of the former two in combination with k and show the latter to be intractable.

3.1   Vertex Cover Number

For our first parameterization, we consider the vertex cover number of the union graph $vc(G^\cup)$. We use a similar approach as Bhore et al. [2] in their parameterization of the problem BOOK THICKNESS. Let C be a minimum vertex cover of $G^\cup$ of size $\varphi := vc(G^\cup)$ and let k be the number of input graphs. For every vertex $v \in V \setminus C$, note that $N_{G^\cup}(v) \subseteq C$, as otherwise there would be an edge in $G^\cup$ not covered by C. We group the vertices of $V(G^\cup) \setminus C$ into *types* based on their neighborhood in all input graphs, i.e., two vertices v_1 and v_2 are of the same type if and only if $N_{G^①}(v_1) = N_{G^①}(v_2)$ for all $i \in [k]$; see Fig. 3. Let $\mathcal{P}$ denote the partition of $V \setminus C$ based on the types of vertices. Let $\mathcal{P}_{\geq 3} \subseteq P$ be the types of $\mathcal{P}$ whose vertices have degree at least 3 in some input graph $G^①$ and let $\mathcal{P}_{\leq 2} := \mathcal{P} \setminus \mathcal{P}_{\geq 3}$ denote the remaining types that have degree at most 2 in every input graph. Our first goal is to bound the number of vertices in $\mathcal{P}_{\geq 3}$. Since every input graph $G^①$ is planar, we can use Lemma 1 to bound the number of vertices of degree at least 3 in $G^①$ linearly in $vc(G^①) \leq \varphi$. The union graph $G^\cup$ can therefore only contain $O(k\varphi)$ vertices that have degree at least 3 in some exclusive graph. Consequently, we obtain the upper bound $|\bigcup \mathcal{P}_{\geq 3}| \in O(k\varphi)$.

It remains to bound the number of vertices contained in $\mathcal{P}_{\leq 2}$. Since these vertices have degree at most 2 in each input graph, there are $O\left(\binom{\varphi}{2}^k\right)$ distinct types in $\mathcal{P}_{\leq 2}$, i.e., $|\mathcal{P}_{\leq 2}| \in O(\varphi^{2k})$. To bound the number of vertices of each type in $\mathcal{P}_{\leq 2}$, we use the following reduction rule; see Fig. 4.

Reduction Rule 1 (*). *If there exists a type $U \in \mathcal{P}_{\leq 2}$ with $|U| > 1$, pick an arbitrary vertex $v \in U$ and reduce the instance to $(G^\cup - v, k)$.*

After exhaustively applying Rule 1, each type of $\mathcal{P}_{\leq 2}$ contains at most one vertex. Because we have $|\mathcal{P}_{\leq 2}| \in O(\varphi^{2k})$, we thus get $|\bigcup \mathcal{P}_{\leq 2}| \in O(\varphi^{2k})$. Using the upper bound $|\bigcup \mathcal{P}_{\geq 3}| \in O(k\varphi)$ from above, we finally obtain a kernel of size $|V| = |\bigcup \mathcal{P}_{\geq 3}| + |\bigcup \mathcal{P}_{\leq 2}| + |C| \in O(\varphi^{2k})$ that can be solved in FPT time by enumerating all subsets of its vertices. Combined with the fact that a minimum vertex cover of size φ can be computed in time $O(1.2738^\varphi + \varphi n)$ [8], we obtain the following result.

Fig. 3. (a) Vertices u and v of different types having different neighborhoods in the two input graphs. $G^①$ is shown with thin black edges, $G^②$ with thick blue edges, overlapping edges belong to G. (b) Vertices u and v of the same type. (Color figure online)

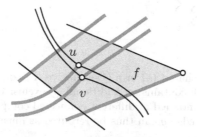

Fig. 4. An example illustrating the proof of Rule 1 for $k = 2$ input graphs with vertices u, v belonging to the same type of $\mathcal{P}_{\leq 2}$. Vertex v can be placed directly next to vertex u in the face f that u is contained in without introducing any crossings in the input graphs.

Theorem 1. *SEFE is FPT with respect to the number of input graphs k plus the vertex cover number φ of the union graph $G^{\cup}$ and admits an $O(\varphi^{2k})$ kernel.*

We note that this result also holds in the non-sunflower case of SEFE.

3.2 Feedback Edge Set Number

In this section, we consider the feedback edge set number $\psi = \mathrm{fes}(G^{\cup})$ of the union graph. We build on ideas of Binucci et al. [3] in their parameterization of the problem STORYPLAN. Given a minimum feedback edge set F of $G^{\cup}$, our goal is to bound the number of vertices of $H := G^{\cup} - F$ using reduction rules. Since F is minimal and we can assume $G^{\cup}$ to be connected, H must be a tree.

Recall that we can even assume the union graph $G^{\cup}$ to be biconnected due to the preprocessing by Bläsius et al. [5]. Since the preprocessing simply decomposes split components around cutvertices of the union graph into independent instances, the feedback edge set number of the graph does not increase. The biconnectivity of $G^{\cup}$ ensures that $G^{\cup}$ does not contain degree-1 vertices and therefore any leaf of H must be incident to an edge of F in $G^{\cup}$. This allows us to bound the number of leaves (and consequently also the number of nodes of degree at least 3) of H linearly in $|F| = \psi$.

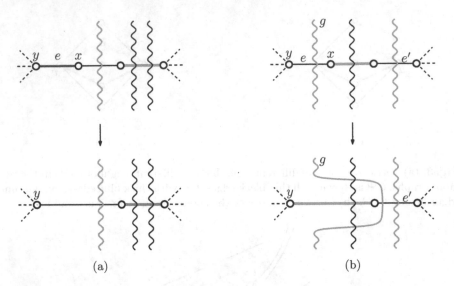

Fig. 5. Contraction of a shared (a) or a ①-exclusive (b) edge e in a path of degree-2 vertices. Because the path contains an additional ①-exclusive edge e' in (b), the endpoints x and y of e are contained in different connected components of all other input graphs. The ②-exclusive edge g can thus be rerouted as illustrated.

It thus only remains to limit the number of degree-2 vertices in H. We do this using a reduction rule that contracts edges in long paths of degree-2 vertices in $G^\cup$. If such a path contains a shared edge or two exclusive edges of the same type, we can shorten the path as outlined in Fig. 5. With this reduction rule, we find a linear kernel for SEFE.

Theorem 2 (*). SEFE *is FPT parameterized by the number of input graphs k plus the feedback edge set number ψ of the union graph $G^\cup$ and admits a kernel of size $O(k\psi)$.*

3.3   Twin Cover Number

Finally, we show that SEFE is para-NP-hard with respect to the twin cover number of the union graph $G^\cup$. To this end, we use the following theorem.

Theorem 3. *When k is part of the input, SEFE is NP-complete, even if the union graph is a complete graph.*

Proof. We start with the instance $G^\cup$ obtained from the reduction by Angelini et al. [1] where the shared graph is a star (the remaining structure of the instance is irrelevant). For every edge e of the complement graph of $G^\cup$ (i.e., for every edge that is missing in $G^\cup$), create a new input graph $G^©$ and add e to $G^©$. Note that $G^©$ only consists of the shared star and the additional edge e. As this does not restrict the rotation of the star, we thus obtain an equivalent instance.   □

Note that, if $G^{\cup}$ is a complete graph, all vertices have the same neighborhood and thus every edge is a twin edge, and consequently $G^{\cup}$ has twin cover number 0.

Theorem 4. *When k is part of the input,* SEFE *is para-NP-hard with respect to the twin cover number of $G^{\cup}$.*

Recall that the FPT algorithms from Sects. 3.1 and 3.2 also require the number of input graphs k as a parameter. In contrast, Theorem 4 only holds for unbounded k, since the reduction requires many input graphs.

4   Parameters of the Shared Graph

We now consider parameters of the shared graph G. In this case, finding safe reduction rules becomes significantly more involved. While we may assume that the union graph is biconnected [5], even isolated vertices or vertices of degree 1 of the shared graph may hold important information due to their connectivity in the union graph as the following result of Angelini et al. [1] shows.

Theorem 5 ([1]). SEFE *is NP-complete for any fixed $k \geq 3$, even if G is a star.*

Since, for a star, the vertex cover number $vc(G)$, the feedback edge set number $fes(G)$, the number of connected components $cc(G)$, and the number of cutvertices $cv(G)$ are all constant, Theorem 5 already implies para-NP-hardness for each of these parameters and for all combinations of them; see Fig. 2. To obtain additional hardness results, one can take the instance resulting from Theorem 5, duplicate every edge of the star, and subdivide each of the new duplication edges once. Since these new edges can be drawn directly next to their original version (or simply be removed from the drawing, if one wants to argue the reverse direction), the resulting instance is equivalent. In this instance, the parameter Δ_1, the maximum number of degree-1 vertices adjacent to a single vertex in G, is zero, and the only cutvertex of G is the center vertex, thus filling additional gaps in Fig. 2. We note that this construction increases $fes(G)$ and $vc(G)$.

Theorem 6. SEFE *is NP-complete for any fixed $k \geq 3$, even if $\Delta_1 = 0$ and $cv(G) = cc(G) = 1$.*

In the remainder of this section, we focus on several degree-related parameters for SEFE. In addition to Δ_1, we consider the maximum degree Δ of the shared graph, the highest maximum degree $\Delta^{\odot}$ among all input graphs, and the maximum degree $\Delta^{\cup}$ of the union graph. Recall that $\Delta_1 \leq \Delta \leq \Delta^{\odot} \leq \Delta^{\cup}$. Theorems 5 and 6 do not prove hardness for the latter three parameters as the shared graph has high degree. To close this gap, we show in Sect. 4.1 that SEFE is NP-complete, even if the shared graph is a tree and the union graph has maximum degree 4. This proves the intractability of many combinations of parameters; see Fig. 2. Finally, we show that SEFE is FPT when parameterized by $vc(G) + \Delta_1$ and by $cv(G) + \Delta$ (Sects. 4.2 and 4.3), which settles the remaining entries of Fig. 2.

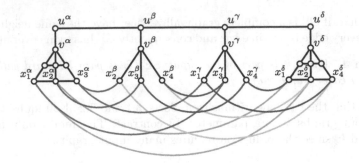

Fig. 6. An example illustrating the reduction of Theorem 7 with $X = \{x_1, x_2, x_3, x_4\}$ and $\mathcal{T} = \{(x_1, x_2, x_3), (x_1, x_2, x_4)\}$. Shared edges are drawn in black, every other color represents one input graph. (Color figure online)

4.1 Maximum Degree

In this section we show that SEFE is NP-complete, even if the instance has bounded degree. Angelini et al. [1] reduce from the NP-complete [15] problem BETWEENNESS, which asks for a linear ordering of a ground set X, subject to a set $\mathcal{T}$ of triplets over X, where each triplet requires an element to appear between two other elements in the ordering. They use the rotation of a star in the shared graph to encode the linear ordering in the SEFE instance, while exclusive edges enforce the triplets. Since this approach leads to a high-degree vertex, we use a different approach. We use small-degree vertices in the shared graph to encode the ordering of every three-element subset of X, while the input graphs ensure compatibility between these orderings. This results in a shared graph with small maximum degree, but the number of input graphs is not constant.

Theorem 7 (*). *When k is part of the input, SEFE is NP-complete, even if the shared graph is connected and has maximum degree 4.*

Proof Sketch. Figure 6 illustrates the SEFE instance resulting from the reduction from BETWEENNESS. For every $\tau \in \binom{X}{3}$, we have a degree-4 vertex v^τ in the shared graph adjacent to vertices representing the three elements of τ. The fourth neighbor u^τ of v^τ acts as a separator that allows us to interpret the circular order of the edges incident to v^τ as a linear order of τ. The embeddings of the shared graph thus correspond to linear orderings of all three-element subsets of X.

It remains to ensure that we only allow linear orderings of these subsets that can be extended to a common linear order of X. For every pair of three-element subsets that share two elements a and b, we add two exclusive edges in a new input graph that ensure that a and b are ordered consistently in the two subsets; see Fig. 6. Since the three-element subsets themselves guarantee transitivity, our SEFE instance thus represents exactly the linear orderings of X. To encode the additional constraint of each triplet in $\mathcal{T}$ into our SEFE instance, we simply fix the vertices of the corresponding three-element subset in a rigid structure in

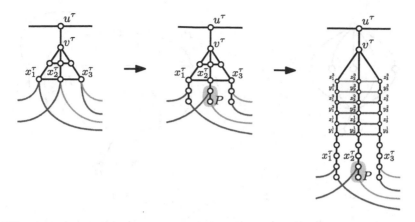

Fig. 7. An example illustrating how the instance obtained from the reduction of Theorem 7 can be modified such that each vertex has degree at most 4 in the union graph (first step) and the shared graph is acyclic (second step).

the shared graph (see triplet (x_1, x_2, x_3) in Fig. 6 for an example). We can show that the resulting SEFE instance is equivalent to the original BETWEENNESS instance and that its shared graph is connected and has maximum degree 4. □

This result can be further improved by eliminating all cycles of the shared graph and bounding the maximum degree of the union graph; see Fig. 7.

Theorem 8 (*). *When k is part of the input, SEFE is NP-complete, even if the shared graph is a tree and the union graph has maximum degree 4.*

4.2 Vertex Cover Number + Maximum Degree-1 Neighbors

Let Δ_1 denote the maximum number of degree-1 vertices adjacent to a single vertex in G and let C denote a minimum vertex cover of G with size $\varphi = \text{vc}(G)$. In this section, we show that SEFE is FPT with respect to $\varphi + \Delta_1$.

Since the shared graph is planar, the number of vertices with degree at least 3 in G is at most 3φ by Lemma 1. Because Δ_1 bounds the number of degree-1 neighbors of each vertex in G, only the number of isolated vertices and degree-2 vertices in the shared graph remains unbounded. Unfortunately, it is difficult to reduce either of these vertex types in the general case. Instead, our goal is to enumerate all suitable embeddings of the shared graph. The SPQR-tree is a data structure of linear size that compactly describes all planar embeddings of a graph by identifying subgraphs which can be flipped and groups of subgraphs that can be arbitrarily permuted [9, 16]. The former are described by so-called *R-nodes*, the latter by *P-nodes* consisting of two *pole* vertices forming a separating pair, their individual split components are the subgraphs that can be arbitrarily permuted; see Fig. 8a. Given the parameter $\varphi + \Delta_1$, the only embedding choices we cannot afford to brute-force involve the P-nodes of G. Consider a set U of

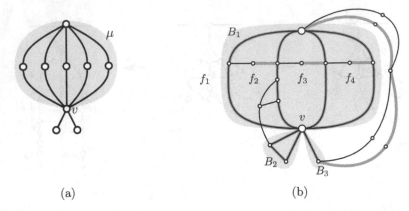

(a)   (b)

Fig. 8. (a) A two-parallel P-node μ with many possible embedding choices at cutvertex v. (b) Three blocks B_1, B_2, and B_3 of the shared graph sharing the same cutvertex v. Block B_2 is a binary block (faces f_2 and f_3) with respect to B_1, while B_3 is a mutable block that can be embedded in all faces.

at least three degree-2 vertices of G that have the same neighborhood $\{u, v\}$. Note that neither of the vertices in U has to be contained in the vertex cover C of G if both u and v are contained in C, thus $|U|$ is not necessarily bounded by a function in $\varphi + \Delta_1$. In the SPQR-tree of G, u and v are the poles of a P-node μ with at least $|U|$ parallel subgraphs and we therefore cannot afford to enumerate all permutations of μ. We call such a P-node *two-parallel*. Even if we fix the embedding of μ, the number of faces incident to its poles is not bounded, leaving too many embedding decisions if one of the poles of μ is a cutvertex in G; see Fig. 8a.

To circumvent these issues, we show that we can limit the number of embedding decisions involving two-parallel P-nodes of G using connectivity information of the union graph, allowing us to brute-force all suitable embeddings of each shared component in FPT-time. Subsequently, we can determine whether $G^{\cup}$ admits a SEFE with the given embeddings using the algorithm by Bläsius and Rutter [7] that solves SEFE if every shared component has a fixed embedding. In other words, we use the bounded search tree technique where the branches of the search tree correspond to embedding decisions in the shared graph. We first branch for all possible embeddings of the blocks in G, and subsequently branch for all possible configurations of blocks around cutvertices.

Theorem 9 (*). *When k is part of the input, SEFE is FPT with respect to the maximum number of degree-1 neighbors Δ_1 and the vertex cover number φ of the shared graph and can be solved in*

$$O(2^{O(\varphi)} \cdot (2(\varphi + \Delta_1))^{(\varphi+\Delta_1)^2 \cdot 3\varphi} \cdot ((\varphi + \Delta_1)!)^{3\varphi} \cdot n^{O(1)}) \ time.$$

Proof Sketch. We give a high-level overview of our strategy. For details, we refer to the full version [11].

As the first step, we embed all blocks of the shared graph G. To this end, we can use the preprocessing steps introduced by Bläsius et al. [5] which essentially guarantee that every P-node of G has a fixed embedding (up to mirroring) due to connectivity information of the union graph and thus (as for R-nodes) only leave a binary embedding choice. Subsequently, we show that the number of P-nodes and R-nodes in G is bounded linearly in φ. We can therefore enumerate all suitable embeddings of all blocks in G using $O(2^\varphi)$ branches.

Given a fixed embedding for all blocks of G, it remains to nest and order the blocks around all cutvertices of G. Let v be a cutvertex and let B_1 and B_2 be two blocks of G containing v. Again, preprocessing steps by Bläsius et al. [5] and the connectivity of the union graph restrict the faces of B_1 that B_2 can potentially be embedded in. If B_2 can be embedded in at most two faces of B_1, we say that B_2 is a *binary block* with respect to B_1, otherwise it is a *mutable block*; see Fig. 8b. First, we branch for each possible assignment of binary blocks to each of their respective admissible faces and subsequently call every face with an assigned block *occupied*. This does not work for mutable blocks as the number of faces they may be embedded in is not necessarily bounded by our parameters. However, we show (roughly speaking) that we can decide whether a mutable block is compatible with a given face f using just connectivity information of the union graph, as long as f is not occupied. If there exists a compatible face that is not occupied, we can assign the block to that face without branching. Otherwise, we successively branch and assign the block to each occupied face, the number of which is bounded by our parameters.

With a fixed nesting of blocks at every cutvertex v of G, it only remains to order blocks that lie in the same face. Since φ bounds the number of vertices of degree at least 3 in G (Lemma 1) and $\varphi + \Delta_1$ bounds the number of blocks in G, we can brute-force all such orderings. In every branch of the search tree, we thus now have a fixed embedding for each component of the shared graph G and can solve the instance using the algorithm by Bläsius and Rutter [7].   □

4.3   Number of Cutvertices + Maximum Degree

In this section, we consider the parameters number of cutvertices $cv(G)$ and maximum degree Δ of the shared graph G. This combination enables us to brute-force all possible orders of incident edges at cutvertices of G, thereby allowing us to effectively treat each biconnected component separately. In the following, we give a brief outline of this approach, details are given in the full version [11].

To solve instances where every connected component of the shared graph is biconnected, Bläsius et al. [5] show that, for a biconnected planar graph G and a set X of vertices of G that share a face in some planar embedding of G, the order of X along any simple cycle in G is fixed up to reversal [5, Lemma 8]. This essentially allows them to decompose the instance at such cycles of the shared graph, as the "interface" the graph provides for the vertices of X is independent of the embedding. We show that the same holds for facial cycles of non-biconnected planar graphs if every cutvertex v has a fixed rotation σ_v. We say that a walk W (i.e., a connected sequence of vertices) of G is *face-embeddable* if there exists a planar embedding of G where W is a facial cycle.

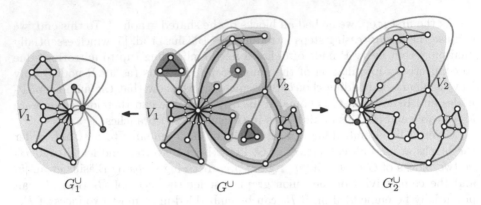

$G_1^{\cup}$ $G^{\cup}$ $G_2^{\cup}$

Fig. 9. Decomposing $G^{\cup}$ at a cutvertex v. The orange edges belong to an extra input graph that enforces the fixed rotation of the cutvertices. (Color figure online)

Lemma 2 (*). *Let G be a planar graph where every cutvertex has a fixed rotation. Let X be a set of vertices that are incident to a common face in some planar embedding of G. Then the order of X in any face-embeddable walk of G containing X is unique up to reversal.*

This allows us to decompose the instance $G^{\cup}$ at cutvertices of the shared graph G; see Fig. 9. We partition the vertices of $G - v$ into two sets V_1 and V_2 such that the edges connecting each of the sets with cutvertex v are consecutive in σ_v and no vertices belonging to the same block of G are separated. We remark that the choice of V_1 and V_2 is in general not unique. Note that the fixed rotation σ_v defines nestings of the split components in V_2. Given this information, we (i) contract in $G^{\cup}$ each set of split components in V_2 that belong to the same nesting (highlighted in red in Fig. 9) into a single vertex that is only connected to v via exclusive edges and (ii) contract every shared component contained in V_2 (highlighted in yellow in Fig. 9) into a single vertex. This yields the graph $G_1^{\cup}$. The construction ensures that all *attachment vertices* of the split components in V_1 that a split component in V_2 is connected to lie on the outer face of the split components in any simultaneous embedding of $G_1^{\cup}$. Symmetrically, we obtain graph $G_2^{\cup}$ by interchanging V_1 and V_2.

To test whether $G^{\cup}$ admits a SEFE, we first test $G_1^{\cup}$ and $G_2^{\cup}$. Note that both graphs contain fewer split components at v, we thus make progress towards biconnectivity. If both graphs admit a SEFE (a necessary condition), we merge their embeddings of the shared graph back together at vertex v and test whether $G^{\cup}$ admits a SEFE with this fixed embedding of the shared graph using the algorithm by Bläsius and Rutter [7]. This is necessary and sufficient as the order of the attachment vertices on the outer face of the split components contained in V_1 (respectively V_2) is the same in any embedding by Lemma 2. Thus, if $G^{\cup}$ admits a simultaneous embedding, it also admits a simultaneous embedding where the shared graph has our chosen embedding.

Theorem 10 (*). *When k is part of the input, SEFE is FPT with respect to the number of cutvertices $cv(G)$ and the maximum degree Δ of G and can be solved in $O((\Delta!)^{cv(G)} \cdot n^{O(1)})$ time.*

5 Conclusion

Our FPT algorithms for the vertex cover number (Sect. 3.1) and the feedback edge set number (Sect. 3.2) of the union graph both additionally require the number of input graphs k as a parameter, raising the question whether similar results can be obtained for unbounded k. In contrast, the reduction showing para-NP-hardness of SEFE with respect to the twin cover number of the union graph (Sect. 3.3) requires k to be unbounded. Is the problem FPT parameterized by the twin cover number plus k?

For the shared graph, we have shown that all eight parameters we considered (and most of their combinations; see Fig. 2) are intractable. This even includes the vertex cover number, one of the strongest graph parameters that upper bounds numerous other metrics like the feedback vertex set number, the treedepth, the pathwidth, and the treewidth. Most of the intractabilities regarding degree-related parameters follow from our hardness reduction for SEFE with maximum degree 4 in Theorem 8. However, Theorem 8 only holds if the number k of input graphs is unbounded, as the reduction inherently requires many input graphs. It is therefore an interesting question whether these parameters become tractable when one additionally considers k as a parameter.

References

1. Angelini, P., Da Lozzo, G., Neuwirth, D.: Advancements on SEFE and partitioned book embedding problems. Theor. Comput. Sci. **575**, 71–89 (2015). https://doi.org/10.1016/j.tcs.2014.11.016
2. Bhore, S., Ganian, R., Montecchiani, F., Nöllenburg, M.: Parameterized algorithms for book embedding problems. J. Graph Algorithms Appl. **24**(4), 603–620 (2020). https://doi.org/10.7155/jgaa.00526
3. Binucci, C., et al.: On the complexity of the storyplan problem. In: Angelini, P., von Hanxleden, R. (eds.) GD 2022. LNCS, vol. 13764, pp. 304–318. Springer, Cham (2022). https://doi.org/10.1007/978-3-031-22203-0_22
4. Bläsius, T., Fink, S.D., Rutter, I.: Synchronized planarity with applications to constrained planarity problems. In: Mutzel, P., Pagh, R., Herman, G. (eds.) Proceedings of the 29th Annual European Symposium on Algorithms (ESA 2021). LIPIcs, vol. 204, pp. 19:1–19:14. Schloss Dagstuhl - Leibniz-Zentrum für Informatik (2021). https://doi.org/10.4230/LIPIcs.ESA.2021.19
5. Bläsius, T., Karrer, A., Rutter, I.: Simultaneous embedding: edge orderings, relative positions, cutvertices. Algorithmica **80**(4), 1214–1277 (2018). https://doi.org/10.1007/s00453-017-0301-9
6. Bläsius, T., Kobourov, S.G., Rutter, I.: Simultaneous embedding of planar graphs. In: Tamassia, R. (ed.) Handbook on Graph Drawing and Visualization, pp. 349–381. Chapman and Hall/CRC (2013)

7. Bläsius, T., Rutter, I.: Disconnectivity and relative positions in simultaneous embeddings. Comput. Geom. **48**(6), 459–478 (2015). https://doi.org/10.1016/j. comgeo.2015.02.002
8. Chen, J., Kanj, I.A., Xia, G.: Improved upper bounds for vertex cover. Theor. Comput. Sci. **411**(40–42), 3736–3756 (2010). https://doi.org/10.1016/j.tcs.2010. 06.026
9. Di Battista, G., Tamassia, R.: On-line maintenance of triconnected components with SPQR-trees. Algorithmica **15**(4), 302–318 (1996). https://doi.org/10.1007/ BF01961541
10. Erten, C., Kobourov, S.G.: Simultaneous embedding of planar graphs with few bends. J. Graph Algorithms Appl. **9**(3), 347–364 (2005). https://doi.org/10.7155/ jgaa.00113
11. Fink, S.D., Pfretzschner, M., Rutter, I.: Parameterized complexity of simultaneous planarity. CoRR abs/2308.11401 (2023). https://doi.org/10.48550/arXiv.2308. 11401
12. Fulek, R., Tóth, C.D.: Atomic embeddability, clustered planarity, and thickenability. In: Chawla, S. (ed.) Proceedings of the 2020 ACM-SIAM Symposium on Discrete Algorithms, (SODA 2020), pp. 2876–2895. SIAM (2020). https://doi.org/ 10.1137/1.9781611975994.175
13. Gassner, E., Jünger, M., Percan, M., Schaefer, M., Schulz, M.: Simultaneous graph embeddings with fixed edges. In: Fomin, F.V. (ed.) WG 2006. LNCS, vol. 4271, pp. 325–335. Springer, Heidelberg (2006). https://doi.org/10.1007/11917496_29
14. Jünger, M., Schulz, M.: Intersection graphs in simultaneous embedding with fixed edges. J. Graph Algorithms Appl. **13**(2), 205–218 (2009). https://doi.org/10.7155/ jgaa.00184
15. Opatrny, J.: Total ordering problem. SIAM J. Comput. **8**(1), 111–114 (1979). https://doi.org/10.1137/0208008
16. Patrignani, M.: Planarity testing and embedding. In: Tamassia, R. (ed.) Handbook on Graph Drawing and Visualization, pp. 1–42. Chapman and Hall/CRC (2013)
17. Rutter, I.: Simultaneous embedding. In: Hong, S.-H., Tokuyama, T. (eds.) Beyond Planar Graphs, pp. 237–265. Springer, Singapore (2020). https://doi.org/10.1007/ 978-981-15-6533-5_13
18. Schaefer, M.: Toward a theory of planarity: Hanani-Tutte and planarity variants. J. Graph Algorithms Appl. **17**(4), 367–440 (2013). https://doi.org/10.7155/jgaa. 00298

The Parametrized Complexity of the Segment Number

Sabine Cornelsen[1]($\boxtimes$)(iD), Giordano Da Lozzo[2](iD), Luca Grilli[3](iD),
Siddharth Gupta[4](iD), Jan Kratochvíl[5](iD), and Alexander Wolff[6](iD)

[1] University of Konstanz, Konstanz, Germany
sabine.cornelsen@uni-konstanz.de
[2] Roma Tre University, Rome, Italy
giordano.dalozzo@uniroma3.it
[3] Università degli Studi di Perugia, Perugia, Italy
luca.grilli@unipg.it
[4] University of Warwick, Coventry, UK
siddharth.gupta.1@warwick.ac.uk
[5] Charles University, Prague, Czech Republic
honza@kam.mff.cuni.cz
[6] Universität Würzburg, Würzburg, Germany

Abstract. Given a straight-line drawing of a graph, a *segment* is a maximal set of edges that form a line segment. Given a planar graph G, the *segment number* of G is the minimum number of segments that can be achieved by any planar straight-line drawing of G. The *line cover number* of G is the minimum number of lines that support all the edges of a planar straight-line drawing of G. Computing the segment number or the line cover number of a planar graph is $\exists\mathbb{R}$-complete and, thus, NP-hard. We study the problem of computing the segment number from the perspective of parameterized complexity. We show that this problem is fixed-parameter tractable with respect to each of the following parameters: the vertex cover number, the segment number, and the line cover number. We also consider colored versions of the segment and the line cover number.

Keywords: Segment number · Line cover number · Vertex cover number · Parameterized complexity · Visual complexity

1 Introduction

A *segment* in a straight-line drawing of a graph is a maximal set of edges that together form a line segment. The *segment number* $\mathrm{seg}(G)$ of a planar graph G is

This research was initiated at the Bertinoro Workshop on Graph Drawing 2023. G.D. was supported, in part, by MUR of Italy (PRIN Project no. 2022ME9Z78 – NextGRAAL and PRIN Project no. 2022TS4Y3N – EXPAND.) L.G. was supported, in part, by the Dipartimento di Ingegneria, Universitá degli Studi di Perugia through grant RICBA21LG. S.G. acknowledges support from Engineering and Physical Sciences Research Council (EPSRC) grant no: EP/V007793/1. J.K. acknowledges support by the Czech Science Foundation through research grant GAČR 23-04949X.

M. A. Bekos and M. Chimani (Eds.): GD 2023, LNCS 14466, pp. 97–113, 2023.
https://doi.org/10.1007/978-3-031-49275-4_7

the minimum number of segments in any planar straight-line drawing of G [10]. The *line cover number* line(G) of a planar graph G is the minimum number of lines that support all the edges of a planar straight-line drawing of G [6]. Clearly, line(G) $\leq$ seg(G) for any graph G [6]. As a side note, we show that seg(G) $\leq$ line^2(G) for any connected graph G. For circular-arc drawings of planar graphs, the *arc number* [28] and *circle cover number* [21] are defined analogously as the segment number and the line cover number, respectively, for straight-line drawings. For an example, see Fig. 1. All these numbers have been considered as meaningful measures of the visual complexity of a drawing of a graph; in particular, for the segment number, a user study has been conducted by Kindermann et al. [20].

 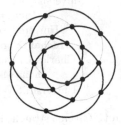

Fig. 1. The dodecahedron graph has 30 edges, segment number 13, line cover number 10, arc number 10, and circle cover number 5 [21].

In general, it is ∃ℝ-complete [22] (and hence NP-hard) to compute the segment number of a planar graph. For the definition of the complexity class ∃ℝ, see [27]. The segment number can, however, be computed efficiently for trees [10], series-parallel graphs of degree at most three [26], subdivisions of outerplanar paths [1], 3-connected cubic planar graphs [10,16], and cacti [14]. Upper and lower bounds for the segment number of various graph classes are known, such as outerplanar graphs, 2-trees, planar 3-trees, 3-connected plane graphs [10], (4-connected) triangulations [11], and triconnected planar 4-regular graphs [14]. Some of the lower and upper bounds are so close that they yield constant-factor approximations, e.g., for outerplanar paths, maximal outerplanar graphs, 2-trees, and planar 3-trees [14]. Segment number and arc number have also been investigated under the restriction that vertices must be placed on a polynomial-size grid [15,19,28]. Also in the setting that a planar graph G comes with a planar embedding (that is, for each face, the cyclic ordering of the edges around it is given and the outer face is fixed), it is NP-hard to compute the segment number of G with respect to the given embedding [12].

This paper focuses on the parametrized complexity of computing the segment number of a graph. A decision problem with input x and parameter $k \in \mathbb{Z}_{\geq 0}$ is *fixed-parameter tractable* (*FPT*) if it can be solved by an algorithm with run time in $\mathcal{O}(f(k) \cdot |x|^c)$ where f is a computable function, $|x|$ is the size of the input x and c is a constant. Given a planar graph G and a parameter $k > 0$, the SEGMENT NUMBER problem is to decide whether seg(G) $\leq k$. This is the *natural parameter* for the problem. By considering additional parameters, we get a more fine-grained picture of the complexity of the problem.

Chaplick et al. [7] showed that computing the line cover number of a planar graph is in FPT with respect to its natural parameter. They observed that, for a given graph G and an integer k, the statement line$(G) \leq k$ can be expressed by a first-order formula about the reals. This observation shows that the problem of deciding whether or not line$(G) \leq k$ lies in $\exists \mathbb{R}$: it reduces in polynomial time to the decision problem for the existential theory of the reals. The algorithm of Chaplick et al. crucially uses the exponential-time decision procedure for the existential theory of the reals by Renegar [23–25] (see Sect. 2). Unfortunately, this procedure does *not* yield a geometric realization. Chaplick et al. even showed that constructing a drawing of a given planar graph that is optimal with respect to the line cover number can be unfeasible since there is a planar graph $G^\star$ [7, Fig. 3b] such that every optimal drawing of $G^\star$ requires irrational coordinates. They argue that any optimal drawing of $G^\star$ contains the Perles configuration. It is known that every realization of the Perles configuration contains a point with an irrational coordinate [3, page 23]. Moreover, any optimal drawing of $G^\star$ admits a cover with ten lines, and each of these lines contains a *single* line segment of the drawing. In other words, every drawing of $G^\star$ that is optimal with respect to the line cover number is also optimal with respect to the segment number.

Our Results. We show that computing the segment number of a graph is FPT with respect to each of the following parameters: the natural parameter, the line cover number (both in Sect. 5), and the *vertex cover number* (in Sect. 4). Recall that the vertex cover number is the minimum number of vertices that have to be removed such that the remaining graph is an independent set. In the parametrized complexity community, the vertex cover number is considered a rather weak graph parameter, but for (geometric) graph drawing problems, FPT results can be challenging to obtain even with respect to the vertex cover number [2,4,5].

We remark that our algorithms use Renegar's decision procedure as a subroutine; hence, when we compute the segment number k of a planar graph G, we do *not* get a straight-line drawing of G that consists of k line segments. Recall, however, that even *specifying* such a drawing is difficult for some graphs such as the above-mentioned graph $G^\star$, which has a point with irrational coordinates in any drawing that is optimal with respect to the segment number.

Motivated by list coloring, we also consider colored versions of the segment and line cover number problems. As a warm up, we provide efficient algorithms for computing the segment number of *banana trees* and *banana cycles*, that is, graphs that can be obtained from a tree or a cycle, respectively, by replacing each edge by a set of paths of length two; see Sect. 3.

Proofs for statements marked with ($\star$) can be found in the full version [9].

2   Preliminaries

In a straight-line drawing of a graph, two incident edges are *aligned* if they are on the same segment. Since the number of segments equals the number of edges minus the number of pairs of aligned edges, we get the following.

Lemma 1. *A planar straight-line drawing has the minimum number of segments if and only if it has the maximum number of aligned edges.*

An *existential first-order formula about the reals* is a formula of the form $\exists x_1 \ldots \exists x_m\ \Phi(x_1, \ldots, x_m)$, where Φ consists of Boolean combinations of order and equality relations between polynomials with rational coefficients over the variables $x_1, \ldots, x_m$. Renegar's result on the existential theory of the reals can be summarized as follows.

Theorem 1 (Renegar [23–25]). *Given any existential sentence Φ about the reals, one can decide whether Φ is true or false in time*

$$(L \log L \log \log L) \cdot (PD)^{O(N)},$$

where N is the number of variables in Φ, P is the number of polynomials in Φ, D is the maximum total degree over the polynomials in Φ, and L is the maximum length of the binary representation over the coefficients of the polynomials in Φ.

The proof of the next lemma follows the approach in [7, Lemma 2.2].

Lemma 2 ($\star$). *Given an n-vertex planar graph G and an integer $k \leq 3n - 6$, there exists a first-order formula Φ about the reals that involves $O(n^2)$ polynomials in $O(n)$ variables (each of constant total degree and with integer coefficients of constant absolute value) such that Φ is satisfiable if and only if $seg(G) \leq k$.*

Lemma 2 and Theorem 1 immediately imply the following.

Corollary 1. *Given an n-vertex planar graph G and an integer k, there exists a $2^{O(n)}$-time algorithm to decide whether $seg(G) \leq k$.*

3   The Segment Number of Banana Trees and Cycles

We first consider some introductory examples. A *banana* is the union of paths that share only their start- and endpoints [29]. In this paper, we additionally insist that all paths have length 2. We say that a banana is a *k-banana* if it consists of k paths (of length 2). We call the joint endpoints of the paths *covering vertices* and the vertices in the interior of the paths *independent vertices*. *Banana graphs*, i.e., graphs in which some edges are replaced by bananas will play an important role in Sect. 4 when we consider the parameter vertex cover.

Lemma 3 ([10]). *A k-banana has segment number $\lfloor 3k/2 \rfloor$.*

Proof. A k-banana has $2k$ edges. There is a drawing in which $\lceil k/2 \rceil$ pairs of edges are aligned. See Fig. 2a. This is optimum: At most one path of length two can be aligned at the independent vertex. If a pair of edges is aligned at one covering vertex then no pair can be aligned at the other covering vertex.   □

Given an integer $\ell > 0$, a *banana path of length* ℓ is a graph that is constructed from a path $\langle v_0, \ldots, v_\ell \rangle$ of length ℓ by replacing, for $i \in \{1, \ldots, \ell\}$, edge $\{v_{i-1}, v_i\}$ of the path by a k_i-banana, called banana i, for some value $k_i > 0$. We say that k_i is the *multiplicity* of edge $\{v_{i-1}, v_i\}$; see Fig. 2b. *Banana trees* and *banana cycles* are defined analogously; see Fig. 3.

Theorem 2. *The segment number of a banana path of length ℓ can be computed in $\mathcal{O}(\ell)$ time and can be expressed explicitly as a function of the multiplicities.*

Proof. Let B be a banana path of length ℓ with multiplicities $k_1, \ldots, k_\ell$. In each banana, align two edges at an independent vertex. At each inner covering vertex v_i with $i \in \{1, \ldots, \ell - 1\}$, align $a_i = \min\{k_i, k_{i+1}\}$ edges from bananas i and $(i + 1)$. In particular, we align the edges from incident bananas that were aligned at their respective independent vertex. Further, let $a_0 = a_\ell = 1$. For each $i \in \{1, \ldots, \ell\}$, let $s_i = \max\{k_i - a_{i-1}, k_i - a_i\}$. Align $\lfloor s_i/2 \rfloor$ pairs of edges of banana i at covering vertex v_{i-1} or v_i, based on where the maximum is assumed. See Fig. 2b. Note that setting $a_0 = a_\ell = 1$ takes into account that one edge is already aligned at an independent vertex. Therefore, it cannot be aligned at a covering vertex at the same time. It is not possible to align more pairs of edges. Thus, the segment number is $\sum_{i=1}^{\ell} (2k_i - 1 - a_i - \lfloor s_i/2 \rfloor)$.   □

The proof of the following theorem is in the full version [9].

Theorem 3. *The segment number of (i) a banana tree (ii) a banana cycle of length at least five where each banana contains at least two independent vertices can be computed in time linear in the number of covering vertices.*

4   Parameter: Vertex Cover Number

In this section, we consider the following parametrized problem.

SEGMENT NUMBER BY VERTEX COVER NUMBER

> *Input:*   A simple planar graph G, an integer s.
> *Parameter:* Vertex cover number k of G.
> *Question:*   Is the segment number of G at most s?

Let $G = (V, E)$ be a simple planar graph, and let $V' \subset V$ be a vertex cover of G with k vertices. In order to compute the segment number of G, we first compute a subgraph of G whose size is in $\mathcal{O}(2^k)$. We vary over all possible alignments of this subgraph and check whether they are geometrically realizable. We finally use an integer linear program (ILP) in order to reinsert the missing parts from G. In the end, we take the best among the thus computed solutions. The details are as follows.

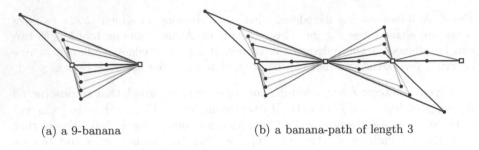

(a) a 9-banana   (b) a banana-path of length 3

Fig. 2. Banana graphs

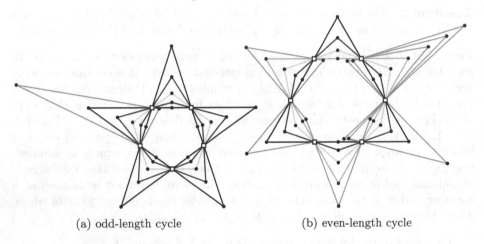

(a) odd-length cycle   (b) even-length cycle

Fig. 3. Segment minimum drawings of banana cycles.

Dividing $V \backslash V'$ into Equivalence Classes. Two vertices of $v \in V \backslash V'$ are *equivalent* if and only if they are adjacent to the same set of vertices in V'. We say that an equivalence class C is a *j-class* if each vertex $v \in C$ is adjacent to exactly j vertices in V'. Observe that the j-classes contain at most two vertices if $j > 2$. Otherwise G would contain a $K_{3,3}$, contradicting that G is planar. Thus, the number of vertices in j-classes, $j > 2$ is bounded by $2 \cdot \sum_{j=3}^{k} \binom{k}{j} \in O(2^k)$.

Reducing the Size of the Graph. Let G_1 be the graph obtained from G by removing all vertices contained in 1-classes. Consider a planar embedding of G_1. Let C be a 2-class and let $v, w \in V'$ be adjacent to all vertices in C. Observe that the edges between v and C do not have to be consecutive in the cyclic order around v (and similarly for w). A *contiguous 2-class* is a maximal subset of C such that the incident edges are consecutive around both v and w. Observe that two contiguous 2-classes are separated by the edge $\{v, w\}$ or by at least a vertex-cover vertex. Thus, a 2-class consists of at most k contiguous 2-classes. Now, for each 2-class C, we remove all but $\min(|C|, k)$ vertices from C. Let the resulting graph be G_2. Observe that the vertices of G_2 are the vertex cover vertices, the vertices of all j-classes, $j > 2$, and at most k vertices from each 2-class. Thus, the number of vertices of G_2 is bounded by $k + 2 \cdot \sum_{j=3}^{k} \binom{k}{j} + k \cdot \binom{k}{2} \in O(2^k)$.

Alignment Within 2-Classes. For each 2-class C, let e_C be the edge connecting the two vertex cover vertices adjacent to the vertices in C. We vary over all subsets $\mathcal{C}$ of 2-classes C for which e_C is not an edge of G_2. There are at most 2^{3k} such subsets. For each $C \in \mathcal{C}$, we add e_C to G_2. These edges represent a pair of aligned edges at an independent vertex. Let the resulting graph be $G_2(\mathcal{C})$.

Varying over all Planar Embeddings $\mathcal{E}$ of $G_2(\mathcal{C})$. For each 2-class C, we replace each contiguous 2-class in $\mathcal{E}$ by four vertices. The resulting four 2-paths represent the boundaries of two consecutive contiguous 2-classes; each of which shall be drawn in a non-convex way, i.e., such that the segment between the two incident vertex cover vertices does not lie in its interior. We call these non-convex quadrilaterals *boomerangs*. Let the thus constructed plane graph be $G_2(\mathcal{C}, \mathcal{E})$.

The reason why we represent each contiguous 2-class by two non-convex quadrilaterals instead of one arbitrary quadrilateral will become clear when we reinsert the 2-classes into the boomerangs; see also Fig. 4b: Let $u \in C$, let v and w be the two vertex cover vertices incident to u and assume that we have opted to draw u into boomerang b. Then we need to ensure that the edges $e_v = \{v, u\}$ and $e_w = \{w, u\}$ meet inside b when we draw e_v starting at v and e_w starting at w independently of each other and with arbitrary slopes.

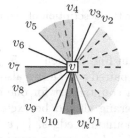
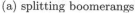
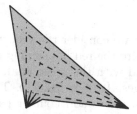

(a) splitting boomerangs   (b) reinserting 2-classes into a boomerang

Fig. 4. We vary over all pairings π of edges and boomerangs around each vertex. (a) The dashed edges are inserted into $G_2(\mathcal{C}, \mathcal{E})$ in order to represent alignment requirements with boomerangs. If the alignment requirements of the thus constructed graph $G_2(\mathcal{C}, \mathcal{E}, \pi)$ can be geometrically realized, we use an ILP to distribute the remaining edges. (b) How to geometrically insert the 2-classes into a boomerang.

Alignment Requirements. We vary over all possible "pairings" π between edges and/or boomerangs incident to a common vertex v that respect the embedding $\mathcal{E}$. If we require that two edges are aligned, that an edge should be aligned with an edge inside a boomerang, or that two edges inside different boomerangs can be aligned, we say that the respective edges and boomerangs are *paired*. E.g.,

Fig. 4a illustrates the following pairing around a vertex v: the yellow boomerang is paired with the blue and the red boomerang as well as with the edge $\{v, v_9\}$. The red and the green boomerangs are also paired. Edges $\{v, v_{10}\}$ and $\{v, v_3\}$ are paired. Edge $\{v_6, v\}$ is not paired with any other edge.

In order to handle pairings within boomerangs, we insert further edges into $G_2(\mathcal{C}, \mathcal{E})$. See the dashed edges in Fig. 4a. Let v be a vertex cover vertex. Let $v_1, \ldots, v_k$ be the neighbors of v in this order around v. Assume that v_1 and v_2 are two independent vertices representing the border of a boomerang b. Assume that we would like to align the edges around v such that $\{v, v_i\}, \ldots, \{v, v_j\}$ should get aligned with some edges in b if any. Then we add up to $j-i+1$ edges incident to v inside the region representing b and require them to be aligned with the respective edges among $\{v, v_i\}, \ldots, \{v, v_j\}$ at v.

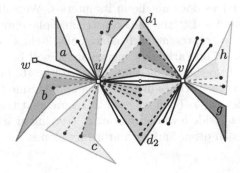

Fig. 5. Example situation for $G_2(\mathcal{C}, \mathcal{E}, \pi)$ as a basis for the description of the ILP.

E.g., when we consider the yellow boomerang b in Fig. 4a then $i = 5$, $j = 9$. Since we require the red boomerang to be partially paired with b (we also want it to be paired with the green one), we only add a counterpart for one of the boundary edges, $\{v, v_5\}$, inside b. The blue boomerang should be exclusively paired with b, so we add counter parts for both boundary edges, $\{v, v_7\}$ and $\{v, v_8\}$. Edge $\{v, v_9\}$ should be aligned with an edge in b, so it also gets a counterpart in b. According to the currently chosen pairing π, the edge $\{v, v_6\}$ does not have to be aligned in v, so we do not add a counterpart for it in b.

We do this operation for each boomerang and each incident vertex cover vertex. The resulting graph $G_2(\mathcal{C}, \mathcal{E}, \pi)$ has at most thrice as many edges as $G_2(\mathcal{C}, \mathcal{E})$.

Testing and Optimally Reinserting the 1- and 2-Class Vertices. We now have a plane graph $G_2(\mathcal{C}, \mathcal{E}, \pi)$ with $\mathcal{O}(2^k)$ vertices. See Fig. 5. Some pairs of edges are required to be aligned. The two 2-paths representing the boundary of a boomerang must bound a quadrilaterial that does not contain the segment between the two vertex cover vertices. We use Renegar's algorithm [23, 24] to test whether these requirements are geometrically realizable.

Lemma 4 ($\star$). *Given a plane graph H with embedding $\mathcal{E}$ and K vertices, a set A of pairs of edges, a set B of 4-cycles of H, and for each 4-cycle Q in B a specified diagonal $d(Q)$, we can decide, in $K^{\mathcal{O}(K)}$ time, whether there exists a planar straight-line drawing of H such that (i) the edge pairs in A are aligned and (ii) for each 4-cycle Q in B, $d(Q)$ does not lie in the interior of Q or on Q.*

We apply Lemma 4 to $H = G_2(\mathcal{C}, \mathcal{E}, \pi)$, setting A according to π, B to the set of all boomerangs, and, for each boomerang b, the specified diagonal $d(b)$ to the segment between the two vertex cover vertices. If the answer is yes, we set up and solve the following ILP to optimally insert 2-class vertices into the boomerangs and to optimally add the 1-class vertices. To this end, we use, among others, the following variables: $x_{v,b,d} = x_{v,d,b}$ expresses how many edges in boomerang b should be aligned to edges in boomerang d at vertex v and $y_{v,d}$ expresses how many edges between v and 1-class vertices should be aligned with edges in boomerang d. We allow that some of the boomerangs remain empty. The ILP uses $\mathcal{O}(2^k)$ variables and constraints and, thus, can be solved in $2^{\mathcal{O}(k2^k)}$ time [13,17,18]. For details, see the full version [9].

We are now ready to prove the main theorem of this section.

Theorem 4. SEGMENT NUMBER BY VERTEX COVER NUMBER *is FPT.*

Proof. The vertex cover number of the input graph G with n vertices can be computed in $\mathcal{O}(kn + 1.274^k)$ time [8]. The number of subsets $\mathcal{C}$ of 2-classes is in $\mathcal{O}(2^{3k})$. Then we vary over all embeddings of $G_2(\mathcal{C})$; see the respective paragraph on Page 7. The number of embeddings $\mathcal{E}$ of the graph $G_2(\mathcal{C})$ with $\mathcal{O}(2^k)$ vertices is in $\mathcal{O}((c2^k)!)$ for some constant c, and the number of possible pairings π is also bounded by a function of k. The plane graph $G_2(\mathcal{C}, \mathcal{E}, \pi)$ has $\mathcal{O}(2^k)$ vertices. Hence, testing whether the alignment and non-convexity requirements of $G_2(\mathcal{C}, \mathcal{E}, \pi)$ can be realized geometrically can be done in $2^{\mathcal{O}(k2^k)}$ time. If the answer is yes, we can solve the ILP in $2^{\mathcal{O}(k2^k)}$ time. In the end we have to determine the minimum over all choices of $\mathcal{C}$, $\mathcal{E}$, π.

In order to prove correctness, consider now a hypothetical geometric realization and an optimum solution of the ILP. Observe that the ILP is always feasible. It remains to show that the 2-class vertices can be inserted into the boomerangs and the 1-class vertices can be added so as to fulfill the alignment requirements prescribed by the ILP. For each triplet (v, b, d), where b and d are two boomerangs incident to a common vertex cover vertex v, we add $x_{v,b,d}$ stubs of aligned edges in a close vicinity of v inside the region reserved for b and d. Further, for each boomerang d incident to a vertex cover vertex v do the following. For each edge e of G_2 that we required to be aligned with an edge in d at v, we add a stub incident to v into d aligned with e. We also add $y_{v,d}$ stubs incident to v into d. If the total number of stubs in all boomerangs representing contiguous subclasses of a 2-class C are less than $|C|$, we add more stubs into arbitrary boomerangs associated with C. We extend the stubs until the respective rays meet. Observe that this intersection point is inside the boomerang and that no crossings are generated; see Fig. 4b. $\qquad\square$

5   Parameters: Segment and Line Cover Number

In this section, we first study the parameterized complexity of computing the segment number of a planar graph with respect to the segment number and the line cover number separately. Recall that, if $\text{seg}(G) \leq k$, then $\text{line}(G) \leq k$. Then, we study the parameterized complexity of colored versions of the segment number and the line cover number with respect to their natural parameters.

We start by reviewing an FPT algorithm for the line cover number [7]. A *path component* (resp., *cycle component*) of a graph is a connected component isomorphic to a path (resp., a cycle). The algorithm first removes all the path components of the input graph G, as they can be placed on a common line. Then each of the remaining connected components of G is reduced by replacing long maximal paths. More precisely, paths of length greater than $\binom{k}{2}$ that contain only vertices of degree at most 2 are replaced by paths of length $\binom{k}{2}$. Indeed, vertices of degree 2 are irrelevant since they can always be reintroduced by subdividing a straight-line segment of the path in a feasible solution of the reduced instance to obtain a feasible solution of the original instance. Note that the number of vertices of degree greater than 2 must be at most $\binom{k}{2}$. This yields an equivalent instance G' with at most $1.1k^4$ vertices and $2k^4$ edges. Next, the algorithm implicitly enumerates all line arrangements of k lines. For each such arrangement, the algorithm enumerates all possible placements of the vertices of G' at crossing points in the arrangement. Observe that the vertices of degree greater than 2 must be placed at the crossing points of the lines. The instance is accepted if at least one of the line arrangements can host G'.

In order to *explicitly* enumerate all line arrangements of k lines (as needed in Algorithm 1), we can proceed as described below. Observe that a line arrangement $\mathcal{A}$ of k lines defines a connected straight-line plane graph $G_{\mathcal{A}}$ as follows. The graph $G_{\mathcal{A}}$ contains, for each crossing point p in $\mathcal{A}$, a vertex u_p, and, for every pair of crossing points p and q that appear consecutively along a line of $\mathcal{A}$, an edge $\{u_p, u_q\}$. Additionally, for each half-line of $\mathcal{A}$ that starts at a crossing point p, the graph $G_{\mathcal{A}}$ contains a leaf adjacent to u_p. The clockwise order of the edges around each vertex u_p of $G_{\mathcal{A}}$ (of degree larger than 1) is naturally inherited from the order of the crossing points of $\mathcal{A}$ around p. By construction, $\mathcal{A}$ defines a *unique* connected straight-line plane graph $G_{\mathcal{A}}$ together with a *unique* covering of the edges of $G_{\mathcal{A}}$ with k edge-disjoint paths starting and ending at leaves. Thus, to enumerate all line arrangements of k lines, we execute the following steps. We consider all possible planar graphs on $\binom{k}{2} + 2k$ vertices containing $2k$ leaves, all their planar embeddings, and all their edge coverings with k edge-disjoint paths, if any. For each such triplet (of a planar graph, an embedding, and an edge covering), we test (using again Renegar's algorithm [23,24]) whether the considered graph admits a straight-line planar drawing preserving the selected embedding, in which each edge-disjoint path of the considered covering is drawn on a straight-line. Each triplet that passes the previous test corresponds to a (combinatorially different) line arrangement of k lines.

To compute the segment number of G, first observe that each path component requires a segment in a drawing of G. Assume that G has p path components.

Then, let $k' = k - p$ be our new parameter and remove the path components from G. Also, observe that the choices above only leave undecided to which line of the arrangement the *hanging paths* of G', i.e., the paths that start at a high degree vertex and end at a degree-1 vertex, are assigned. Clearly, the number of such choices are also bounded by a function of k. Therefore, for every line arrangement and every placement of the vertices of G' to the crossing points of the arrangement, we consider all possible ways to assign the hanging paths to the lines and compute the actual number of segments determined by these choices. We conclude that $\text{seg}(G) \leq k$ if and only if we encountered a line arrangement and a placement of vertices of G' in this line arrangement that determines at most k' segments. This immediately implies the following theorem.

Theorem 5. SEGMENT NUMBER *is FPT with respect to the natural parameter.*

Note that the line cover number is lower or equal to the segment number. However, assuming no path component exists, we can show that the segment number is polynomially bounded by the line cover number (see the full version [9]). This and Theorem 5 yield the following theorem.

Theorem 6 ($\star$). *Computing the segment number is FPT parameterized by the line cover number of the input graph.*

Motivated by list coloring, we generalize both SEGMENT NUMBER and LINE COVER NUMBER by prescribing admissible segments or lines for certain edges.

LIST-INCIDENCE SEGMENT NUMBER (LINE COVER NUMBER)

Input: A planar graph G and, for each $e \in E(G)$, a list $L(e) \subseteq [k]$.
Parameter: An integer k.
Question: Does there exist a planar straight-line drawing of G and a labeling $\ell_1, \ell_2, \ldots, \ell_k$ of the segments (supporting lines) of this drawing such that, for every $e \in E(G)$, e is drawn on a segment (line) in $\{\ell_i : i \in L(e)\}$?

Theorem 7. *The problems* LIST-INCIDENCE LINE COVER NUMBER *and* LIST-INCIDENCE SEGMENT NUMBER *are FPT with respect to the natural parameter.*

Note that Theorem 7 generalizes Theorem 5, but the algorithm for SEGMENT NUMBER, which yields Theorem 5, is faster than the algorithm for LIST-INCIDENCE SEGMENT NUMBER. Our main tool to prove Theorem 7 is the next lemma, which proves the theorem assuming that the input does not contain any path or cycle components. In the full version [9], we show how to solve the remaining technicalities.

Lemma 5. *The problems* LIST-INCIDENCE LINE COVER NUMBER *and* LIST-INCIDENCE SEGMENT NUMBER *are FPT with respect to the natural parameter k for instances containing no path or cycle components.*

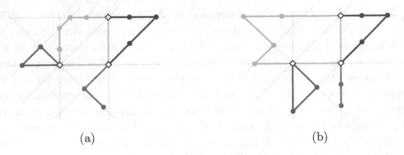

(a)   (b)

Fig. 6. Two routings of the light paths of G on the same pair of a line arrangement and a vertex-placement of $G_{>2}$. The line arrangement is shown by dotted lines, the vertices of $G_{>2}$ are drawn as diamonds, and other vertices are drawn as circles.

Proof. We first consider the LIST-INCIDENCE LINE COVER NUMBER problem. Note that a feasible instance can have at most $\binom{k}{2}$ vertices of degree greater than 2. Let $G_{>2}$ be the subgraph of G induced by vertices of degree greater than 2. We enumerate all possible arrangements of k lines and all possible placements of vertices of G of degrees greater than 2 in the crossing points of the lines. The placement of vertices determines a straight-line drawing of $G_{>2}$, (see green edges in Fig. 6). We continue with the considered arrangement/vertex-placement pair only if the drawing of $G_{>2}$ is *proper*, i.e., it is planar, and each edge of G_2 belongs to a line of the arrangement and does not pass through another vertex of degree greater than 2. Sections of lines between consecutive crossing points not occupied by edges of $G_{>2}$ are called *intervals*. Let a and b be two intervals. a is *aligned* to b (in this order) if they belong to the same line and the ending point of a is the starting point of b. a is *adjacent* to b (in this order) if they do not belong to the same line and the ending point of a is the starting point of b.

Now consider a path $P = \langle u_0, u_1, \ldots, u_t \rangle$ in G such that $\deg(u_0) > 2$, i.e., $u_0 \in V(G_{>2})$, $\deg(u_1) = \deg(u_2) = \cdots = \deg(u_{t-1}) = 2$, and $\deg(u_t)$ is either 1 or greater than 2. We call such a path *light*. Consider the representation of a light path in a solution. The path leaves the vertex u_0 along some line, then turns to another line, possibly several times, until it either ends in a crossing point, where vertex u_t had been placed (when $\deg(u_t) > 2$) or in an inner point of some interval (when $\deg(u_t) = 1$, note that w.l.o.g. we may assume that vertices of degree 1 are not placed in crossing points). This describes the *routing* of P. Every maximal sequence of consecutive intervals that belong to the same line will be called a *superinterval* in the routing. In particular, we can represent a routing R of P by the sequence $\langle S_1, S_2, \ldots, S_h \rangle$ of superintervals such that S_i and S_{i+1} are consecutive. Every two consecutive intervals within a superinterval are aligned, while for two consecutive superintervals, the last interval of the first one is adjacent to the first interval of the second one (see Fig. 6).

We use the following strategy. For every light path, we select its routing as a sequence of superintervals. Moreover, we enumerate over all possible $k!$ namings of the lines of the arrangement with labels in $\ell_1, \ell_2, \ldots, \ell_k$ and proceed only with those namings that determine a *compatible drawing* of $G_{>2}$, i.e., one in which, for each edge e of $G_{>2}$, it holds that e lies on a line in $L(e)$.

Let $\mathcal{R}$ be a collection of routings for the light paths. We check if $\mathcal{R}$ is *feasible*, i.e., any two routings in $\mathcal{R}$ are non-crossing and internally disjoint, and each of them is internally disjoint with the drawing of $G_{>2}$. For every light path, we check if the lists $L(e)$ assigned to each of its edge e allow its allocation along this routing. The first superinterval S_1 of a routing must start in vertex u_0. The last one must end in u_t if $\deg(u_t) > 2$. (If $\deg(u_t) = 1$, there is no restriction on S_h.)

Algorithm 1: LISTLINECOVERNUMBER(G, k, L)

Input   : A planar graph G, an integer k, and for each $e \in E(G)$, a list $L(e) \subseteq [k]$.

Output: true if and only if there exists a straight-line drawing of G and a labelling $\ell_1, \ell_2, \ldots, \ell_k$ of the supporting lines of this drawing such that, for every $e \in E(G)$, e is drawn on a line in $\{\ell_i : i \in L(e)\}$.

foreach arrangement $\mathcal{A}$ of k lines **do**
  foreach placement $\mathcal{P}$ of vertices of degree greater than 2 into crossing points of $\mathcal{A}$ **do**
    if $\mathcal{P}$ determines a proper drawing of $G_{>2}$ **then**
      foreach collection of routings $\mathcal{R}$ of light paths of G **do**
        foreach naming of the lines in $\mathcal{A}$ with labels in $\ell_1, \ell_2, \ldots, \ell_k$ **do**
          if $\mathcal{R}$ is feasible and the drawing of $G_{>2}$ is compatible **then**
            $X \leftarrow$ true;
            foreach light path P and its routing $R_P \in \mathcal{R}$ **do**
              $X \leftarrow X \wedge$ CHECKROUTING(P, R_P, L)
            if $X =$ true **then**
              return true

return false

The strategy described above is summarized in Algorithm 1. To check whether the lists assigned to the edges of a path allow their allocation along a given routing, Algorithm 1 exploits the procedure CHECKROUTING (see Algorithm 2), which works as follows. For a light path $P = \langle u_0, u_1, \ldots, u_t \rangle$ and a routing $R = \langle S_1, S_2, \ldots, S_h \rangle$, we fill in a table $CR(i, j) \in \{$true,false$\}$, $i = 1, \ldots, t$, $j = 1, \ldots, h$ via dynamic programming. Its meaning is that $CR(i, j) =$ true if and only if the subpath $P_i = P[u_0, \ldots, u_i]$ is *realizable* along $R_j = \langle S_1, \ldots, S_j \rangle$, i.e., it admits a planar geometric realization such that all the vertices $u_1, u_2, \ldots, u_i$ are placed inside the superintervals of R_j and every edge $e \in P_i$ lies on a line in the list $L(e)$.

The correctness of Algorithm 1 follows from the fact that we are performing an exhaustive exploration of the solution space. The correctness of Algorithm 2 follows from the fact that $P_i = \langle u_0, \ldots, u_i \rangle$ can be realized along the initial part $\langle S_1, \ldots, S_j \rangle$ of R so that vertex u_i is placed inside the superinterval S_j if and only if P_i can be realized so that u_i is placed on the crossing point that is the last point of S_j. The former is used (and needed) when updating $CR(i, j)$ to $CR(i + 1, j)$, the latter for updating $CR(i, j)$ to $CR(i + 1, j + 1)$.

The following claim states that Algorithm 1 is FPT with respect to k.

Claim 1 ($\star$) *Let $F(k)$ be the time needed to enumerate all non-isomorphic line arrangements of k lines. Then Algorithm 1 runs in time $O(F(k)k^{k^6+3k})n$.*

We now consider the LIST-INCIDENCE SEGMENT NUMBER problem. As already mentioned, any realization with k segments is itself a realization with at most k lines. Thus, we can use Algorithm 1, with just two slight modifications. Observe that, in the third *for loop* of the algorithm, the considered collection of routings of the light paths together with the drawing of $G_{>2}$ uniquely determines the segments of the representation. Therefore, we only proceed to the fourth *for loop* with those collections that determine at most k segments. However, if we proceed, in the fourth *for loop*, we consider all possible namings of these segments instead of considering all possible namings of the lines of the current arrangement. Clearly, the notion of compatible drawing of $G_{>2}$ is naturally modified using the segment labels. Finally, note that counting the number of segments affects the running time by a factor that depends solely on k.   □

Algorithm 2: CHECKROUTING(P, R_P, L)

Input : A light path P, a routing R_P, and for each $e \in E(G)$, a list $L(e) \subseteq [k]$.
Output: true if and only if P is realizable along R_P.
// *Initialization of the table*
for $i = 1$ to t do
 | for $i = 1$ to h do
 | | $CR(i,j) \leftarrow$ false;

$\ell_x \leftarrow$ the supporting line of S_1;
if $x \in L(\{u_0, u_1\})$ then
 | $CR(1,1) \leftarrow$ true;
// *Update of the table*
for $i = 1$ to $t - 1$ do
 | for $i = 1$ to h do
 | | $\ell_x \leftarrow$ the supporting line of S_j;
 | | if $CR(i,j)$ and $x \in L(\{u_i, u_{i+1}\})$ then
 | | | $CR(i+1, j) \leftarrow$ true;
 | | if $j < h$ then
 | | | $\ell_y \leftarrow$ the supporting line of S_{j+1};
 | | | if $CR(i,j)$ and $y \in L(\{u_i, u_{i+1}\})$ then
 | | | | $CR(i+1, j+1) \leftarrow$ true;

// *Return feasibility*
return $CR(t, h)$;

---

6 Open Problems

We have shown that segment number parameterized by segment number, line cover number, and vertex cover number is fixed parameter tractable. Another interesting parameter would be the treewidth. So far, even for treewidth 2, efficient optimal algorithms are only known for some subclasses [1,14,26].

The *cluster deletion number* of a graph is the minimum number of vertices that have to be removed such that the remainder is a union of disjoint cliques. Clearly, the cluster deletion number is always upper-bounded by the vertex cover number. Is the segment number problem FPT w.r.t. the cluster deletion number?

References

1. Adnan, M.A.: Minimum Segment Drawings of Outerplanar Graphs. Master's thesis, Department of Computer Science and Engineering, Bangladesh University of Engineering and Technology (BUET), Dhaka (2008). http://lib.buet.ac.bd:8080/xmlui/bitstream/handle/123456789/1565/Full%20%20Thesis%20.pdf?sequence=1&isAllowed=y

2. Balko, M., Chaplick, S., Gupta, S., Hoffmann, M., Valtr, P., Wolff, A.: Bounding and computing obstacle numbers of graphs. In: Chechik, S., Navarro, G., Rotenberg, E., Herman, G. (eds.) Proceedings of 30th European Symposium Algorithms (ESA 2022). LIPIcs, vol. 244, pp. 11:1–13. Schloss Dagstuhl - Leibniz-Zentrum für Informatik (2022). https://doi.org/10.4230/LIPIcs.ESA.2022.11

3. Berger, M.: Geometry Revealed: A Jacob's Ladder to Modern Higher Geometry. Springer, Cham (2010). https://doi.org/10.1007/978-3-540-70997-8

4. Bhore, S., Ganian, R., Montecchiani, F., Nöllenburg, M.: Parameterized algorithms for book embedding problems. J. Graph Algorithms Appl. **24**(4), 603–620 (2020). https://doi.org/10.7155/jgaa.00526

5. Bhore, S., Ganian, R., Montecchiani, F., Nöllenburg, M.: Parameterized algorithms for queue layouts. J. Graph Algorithms Appl. **26**(3), 335–352 (2022). https://doi.org/10.7155/jgaa.00597

6. Chaplick, S., Fleszar, K., Lipp, F., Ravsky, A., Verbitsky, O., Wolff, A.: Drawing graphs on few lines and few planes. J. Comput. Geom. **11**(1), 433–475 (2020). https://doi.org/10.20382/jocg.v11i1a17

7. Chaplick, S., Fleszar, K., Lipp, F., Ravsky, A., Verbitsky, O., Wolff, A.: The complexity of drawing graphs on few lines and few planes. J. Graph Algorithms Appl. **27**(6), 459–488 (2023). https://doi.org/10.7155/jgaa.00630

8. Chen, J., Kanj, I.A., Xia, G.: Improved upper bounds for vertex cover. Theor. Comput. Sci. **411**(40–42), 3736–3756 (2010). https://doi.org/10.1016/j.tcs.2010.06.026

9. Cornelsen, S., Da Lozzo, G., Grilli, L., Gupta, S., Kratochvíl, J., Wolff, A.: The parametrized complexity of the segment number. Technical report. arXiv:2308.15416, Cornell University Library (2023). https://doi.org/10.48550/arXiv.2308.15416

10. Dujmović, V., Eppstein, D., Suderman, M., Wood, D.R.: Drawings of planar graphs with few slopes and segments. Comput. Geom. **38**, 194–212 (2007). https://doi.org/10.1016/j.comgeo.2006.09.002

11. Durocher, S., Mondal, D.: Drawing plane triangulations with few segments. Comput. Geom. **77**, 27–39 (2019). https://doi.org/10.1016/j.comgeo.2018.02.003

12. Durocher, S., Mondal, D., Nishat, R., Whitesides, S.: A note on minimum-segment drawings of planar graphs. J. Graph Algorithms Appl. **17**(3), 301–328 (2013). https://doi.org/10.7155/jgaa.00295

13. Frank, A., Tardos, É.: An application of simultaneous Diophantine approximation in combinatorial optimization. Combinatorica **7**(1), 49–65 (1987). https://doi.org/10.1007/BF02579200

14. Goeßmann, I., et al.: The segment number: algorithms and universal lower bounds for some classes of planar graphs. In: Bekos, M.A., Kaufmann, M. (eds.) WG 2022. LNCS, vol. 13453, pp. 271–286. Springer, Cham (2022). https://doi.org/10.1007/978-3-031-15914-5_20

15. Hültenschmidt, G., Kindermann, P., Meulemans, W., Schulz, A.: Drawing planar graphs with few geometric primitives. J. Graph Algorithms Appl. **22**(2), 357–387 (2018). https://doi.org/10.7155/jgaa.00473

16. Igamberdiev, A., Meulemans, W., Schulz, A.: Drawing planar cubic 3-connected graphs with few segments: algorithms & experiments. J. Graph Algorithms Appl. **21**(4), 561–588 (2017). https://doi.org/10.7155/jgaa.00430

17. Lenstra, J., H.W.L.: Integer programming with a fixed number of variables. Math. Oper. Res. **8**(4), 538–548 (1983). https://doi.org/10.1287/moor.8.4.538

18. Kannan, R.: Minkowski's convex body theorem and integer programming. Math. Oper. Res. **12**(3), 415–440 (1987). https://doi.org/10.1287/moor.12.3.415

19. Kindermann, P., Mchedlidze, T., Schneck, T., Symvonis, A.: Drawing planar graphs with few segments on a polynomial grid. In: Archambault, D., Tóth, C.D. (eds.) GD 2019. LNCS, vol. 11904, pp. 416–429. Springer, Cham (2019). https://doi.org/10.1007/978-3-030-35802-0_32

20. Kindermann, P., Meulemans, W., Schulz, A.: Experimental analysis of the accessibility of drawings with few segments. J. Graph Algorithms Appl. **22**(3), 501–518 (2018). https://doi.org/10.7155/jgaa.00474

21. Kryven, M., Ravsky, A., Wolff, A.: Drawing graphs on few circles and few spheres. J. Graph Algorithms Appl. **23**(2), 371–391 (2019). https://doi.org/10.7155/jgaa.00495

22. Okamoto, Y., Ravsky, A., Wolff, A.: Variants of the segment number of a graph. In: Archambault, D., Tóth, C.D. (eds.) GD 2019. LNCS, vol. 11904, pp. 430–443. Springer, Cham (2019). https://doi.org/10.1007/978-3-030-35802-0_33

23. Renegar, J.: On the computational complexity and geometry of the first-order theory of the reals. Part I: Introduction. Preliminaries. The geometry of semi-algebraic sets. The decision problem for the existential theory of the reals. J. Symb. Comput. **13**(3), 255–299 (1992). https://doi.org/10.1016/S0747-7171(10)80003-3

24. Renegar, J.: On the computational complexity and geometry of the first-order theory of the reals. Part II: The general decision problem. Preliminaries for quantifier elimination. J. Symb. Comput. **13**(3), 301–327 (1992). https://doi.org/10.1016/S0747-7171(10)80004-5

25. Renegar, J.: On the computational complexity and geometry of the first-order theory of the reals. Part III: Quantifier elimination. J. Symb. Comput. **13**(3), 329–352 (1992). https://doi.org/10.1016/S0747-7171(10)80005-7

26. Samee, M.A.H., Alam, M.J., Adnan, M.A., Rahman, M.S.: Minimum segment drawings of series-parallel graphs with the maximum degree three. In: Tollis, I.G., Patrignani, M. (eds.) GD 2008. LNCS, vol. 5417, pp. 408–419. Springer, Cham (2008). https://doi.org/10.1007/978-3-642-00219-9_40

27. Schaefer, M.: Complexity of some geometric and topological problems. In: Eppstein, D., Gansner, E.R. (eds.) GD 2009. LNCS, vol. 5849, pp. 334–344. Springer-Verlag, Cham (2010). https://doi.org/10.1007/978-3-642-11805-0_32

28. Schulz, A.: Drawing graphs with few arcs. J. Graph Algorithms Appl. **19**(1), 393–412 (2015). https://doi.org/10.7155/jgaa.00366
29. Scott, A., Seymour, P.: Induced subgraphs of graphs with large chromatic number. VI. Banana trees. J. Combin. Theory Ser. B. **145**, 487–510 (2020). https://doi.org/10.1016/j.jctb.2020.01.004

Planar Graphs

A Schnyder-Type Drawing Algorithm
for 5-Connected Triangulations

Olivier Bernardi[1] [ID], Éric Fusy[2(✉)] [ID], and Shizhe Liang[1] [ID]

[1] Department of Mathematics, Brandeis University, Waltham, MA, USA
{bernardi,shizhe1011}@brandeis.edu
[2] CNRS/LIGM, Université Gustave Eiffel, Champs-sur-Marne, France
eric.fusy@univ-eiffel.fr

Abstract. We define some Schnyder-type combinatorial structures on a class of planar triangulations of the pentagon which are closely related to 5-connected triangulations. The combinatorial structures have three incarnations defined in terms of orientations, corner-labelings, and woods respectively. The wood incarnation consists in 5 spanning trees crossing each other in an orderly fashion. Similarly as for Schnyder woods on triangulations, it induces, for each vertex, a partition of the inner triangles into face-connected regions (5 regions here). We show that the induced barycentric vertex-placement, where each vertex is at the barycenter of the 5 outer vertices with weights given by the number of faces in each region, yields a planar straight-line drawing.

Keywords: Schnyder structures · barycentric drawing · 5-connected triangulations

1   Introduction

In 1989, Walter Schnyder showed that planar triangulations can be endowed with remarkable combinatorial structures, which now go by the name of *Schnyder woods* [24]. A Schnyder wood of a planar triangulation is a partition of its inner edges into three trees, crossing each other in an orderly manner; see Fig. 1(a) for an example. In [25], Schnyder used his structures to define an elegant algorithm to draw planar triangulations with straight edges [25].

In this article we study an analogue of Schnyder woods for triangulations of the pentagon, which we call *5c-woods*, and we show that such structures exist if and only if cycles of length less than 5 have no vertex in their interior (a property closely related to 5-connectedness). We then use these structures to define a graph-drawing algorithm in the spirit of Schnyder's algorithm.

OB was partially supported by NSF Grant DMS-2154242. EF was partially supported by the project ANR19-CE48-011-01 (COMBINÉ), and the project ANR-20-CE48-0018 (3DMaps).

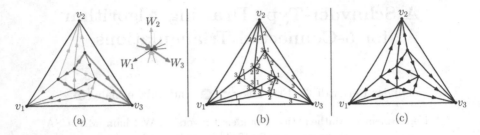

Fig. 1. (a) A Schnyder wood (with the local condition at inner vertices), (b) the corresponding corner labeling, (c) the corresponding 3-orientation.

A disadvantage of our algorithm compared to Schnyder's original algorithm (and to algorithms such as [7,8,11,15,16,19,21] using an underlying combinatorial structure or shelling order to define a vertex-placement) is that it does not yield a *grid-drawing* algorithm, that is, a vertex placement where the coordinates are integers bounded by a linear function of the number of vertices of the graph. Nevertheless, we stress some nice features of our algorithm: it can be implemented in linear time, it respects rotational symmetries, and the worst-case vertex resolution (minimal distance between vertices) is better than in Schnyder's drawing. On the examples we have tested, our algorithm seems to output aesthetically pleasant drawings; see Fig. 7 for an example.

Our article is organized as follows. All the combinatorial results are presented in Sect. 2. We start by defining the 5c-structures for triangulations of the pentagon (see Fig. 2 for an example, which parallels Fig. 1). We give three different incarnations of these structures (5c-woods, 5c-labelings, 5c-orientations), we explain the equivalence between them, state the exact condition for their existence, and point to the appendix of the full version [4] for detailed proofs and the description of a linear-time construction algorithm. The drawing algorithm is then presented in Sect. 3. Together with the proof of planarity of the drawing, this section includes a discussion of the drawing properties mentioned above, and some open questions.

Let us mention that the 5c-structures presented here are closely related to the *quasi-Schnyder structures* which we introduced in [3]. The quasi-Schnyder structures are a far-reaching generalization of Schnyder woods (encompassing regular edge labelings [18], and Schnyder decompositions considered in [1]), and in the case of triangulations of the pentagon, these structures can be identified with the 5c-structures which are the focus of the present article. Focusing on triangulations of the pentagon allows us to provide a simplified presentation, both in terms of definitions and proofs, and also to define an additional incarnation in terms of woods. Let us add that the quasi-Schnyder structures for triangulations of the square coincide (after simplifications) with *regular edge labelings* [18] (a.k.a. *transversal structures* [16]), and the quasi-Schnyder structures for triangulations coincide (after simplifications) with the classical Schnyder woods [24]. Another combinatorial structure which bears some resemblance to 5c-structures are the five color forests introduced in [13] that are related to pentagon contact representations. As we will explain in Remark 1, one can easily construct a five color forest from a 5c-structure (but the opposite is not true).

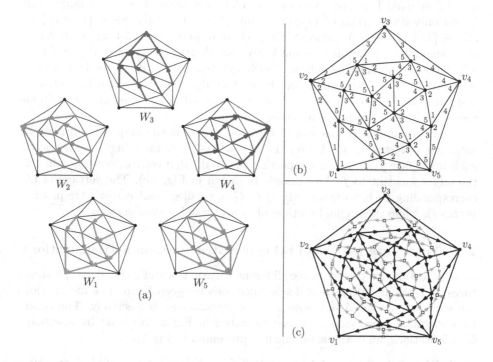

Fig. 2. The three incarnations of 5c-structures. (a) A 5c-wood $\mathcal{W} = (W_1, \ldots, W_5)$. (b) The 5c-labeling $\mathcal{L} = \Theta^{-1}(\mathcal{W})$. (c) The 5c-orientation $\mathcal{O} = \Phi(\mathcal{L})$.

2   Schnyder Structures for Triangulations of the Pentagon

2.1   Definitions About Triangulations

A *plane map* is a connected planar graph G with a fixed planar embedding. The *faces* of G are the connected components of $\mathbb{R}^2 \backslash G$. The *outer face* is the unique unbounded face, the other faces are called *inner*. An *arc* of G is an edge with a choice of direction. We use the notation $\{u, v\}$ for an edge connecting the vertices u and v, and (u, v) for an arc oriented from u to v. The two arcs on an edge are called *opposite*. A *corner* is a sector delimited by two consecutive edges around a vertex. Corners, vertices, edges and arcs are called *outer* when they are incident to the outer face, and *inner* otherwise.

A *triangulation of the pentagon*, or *5-triangulation* for short, is a plane map such that the inner faces have degree 3 and the outer face contour is a simple cycle of length 5. The outer vertices of a 5-triangulation are denoted by $v_1, v_2, \ldots, v_5$

in clockwise order around the outer face; see Fig. 2. A *5c-triangulation* is a 5-triangulation such that every cycle with at least one vertex in its interior has length at least 5. Note that 5c-triangulations have no loops nor multiple edges. In the terminology of [3], 5c-triangulations are the 5-triangulations which are *quasi 5-adapted*. It is easy to check that a 5-triangulation G is a 5c-triangulation if and only if the graph G' obtained from G by adding the edges $\{v_i, v_{i+2}\}$ for all $i \in \{1, 2, \ldots, 5\}$ is 5-connected (in particular 5-connected 5-triangulations are 5c-triangulations). Let us mention lastly that deleting a vertex of degree 5 from a 5-connected planar triangulation (such a vertex necessarily exists, by the Euler relation) yields a 5c-triangulation; hence the algorithm presented in Sect. 3 for 5c-triangulations gives a way to draw 5-connected planar triangulations (upon seeing the deleted vertex as a vertex at infinity).

The *primal-dual completion* of a plane map G is the map G^+ obtained by inserting a vertex v_e in the middle of each edge e, inserting a vertex v_f in each inner face f, and then connecting v_f to all edge-vertices corresponding to the edges incident to f. An example is shown in Fig. 2(c). The vertices of G^+ corresponding to faces (resp. edges) of G are called *dual vertices* (resp. *edge-vertices*), while the original vertices of G are called *primal vertices*.

2.2   Three Incarnations of Schnyder Structures on 5-Triangulations

In this section we present three different incarnations of Schnyder-type structures on 5-triangulations and define bijections between them. The incarnations are called *5c-woods*, *5c-labelings* and *5c-orientations* respectively. The conditions defining these structures are indicated in Fig. 3. We start by discussing 5c-orientations, an example of which is presented in Fig. 2(c).

Definition 2.1 *Given a 5-triangulation G, a 5c-orientation of G is an orientation of the inner edges of the primal-dual completion G^+ of G satisfying the following conditions:*

(O0) The outer primal vertices have outdegree 0.
(O1) The inner primal vertices have outdegree 5, the dual vertices have outdegree 2, and the edge-vertices (including the outer ones) have outdegree 1.

The definition of 5c-orientations is illustrated in the top row of Fig. 3.

Next, we define *5c-labelings*, an example of which is presented in Fig. 2(b). A *corner labeling* of a 5-triangulation G is an assignment of a *label* in $[1 : 5] := \{1, 2, 3, 4, 5\}$ to each inner corner of G. For two corners c and c' with labels i and i' respectively, we call *label jump* from c to c' the integer $\delta \in \{0, 1, 2, 3, 4\}$ such that $i + \delta \equiv i' \pmod 5$.

Definition 2.2 *Given a 5-triangulation G, a 5c-labeling of G is a corner labeling of G satisfying the following conditions:*

(L0) For all $i \in [1 : 5]$, every inner corner incident to v_i has label i.
(L1) Around every inner vertex, the incident corners form 5 non-empty intervals I_1, I_2, I_3, I_4, I_5 in clockwise order, with all corners in I_i having label i .

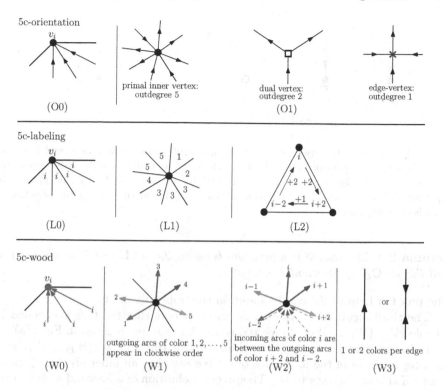

Fig. 3. Definition of 5c-structures. Top row: Conditions defining 5c-orientations. Middle row: Conditions defining 5c-labelings. Bottom row: Conditions defining 5c-woods.

(L2) Around every inner face, in clockwise order, there are two label jumps equal to 2 and one label jump equal to 1.

Conditions (L0-L2) are illustrated in Fig. 3 (middle row).

Next, we define a bijection Φ between 5c-labelings and 5c-orientations. The mapping Φ is represented in Fig. 4(a).

Definition 2.3 *Given a 5c-labeling $\mathcal{L}$ of G, we define an orientation $\Phi(\mathcal{L})$ on G^+ as follows. First we note that there is a one-to-one correspondence between the inner corners of G and the inner faces of G^+, hence we interpret $\mathcal{L}$ as a labeling of the inner faces of G^+. Let $e = \{v, x\}$ be an inner edge of G^+, where v is either a primal or dual vertex, and x is an edge-vertex.*

- *If v is a primal vertex, and f^- and f^+ are the faces incident to e in G^+ in clockwise order around v, then e is oriented toward v if and only if the label jump from f^- to f^+ is 0 (i.e., f^- and f^+ have the same label).*
- *If v is a dual vertex, and f^- and f^+ are the faces incident to e in G^+ in clockwise order around v, then e is oriented toward v if and only if the label jump from f^- to f^+ is 1.*

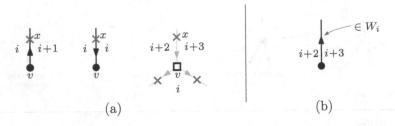

Fig. 4. Rules giving the bijections between 5c-orientations, 5c-labelings, and 5c-wood. Vertices of G are represented by solid black dots, whereas dual vertices are represented by squares, and edge-vertices are represented by crosses. (a) Local rule giving the bijection Φ from 5c-labelings to 5c-orientations. (b) Local rule giving the bijection Θ from 5c-labelings to 5c-woods.

Lemma 2.4 *The map Φ is a bijection between the set $\mathbf{L}_G$ of 5c-labelings on G and the set $\mathbf{O}_G$ of 5c-orientations on G.*

The proof of Lemma 2.4 can be found in the appendix of [4].

The third incarnation is called *5c-woods*, where the structure is encoded by a tuple $\mathcal{W} = (W_1, ..., W_5)$ of directed trees. An example is given in Fig. 2(a).

It is sometimes convenient to think of the tuple $\mathcal{W} = (W_1, ..., W_5)$ as a coloring in $[1:5]$ of the inner arcs of G. We say that an inner arc a of G has a color i if this arc belongs to W_i. The precise definition of a 5c-wood is as follows.

Definition 2.5 *Given a 5-triangulation G, a 5c-wood of G is a tuple $\mathcal{W} = (W_1, ..., W_5)$ of disjoint subsets of inner arcs, satisfying the following conditions:*

(W0) No arc in $\mathcal{W}$ starts at an outer vertex, and those ending at the outer vertex v_i have color i for all $i \in [1:5]$.

(W1) Every inner vertex v has a unique outgoing arc of color i, for $i \in [1:5]$, and these arcs appear in clockwise order around v.

(W2) Let v be an inner vertex with incident outgoing arcs $a_1, ..., a_5$ of colors $1, ..., 5$, respectively. Any arc a of color i having terminal vertex v appears weakly between a_{i+2} and a_{i+3} in clockwise order around v (weakly means that a may be the arc opposite to a_{i+2} or a_{i+3}).

(W3) Every inner edge has at least one color.

Conditions (W0-W3) are illustrated in the bottom row of Fig. 3. We will prove later (see Proposition 3.1) that for any 5c-wood $\mathcal{W} = (W_1, ..., W_5)$ the set of arcs W_i is acyclic for all i. Given that every inner vertex has outdegree 1 in W_i, it follows that W_i is a tree spanning all the inner vertices and the outer vertex v_i, and oriented toward the root-vertex v_i (that is, every arc in W_i is oriented from child to parent when v_i is taken as the root of the tree W_i).

Now we define a bijection Θ between 5c-labelings and 5c-woods. The mapping Θ is represented in Fig. 4(b).

Definition 2.6 *Given a 5c-labeling $\mathcal{L}$ of G, we define a tuple $\Theta(\mathcal{L}) = (W_1, ..., W_5)$ of subsets of inner arcs of G (interpreted as a partial arc coloring) as follows: an inner arc $a = (u, v)$ receives color i if the labels of the corners on the left and on the right of a (at u) are $i + 2$ and $i + 3$, respectively; the arc a has no color if the two labels are equal.*

Lemma 2.7 *The map Θ is a bijection between the set $\mathbf{L}_G$ of 5c-labelings on G and the set $\mathbf{W}_G$ of 5c-woods on G.*

The proof of Lemma 2.7 can be found in the appendix of [4].

Remark 1. We can give a fourth incarnation of 5c-structures, as a representation of the 5-triangulation by a contact system of "soft pentagons" as indicated in Fig. 5. Let G be a 5-triangulation. A *soft pentagon contact representation* of G is a collection of interior-disjoint "soft pentagons" (the sides of which are curves rather than straight-line segments) inside a regular "outer pentagon", with vertices of each pentagon labeled 1 to 5 in clockwise order such that

- the side $[i + 2, i + 3]$ of the outer pentagon corresponds to the outer vertex v_i of G, for $i \in [1 : 5]$,
- the soft pentagons correspond to the inner vertices of G,
- contacts between pentagons correspond to the inner edges of G,
- all the pentagon contacts occur between a soft pentagon vertex of label $i \in [1 : 5]$ and either the side labeled $[i + 2, i + 3]$ of another soft pentagon (vertex-to-side contact), or the vertex $i + 2$ or $i + 3$ of another soft pentagon (vertex-to-vertex contact), or the side labeled $[i + 2, i + 3]$ of the outer pentagon (outer contact, always vertex-to-side).

A soft pentagon contact representation is *complete* if every vertex of every soft pentagon has a contact. It is easy to see that the 5c-woods of G are in bijection with its complete soft pentagon contact representations. This is illustrated in Fig. 5. This topological interpretation of 5c-woods allows us to compare them to the *five color forests* introduced in [13]. For a 5-triangulation G these structures correspond to general (potentially incomplete) soft pentagon contact representations without vertex-to-vertex contact. In particular, one can easily turn any 5c-wood into a five color forest, upon resolving each vertex-to-vertex contact into a vertex-to-side contact. At the level of the 5c-wood (interpreted as a collection of arcs colored in [1 : 5]), this amounts to deleting one colored arc for each edge of G bearing 2 colored arcs. Note however that it is usually not possible to turn a five color forest into a 5c-wood. Indeed, five color forests exist for any simple 5-triangulation, while (as stated below) 5c-woods only exist for 5c-triangulations.

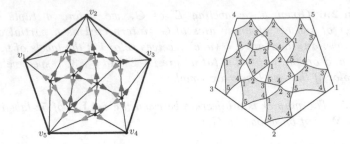

Fig. 5. Left: A 5c-wood. Right: The corresponding (complete) soft pentagon contact representation.

2.3   Existence and Computation of Schnyder Structures

We now state our main result on 5c-structures (where we include the already obtained bijective statements).

Theorem 2.8 *A 5-triangulation G admits a 5c-wood (resp. 5c-labeling, 5c-orientation) if and only if G is a 5c-triangulation. In this case, the sets of 5c-woods, 5c-labelings and 5c-orientations of G are in bijection. Moreover, a 5c-wood (resp. 5c-labeling, 5c-orientation) of a 5c-triangulation can be computed in linear time in the number of vertices.*

We have already stated in Sect. 2.2 that, for any 5-triangulation G, the sets of 5c-woods, 5c-labelings and 5c-orientations of G are in bijection. Moreover, it is clear that these bijections can be performed in linear time in the number of vertices. It thus remains to show that a 5-triangulation admits a 5c-orientation if and only if it is a 5c-triangulation, and that, in this case, a 5c-orientation can be computed in time linear. We already proved this existence (and algorithmic) result in [3] within the larger framework of *quasi-Schnyder structures*. Since there are many layers to the proof given in [3], we sketch a more direct proof (and construction algorithm) in the appendix of [4].

3   Graph Drawing Algorithm for 5c-Triangulations

3.1   Paths and Regions

In this section we explain how a 5c-wood gives rise to paths and regions associated to each vertex of a 5c-triangulation; see Fig. 6(b).

Let G be an undirected graph. A *biorientation* of G is an arbitrary subset of arcs of G (so that each edge of G is oriented in either 0, 1 or 2 directions). A biorientation is *acyclic* if it contains no simple directed cycle, with the convention that two opposite arcs (coming from the same edge oriented in 2 directions) do not constitute a simple cycle. Suppose now that G is a 5c-triangulation, and let $\mathcal{W} = (W_1, \ldots, W_5)$ be a 5c-wood. For $i \in [1 : 5]$, we denote by W_i^- the set of arcs obtained by reversing the arcs in W_i (that is, taking the opposite arcs).

Proposition 3.1 *Let* $\mathcal{W} = (W_1, \ldots, W_5)$ *be a 5c-wood of* G. *For all* $i \in [1:5]$, *the biorientation* $\mathcal{O}_i = W_i \cup W_{i-1} \cup W_{i+1} \cup W_{i-2}^- \cup W_{i+2}^-$ *is acyclic. Consequently, for each pair* $j, k \in [1:5]$, *the biorientation* $W_j \cup W_k^-$ *is acyclic.*

The proof of Proposition 3.1 can be found in the appendix of [4]. An example is shown in Fig. 6(a).

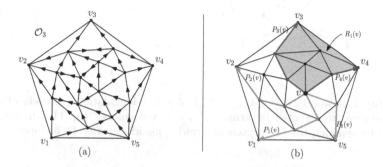

Fig. 6. (a) The biorientation $\mathcal{O}_3$ for the 5c-wood represented in Fig. 2. (b) The paths $P_1(v), \ldots, P_5(v)$ and the region $R_1(v)$ for the 5c-wood in Fig. 2.

As mentioned in Sect. 2, Proposition 3.1 implies that in any 5c-wood $\mathcal{W} = (W_1, \ldots, W_5)$ of G, each set W_i is a tree directed toward the outer vertex v_i. For an inner vertex v of G, we denote by $P_i(v)$ the directed path from v to v_i in W_i. These paths are indicated in Fig. 6(b). Proposition 3.1 implies that the paths $P_1(v), \ldots, P_5(v)$ have no vertex in common except v (since if $P_j(v)$ and $P_k(v)$ had a vertex in common, the biorientation $W_j \cup W_k^-$ would contain a directed cycle). We denote by $R_i(v)$ the region enclosed by the simple cycle made of the outer edge (v_{i+2}, v_{i-2}) together with the paths $P_{i-2}(v)$ and $P_{i+2}(v)$. See Fig. 6(b) for an example.

3.2 Graph Drawing Algorithm

For convenience, we extend the definition of $R_i(v)$ for v an outer vertex by declaring $R_i(v_i) = G$ for all $i \in [1:5]$, and $R_i(v_j) = \{v_j\}$ for all $j \neq i$. For a vertex v of G and for $i \in [1:5]$, the *size* of $R_i(v)$, denoted by $|R_i(v)|$, is the number of inner faces in $R_i(v)$. Letting n be the number of vertices of G, the Euler relation implies that G has $2n - 7$ inner faces. Let $\alpha_i(v) := |R_i(v)|/(2n-7)$. Since the inner faces of G are partitioned among the 5 regions, we have $\sum_{i=1}^{5} \alpha_i(v) = 1$.

The *5c-barycentric drawing algorithm* then consists of the following steps:

1. Draw a regular pentagon $(v_1, v_2, v_3, v_4, v_5)$ (in clockwise order).
2. Place each vertex v at the barycenter $\sum_{i=1}^{5} \alpha_i(v)v_i$.
3. Draw each edge of G as a segment connecting the corresponding points.

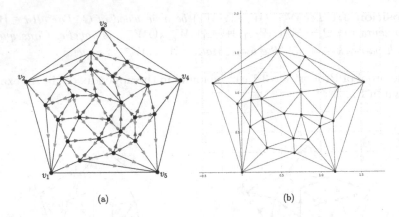

(a)   (b)

Fig. 7. (a) A 5c-wood $\mathcal{W} = (W_1, \ldots, W_5)$ of a 5c-triangulation (the sets of arcs in $W_1, \ldots, W_5$ are represented by arrows of 5 different colors). (b) The drawing of G obtained by applying the 5c-barycentric drawing algorithm with $\mathcal{W}$ as input.

An output of this algorithm is represented in Fig. 7. In all our drawings, the outer pentagon is drawn with the edge $\{v_1, v_5\}$ as an horizontal bottom segment.

Theorem 3.2 *For each 5c-triangulation G with n vertices, the 5c-barycentric drawing algorithm yields a planar straight-line drawing of G, which can be computed in $O(n)$ operations.*

Remark 2. Compared to Schnyder's algorithm [25] and to Tutte's spring embedding [27], it is important here that the 5 outer vertices are placed so as to form a regular pentagon, and we use this property in our proof of planarity.

We will prove the planarity of the 5c-barycentric drawing in the next section. The time-complexity analysis can be found in the appendix of [4].

3.3   Proof of Planarity

The proof of planarity in Theorem 3.2 crucially relies on a property (illustrated in the right drawing of Fig. 9 and to be established in Lemma 3.4) stating that the directions of the arcs of each color from an inner vertex are constrained to be in certain cones. As a first step we show the following "half-plane property", illustrated in Fig. 8.

Lemma 3.3 *Let G be a 5c-triangulation endowed with a 5c-wood. In the associated 5c-barycentric drawing, for $i \in [1:5]$, let $\overrightarrow{V_i}$ be the vector connecting the center of the outer pentagon (i.e. the center of the circle circumscribed to that pentagon) to the outer vertex v_i. Then, for any vertices $u \neq v$ of G such that $u \in R_i(v)$, we have $\overrightarrow{V_i} \cdot \overrightarrow{vu} < 0$, where $\overrightarrow{vu}$ is the vector connecting v to u in the 5c-barycentric drawing.*

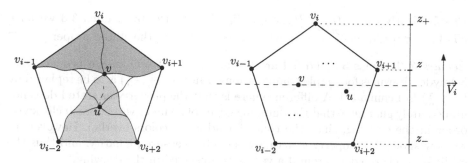

Fig. 8. Left: Situation for two vertices $u \neq v$ such that $u \in R_i(v)$, where the interior of $R_i(v)$ is shown in yellow, and the interior of $R_{i-2,i+2}(v)$ is shown in blue. Right: In the 5c-barycentric drawing, u is in the half-plane "below" v upon rotating the drawing such that $\vec{V_i}$ points upward.

Proof. For each vertex w and each subset S of $[1:5]$, we use the extended notation $R_S(w) = \cup_{i \in S} R_i(w)$, and $\alpha_S(w) = \sum_{i \in S} \alpha_i(w)$. We then observe the following containment relations (see Fig. 8):

$$R_i(v) \supset R_i(u), \quad R_{i-2,i+2}(v) \subset R_{i-2,i+2}(u).$$

The second relation is due to the fact (which follows from Condition (W2) of 5c-woods) that, whenever $P_{i-1}(u)$ (resp. $P_{i+1}(u)$) leaves the region $R_i(v)$, it occurs just after visiting a vertex on $P_{i-2}(v)$ (resp. on $P_{i+2}(v)$).

Defining $z_- < z < z_+$ as

$$z_- = \vec{V_i} \cdot \vec{V_{i-2}} = \vec{V_i} \cdot \vec{V_{i+2}}, \quad z = \vec{V_i} \cdot \vec{V_{i-1}} = \vec{V_i} \cdot \vec{V_{i+1}}, \quad z_+ = \vec{V_i} \cdot \vec{V_i},$$

and letting $r = \alpha_{i-2,i+2}(u) - \alpha_{i-2,i+2}(v)$, $s = \alpha_{i-1,i+1}(u) - \alpha_{i-1,i+1}(v)$, and $t = \alpha_i(u) - \alpha_i(v)$, we then have

$$\vec{vu} \cdot \vec{V_i} = \sum_{k=1}^{5} (\alpha_k(u) - \alpha_k(v)) \vec{V_k} \cdot \vec{V_i}$$

$$= r\, z_- + s\, z + t\, z_+$$

$$= r\, z_- + (-r-t)z + t\, z_+ = r\, (z_- - z) + t\, (z_+ - z),$$

where from the 2nd to 3rd line we use the identity $r + s + t = 1 - 1 = 0$. The above containment relations ensure that $t < 0$ and $r > 0$, so that $\vec{vu} \cdot \vec{V_i} < 0$. $\square$

Lemma 3.4 *Let G be a 5c-triangulation endowed with a 5c-wood. For v an inner vertex of G, and for $i \in [1:5]$, let u be the terminal vertex of the arc of color i from v. Then, in the 5c-barycentric drawing of G, the angle between the vector $\vec{V_i}$ (defined as in Lemma 3.3) and the vector $\vec{vu}$ is in the open interval $\left(-\frac{3\pi}{10}, \frac{3\pi}{10}\right)$ (see Fig. 9 right).*

Proof. Clearly $u \in P_i(v) = R_{i-2}(v) \cap R_{i+2}(v)$. Hence, by Lemma 3.3 we have $\overrightarrow{vu} \cdot \overrightarrow{V_{i-2}} < 0$ and $\overrightarrow{vu} \cdot \overrightarrow{V_{i+2}} < 0$, which is equivalent to the stated property. □

Remark 3. Lemmas 3.3 and 3.4 are the analogues of well-known properties of Schnyder drawings for simple triangulations, where $(-\frac{3\pi}{10}, \frac{3\pi}{10})$ is to be replaced by $(-\frac{\pi}{6}, \frac{\pi}{6})$ in Lemma 3.4. A difference here is that the property as stated does not immediately guarantee that the outgoing edges of an inner vertex are in clockwise order in the drawing, since the cones for adjacent colors overlap. However, as we now show, the property is actually enough to ensure planarity, and thus the cyclic ordering of edges around a vertex is preserved in the drawing.

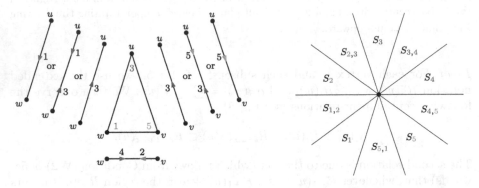

Fig. 9. Left: Possible cases for the colors of arcs around an inner face with corner labels $1, 3, 5$. Right: Conditions of Lemma 3.4 (a sector S_i can only have the arc of color i from the shown vertex, a sector $S_{i,j}$ can only have the arcs of colors in $\{i, j\}$).

To show that the drawing is planar, by a known topological argument (see e.g. [9, Lem.4.4]), it suffices to show that each inner face is *properly represented*, that is, represented as a non-degenerated triangle that is not flipped (indeed, this condition ensures that the function mapping a point p inside the enclosing pentagon to the number $n(p)$ of triangles covering it is locally constant, hence has to be equal to 1 everywhere). In other words, if the corner labels are $i, i+2, i+3$ in clockwise order around an inner face, we have to show that these corners are seen in the same cyclic order in the triangle representing the face. It is clear that the 5 inner faces incident to the 5 outer edges are properly represented, so we can focus on inner faces incident to 3 inner edges. Let us treat the case $i = 3$ (the other cases can be treated symmetrically), where the vertices at the corners $3, 5, 1$ are respectively denoted u, v, w. Given the conditions defining corner-labelings, and the bijection between 5c-woods and 5c-labelings, it is easy to check that the following holds (see Fig. 9 left):

– The arc (v, w) has color 2, and the arc (w, v) has color 4.
– The arc (v, u) has color 3 if colored, the arc (u, v) has color 5 if colored, and at least one of these two arcs is colored.

– The arc (w, u) has color 3 if colored, the arc (u, w) has color 1 if colored, and at least one of these two arcs is colored.

By Lemma 3.4, and using the notation of Fig. 9 right, the arc (v, w) is in sector $S_2 \cup S_{1,2}$ from v, and the arc (w, v) is in sector $S_4 \cup S_{4,5}$ from w (the fact that $\{v, w\}$ is a straight segment in the drawing excludes sectors $S_{2,3}$ and $S_{3,4}$ respectively). If the arc (v, u) is colored (with color 3), then it is in sector $S_{2,3} \cup S_3 \cup S_{3,4}$ from v, hence the directed angle from (v, u) to (v, w) around v is in the open interval $(0, \pi)$, so that the face (u, v, w) is properly represented. Similarly, if the arc (w, u) is colored (with color 3), then the directed angle from (w, v) to (w, u) around w is in the open interval $(0, \pi)$, so that the face (u, v, w) is properly represented. Finally, if none of the arcs (v, u) or (w, u) is colored, then the arc (u, w) has color 1, and the arc (u, v) has color 5. Assume for contradiction that the face (u, v, w) is not properly represented. Then the directed angle from (u, w) to (u, v) around u is not in the open interval $(0, \pi)$. Given Lemma 3.4, these two arcs have to be in $S_{5,1}$ from u, that is, the arc (w, u) is in S_3 from w and the arc (v, u) is in S_3 from v. But then, the angle from (w, v) to (w, u) around w is in the open interval $(\pi/5, 4\pi/5)$, so that the face (u, v, w) is properly represented. This concludes the proof of planarity in Theorem 3.2.

3.4 Variations and Other Properties

We discuss here some variants (weighted faces, vertex-counting) and aesthetic properties of the drawing, and end with some open questions.

Variants of the Embedding Algorithm. As in the case of Schnyder's algorithm [14], one can give a *weighted version* of the 5c-barycentric drawing algorithm. In this version, each inner face is assigned a positive weight, and $\alpha_i(v)$ is the total weight of inner faces in $R_i(v)$, divided by the total weight of inner faces. The proof of Lemma 3.4 and the proof of planarity extend verbatim. Moreover, as in [25, Sec.7], a vertex-counting variant can also be given, upon changing $|R_i(v)|$ to be the number of vertices in $R_i(v) \backslash P_{i-2}(v)$, and using $\alpha_i(v) = |R_i(v)|/(n-1)$, with n the number of vertices of G. The relevant inequalities to prove Lemma 3.3 become, for $u \neq v$ such that $u \in R_i(v)$,

$$|R_i(v)| \geq |R_i(u)|, \quad |R_{i-2,i+2}(v)| < |R_{i-2,i+2}(u)|.$$

(Note that, even if $R_i(v) \supset R_i(u)$, here we may have $|R_i(v)| = |R_i(u)|$ when u, v are part of a triangular face (v, u, w) such that u is the end of the arc of color $i - 2$ and w is the end of the arc of color $i + 2$ from v.) Hence, Lemma 3.4 still holds, which as before implies planarity.

Symmetries. Compared to Tutte's spring embedding [27] (and similarly as for Schnyder's drawing), the drawing algorithm (either in its face-counting or vertex-counting version) can display rotational symmetries, but not mirror symmetries. Precisely, there is a canonical 5c-wood corresponding to the *minimal 5c-orientation*, that is, the unique 5c-orientation with no counterclockwise directed cycle. If G is invariant by a rotation of order 5, then so is the minimal 5c-orientation, and so is the corresponding 5c-wood under a shift of the arc colors (and indices of the outer vertices) by 1. Thus, the drawing obtained using this canonical 5c-wood displays the rotational symmetry.

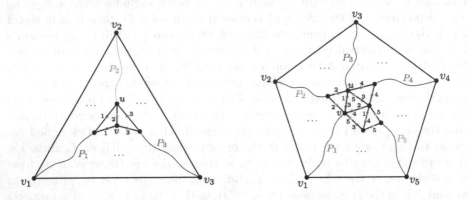

Fig. 10. The configuration (up to rotation and mirror) of vertex-pairs (u, v) of smallest possible distance in Schnyder's drawing (left) and in our drawing (right).

Quality of the Drawing. Besides the fact that the family of triangulations is more restricted, an obvious disadvantage of our algorithm compared to Schnyder's one (and also to straight-line drawings using transversal structures [16], or using shelling orders [8,15,19,21,22]) is that we can not use an affine transformation to turn the drawing into one on a regular square-grid of linear width and height[1]. Let us however discuss a parameter (vertex resolution) for which our algorithm brings some improvement. For $n \geq 1$, we let $\mu_3(n)$ (resp. $\mu_5(n)$) denote the smallest possible distance between vertices over Schnyder's drawings on simple triangulations with n vertices (resp. over our 5c-barycentric drawings on 5-connected triangulations with n vertices), assuming the drawing is normalized to have circumscribed circle of radius 1. Below we use complex numbers to represent points in the plane. Let $\omega_3 = e^{2i\pi/3}$ (resp. $\omega_5 = e^{2i\pi/5}$). For a vector $\delta = (\delta_1, \delta_2, \delta_3) \in \mathbb{Z}^3$ (resp. $\delta = (\delta_1, \delta_2, \delta_3, \delta_4, \delta_5) \in \mathbb{Z}^5$) whose components add up to 0, we define the *modulus* $\|\delta\|$ of δ as the modulus $|z|$ of the complex number $z = \sum_{i=1}^{3} \delta_i \omega_3^i$ (resp. $z = \sum_{i=1}^{5} \delta_i \omega_5^i$). The vector δ is called *possible* if there is a vertex pair (u, v) in a Schnyder's drawing (resp. in our drawing) such that the number of faces in $R_i(u)$ minus the number of faces in $R_i(v)$ equals δ_i for $i \in \{1, 2, 3\}$ (resp. for $i \in [1 : 5]$), in which case the distance between u and v in the drawing is $\|\delta\|/(2n-5)$ (resp. is $\|\delta\|/(2n-9)$).

As shown in Fig. 10 (and following an easy case inspection), the possible δ of smallest modulus (up to dihedral permutation of the entries) for Schnyder's drawing is $(-1, 2, -1)$, of modulus $d_3 = 3$, and for our drawing is

---

[1] If we use an affine transformation to have v_1, v_2 and v_5 placed at $(0,0)$, $(0,1)$ and $(1,0)$ respectively, then we get a drawing with vertex-coordinates in $\mathbb{Q}(\sqrt{5})$.

$(-2, -1, 3, 1, -1)$, of modulus $d_5 \approx 5.97$. Thus (for n large enough) we have $\mu_3(n) = \frac{d_3}{2n-5}$, and $\mu_5(n) = \frac{d_5}{2n-9}$. Hence, the vertices are better kept away from each other in our drawing, the worst-case distance being increased by a factor $d_5/d_3 \approx 1.99$ (to have a comparison over the same objects, we note that $\mu_3(n)$ is also attained by drawings of 5-connected triangulations with n vertices, for n large enough).

In the vertex-counting variant, the possible δ of smallest modulus (up to cyclic permutation of the entries) in Schnyder's algorithm is now $(0, 1, -1)$, of modulus $d'_3 = \sqrt{3} \approx 1.73$, and in our drawing is $(-1, -1, 1, 1, 0)$, of modulus $d'_5 \approx 3.08$ (these are again attained by the patterns shown in Fig. 10). Then we have (for n large enough) $\mu_3(n) = \frac{d'_3}{n-1}$ and $\mu_5(n) = \frac{d'_5}{n-2}$, so that the worst-case distance is increased by a factor $d'_5/d'_3 \approx 1.78$ (in both algorithms the vertex-counting variant brings a slight improvement over the face-counting version).

Open Questions: We conclude with some open questions:

Question 1. Can 5c-woods, and the 5c-barycentric algorithm, be generalized to 5-connected plane graphs in the spirit of the extension of Schnyder woods, and drawing algorithm, to 3-connected planar graphs [7,10, 11,20]?

Question 2. Can 5c-woods be used to define a graph drawing algorithm for the dual of 5c-triangulations (possibly with bent edges but restrictions on the directions of the edge segments). Such algorithms are known for the dual of triangulations [12,20], the dual of irreducible triangulations of the 4-gon [18], and the dual of quadrangulations [1,5,26].

Question 3. Is there a shelling procedure on 5c-triangulations to output a 5c-wood (perhaps by adapting the 5-canonical decomposition introduced in [23])?

Question 4. Can 5c-orientations be used (e.g. using the framework of [2]) to have a bijective derivation of the generating function of 5-connected triangulations, expressed in [17]?

Question 5. Is there a nice counting formula for the total number of 5c-woods on 5c-triangulations with n inner vertices? (For Schnyder woods such a formula is $\frac{6(2n)!(2n+2)!}{n!(n+1)!(n+2)!(n+3)!}$ [6].)

References

1. Bernardi, O., Fusy, É.: Schnyder decompositions for regular plane graphs and application to drawing. Algorithmica **62**(3), 1159–1197 (2012)
2. Bernardi, O., Fusy, É.: Unified bijections for maps with prescribed degrees and girth. J. Comb. Theory, Ser. A **119**, 1351–1387 (2012)
3. Bernardi, O., Fusy, É., Liang, S.: Grand Schnyder woods. arXiv preprint: arXiv:2303.15630 (2023)
4. Bernardi, O., Fusy, É., Liang, S.: A Schnyder-type drawing algorithm for 5-connected triangulations. arXiv preprint: arXiv:2305.19058v2 (2023)
5. Biedl, T., Kant, G.: A better heuristic for orthogonal graph drawings. Comput. Geom. Theory Appl **9**, 159–180 (1998)

6. Bonichon, N.: A bijection between realizers of maximal plane graphs and pairs of non-crossing Dyck paths. Discrete Math. **298**, 104–114 (2005)
7. Bonichon, N., Felsner, S., Mosbah, M.: Convex drawings of 3-connected plane graphs. Algorithmica **47**(4), 399–420 (2007)
8. Chrobak, M., Kant, G.: Convex grid drawings of 3-connected planar graphs. Int. J. Comput. Geometry Appl. **7**(03), 211–223 (1997)
9. Colin de Verdière, E.: Shortening of curves and decomposition of surfaces. Ph.D. thesis, Université Paris 7 (2003)
10. Di Battista, G., Tamassia, R., Vismara, L.: Output-sensitive reporting of disjoint paths. Algorithmica **23**(4), 302–340 (1999)
11. Felsner, S.: Convex drawings of planar graphs and the order dimension of 3-polytopes. Order **18**, 19–37 (2001)
12. Felsner, S.: Geodesic embeddings and planar graphs. Order **20**, 135–150 (2003)
13. Felsner, S., Schrezenmaier, H., Steiner, R.: Pentagon contact representations. Electr. J. Comb. **25**(3), P3.39 (2018)
14. Felsner, S., Zickfeld, F.: Schnyder woods and orthogonal surfaces. Discrete Comput. Geometry **40**(1), 103–126 (2008)
15. de Fraysseix, H., Pach, J., Pollack, R.: Small sets supporting Fáry embeddings of planar graphs. In: Proceedings of STOC, pp. 426–433. ACM (1988)
16. Fusy, É.: Transversal structures on triangulations: a combinatorial study and straight-line drawings. Discrete Math. **309**, 1870–1894 (2009)
17. Gao, Z., Wanless, I., Wormald, N.: Counting 5-connected planar triangulations. J. Graph Theory **38**(1), 18–35 (2001)
18. He, X.: On finding the rectangular duals of planar triangulated graphs. SIAM J. Comput. **22**, 1218–1226 (1993)
19. He, X.: Grid embedding of 4-connected plane graphs. Discrete Comput. Geometry **17**(3), 339–358 (1997)
20. Miller, E.: Planar graphs as minimal resolutions of trivariate monomial ideals. Doc. Math. **7**, 43–90 (2002)
21. Miura, K.: Grid drawings of five-connected plane graphs. IEICE Trans. Fundam. Electron. Commun. Comput. Sci. **105**(9), 1228–1234 (2022)
22. Miura, K., Nakano, S., Nishizeki, T.: Grid drawings of 4-connected plane graphs. Discret. Comput. Geom. **26**(1), 73–87 (2001)
23. Nagai, S., Nakano, S.: A linear-time algorithm to find independent spanning trees in maximal planar graphs. In: Graph-Theoretic Concepts in Computer Science: WG'2000, pp. 290–301 (2000)
24. Schnyder, W.: Planar graphs and poset dimension. Order **5**(4), 323–343 (1989)
25. Schnyder, W.: Embedding planar graphs in the grid. In: Symposium on Discrete Algorithms (SODA), pp. 138–148 (1990)
26. Tamassia, R.: On embedding a graph in a grid with the minimum number of bends. SIAM J. Comput. **16**, 421–444 (1987)
27. Tutte, W.: How to draw a graph. Proc. Lond. Math. Soc. **3**(1), 743–767 (1963)

A Logarithmic Bound for Simultaneous Embeddings of Planar Graphs

Raphael Steiner[(✉)] [iD]

Institute of Theoretical Computer Science, Department of Computer Science,
ETH Zürich, Zürich, Switzerland
raphaelmario.steiner@inf.ethz.ch

Abstract. A set $\mathcal{G}$ of planar graphs on the same number n of vertices is called *simultaneously embeddable* if there exists a set P of n points in the plane such that every graph $G \in \mathcal{G}$ admits a (crossing-free) straight-line embedding with vertices placed at points of P. A *conflict collection* is a set of planar graphs of the same order with no simultaneous embedding. A well-known open problem from 2007 posed by Brass, Cenek, Duncan, Efrat, Erten, Ismailescu, Kobourov, Lubiw and Mitchell, asks whether there exists a conflict collection of size 2. While this remains widely open, we give a short proof that for sufficiently large n there exists a conflict collection consisting of at most $(3 + o(1)) \log_2(n)$ planar graphs on n vertices. This significantly improves the previous exponential bound of $O(n \cdot 4^{n/11})$ for the same problem which was recently established by Goenka, Semnani and Yip.

We also give a computer-free proof that there exists a conflict collection of size 30, improving on the previously smallest known conflict collection of size 49 which was found using heavy computer assistance.

Keywords: Planar graphs · Simultaneous embedding · Conflict collection · Probabilistic method

1 Introduction

Given a planar graph G, a *straight-line embedding* of G in the plane is an injective mapping π from the vertex-set V of G to $\mathbb{R}^2$ such that adding in all the straight-line segments with endpoints $\pi(u)$ and $\pi(v)$ for every edge uv in G yields a crossing-free drawing of G in the plane. Given a planar graph G and a point set $P \subseteq \mathbb{R}^2$ in the plane, we say that G *has a straight-line embedding on P* or equivalently that it *embeds straight-line on P* if there exists a straight-line embedding of G in the plane in which every vertex of G is mapped to a distinct point of P. If G is a *labelled planar graph* with vertices numbered as $\{v_1, \ldots, v_n\}$, and if P is a *labelled point set*, that is, its elements are numbered as $P = \{p_1, \ldots, p_n\}$, then we say that G has a *label-preserving straight-line embedding on P* if the bijection $\pi : V(G) \to P$, $\pi(v_i) := p_i$, forms a straight-line embedding of G.

The study of straight-line embeddings of planar graphs is a classical area in graph drawing. For instance, one of the most fundamental results on this

© The Author(s), under exclusive license to Springer Nature Switzerland AG 2023
M. A. Bekos and M. Chimani (Eds.): GD 2023, LNCS 14466, pp. 133–140, 2023.
https://doi.org/10.1007/978-3-031-49275-4_9

topic, the Fáry-Wagner-Theorem [6], states that every planar graph G admits a straight-line embedding in the plane on *some* point set. However, it is not true that a planar graph on n vertices can be embedded on any given point set of size n. In fact, for a fixed planar graph G on n vertices, only a small fraction of all potential n-point sets may allow a straight-line embedding of G. Thus, many interesting questions in graph drawing arise from considering the embeddability of (restricted classes of) planar graphs on (restricted types of) point sets. One of the biggest branches of research in this direction concerns *simultaneous embeddings* of sets of planar graphs, we refer to [3] for a survey on this topic. Given a set $\mathcal{G}$ of planar graphs and a point set P in the plane, we say that $\mathcal{G}$ is *simultaneously embeddable on* P if every member $G \in \mathcal{G}$ admits a straight-line embedding on P. If $n \in \mathbb{N}$ and $\mathcal{G}$ is a set of planar graphs, each on n vertices, we say that $\mathcal{G}$ is *simultaneously embeddable* (without mapping) if there exists a point set $P \subseteq \mathbb{R}^2$ of size n such that $\mathcal{G}$ is simultaneously embeddable on P. Brass et al. [4] initiated a systematic study of the simultaneous embeddability of small sets $\mathcal{G}$ of planar graphs. In particular, they raised the following intriguing open problem, which remains unsolved.

Problem 1 (cf. [4]). Is there a set $\mathcal{G} = \{G_1, G_2\}$ consisting of two planar graphs of the same order such that $\mathcal{G}$ is not simultaneously embeddable?

Following the terminology of [5,7,9], let us call a set $\mathcal{G}$ of planar graphs, all of the same order $n \in \mathbb{N}$, a *conflict collection* if $\mathcal{G}$ is not simultaneously embeddable. Addressing small values of n, Cardinal et al. [5] proved that for $n \leq 10$, there exists no conflict collection consisting of n-vertex planar graphs. In contrast, they showed that for every $n \geq 15$ a conflict collection *does* exist.

Motivated by Problem 1, it is natural to study the value $\sigma(n)$, defined as the smallest size of a conflict collection of n-vertex planar graphs (if such a collection exists). By Fáry's theorem, we have $\sigma(n) \geq 2$ for every n, and Problem 1 is equivalent to the question whether there exists some n such that $\sigma(n) = 2$. Approaching this question, Cardinal et al. [5] constructed a relatively small conflict collection on 35-vertex graphs, proving that $\sigma(35) \leq 7393$. A significantly smaller conflict collection consisting of 11-vertex graphs was found by Scheucher et al. [9], showing that $\sigma(11) \leq 49$. Regarding the general asymptotic bounds on the function $\sigma(n)$, it was recently proved by Goenka et al. [7] that $\sigma(n) \leq O(n \cdot 4^{n/11}) < O(1.135^n)$, by an explicit general construction of a conflict collection on n-vertex planar graphs. While this bound is exponential in n, in this paper we give a short probabilistic proof that $\sigma(n) \leq (3 + o(1)) \log_2(n)$, and thus for large n much smaller conflict collections of n-vertex graphs of only logarithmic size in n exist.

Theorem 1. *For every $\varepsilon > 0$ there exists $n_0 = n_0(\varepsilon) \in \mathbb{N}$ such that for every integer $n \geq n_0$ there exists a set $\mathcal{G}_n$ of planar n-vertex graphs with $|\mathcal{G}_n| \leq \lceil (3 + \varepsilon) \log_2(n) \rceil$ such that $\mathcal{G}_n$ is not simultaneously embeddable.*

The rest of this note is devoted to presenting our proof of Theorem 1.

2   Order Types and Straight-Line Embeddings

Usually when working on a geometric problem relating to point sets that has a combinatorial flavour, small perturbations of the point set at hand do not change the behavior of the problem. In fact, often times one can reduce the infinite set of potential point sets P in question to a finite set of "types", such that all point sets of the same type have identical behavior for the considered problem. This is also the case here, if one groups the point sets consisting of n points in the plane according to their so-called *order type*.

In the following definition, formally the *orientation* of a triple abc of three distinct points $a = (a_1, a_2)$, $b = (b_1, b_2)$, $c = (c_1, c_2) \in \mathbb{R}^2$ in the plane is defined as the sign of the determinant of the matrix

$$\begin{pmatrix} 1 & 1 & 1 \\ a_1 & b_1 & c_1 \\ a_2 & b_2 & c_2 \end{pmatrix}.$$

It can be seen quite easily that the orientation of abc is 1 if the triangle abc is oriented counterclockwise, it is 0 if a, b, c are collinear and it is -1 if the orientation of abc is clockwise.

Definition 1 (cf. [1,2,5,8,9]).

- Given two finite point sets P and Q in the plane, we say that P and Q are *combinatorially equivalent* if there exists a bijection $f : P \to Q$ with the following property: For every ordered triple abc consisting of three distinct points of P, the *orientation* of the triple abc is the same as the orientation of the triple $f(a)f(b)f(c)$.
- Given two *labelled* point sets $P = \{p_1, \ldots, p_n\}$ and $Q = \{q_1, \ldots, q_n\}$ in the plane, we say that they are *isomorphic* if for every choice of $1 \leq i, j, k \leq n$ the orientations of the point triples $p_i p_j p_k$ and $q_i q_j q_k$ are identical.

It is readily verified that combinatorial equivalence and isomorphy form equivalence relations on the sets of unlabelled and labelled n-point sets in the plane, respectively. Accordingly, one can partition the set of n-point sets into equivalence classes w.r.t. combinatorial equivalence, which are called *order types*. Similarly, one can partition the set of labelled n-point sets into equivalence classes w.r.t. isomorphy, and these are called *labelled order types*. Note that by definition, two unlabelled point sets are combinatorially equivalent if and only if they admit labellings which are isomorphic. The following observation, which is folklore and is explained, for instance, in [1,5,9], shows that with regards to straight-line embeddability of planar graphs, it does not make a difference which among several combinatorially equivalent point sets we consider.

Observation 1 (cf. [1,5,9]).

(i) If two point sets P and Q in the plane are combinatorially equivalent, then for every planar graph G it holds that G embeds straight-line on P if and only if it embeds straight-line on Q.

(ii) If G is a labelled planar graph with $V(G) = \{v_1, \ldots, v_n\}$ and $P = \{p_1, \ldots, p_n\}$, $Q = \{q_1, \ldots, q_n\}$ are isomorphic labelled point sets, then G admits a label-preserving straight-line embedding on P if and only if it admits a label-preserving straight-line embedding on Q.

For our purposes, it will be important to have a limit on how many labelled order types exist, and luckily, Alon [2] provided an asymptotically precise answer to this question in 1986. In contrast, the precise asymptotics of the number of unlabelled order types remains currently unknown, see the discussion in [8].

Theorem 2 (cf. [2], Corollary 4.2). Let $t(n, 2)$ denote the number of distinct labelled order types of n-point sets in the plane. Then

$$t(n, 2) = n^{(4+o(1))n}.$$

The following consequence of Observation 1 and Theorem 2 will be useful later.

Corollary 1. *For every $n \in \mathbb{N}$ there exists a collection $\mathcal{P}_n$ consisting of labelled n-point sets in $\mathbb{R}^2$ such that $|\mathcal{P}_n| \leq n^{(4+o(1))n}$ and such that the following holds.*
  For every collection $\mathcal{G} = \{G_1, G_2, \ldots, G_k\}$ of labelled n-vertex planar graphs, if $\mathcal{G}$ is simultaneously embeddable, then there exists $P \in \mathcal{P}_n$ such that

- *G_1 admits a label-preserving straight-line embedding on P, and*
- *each of $G_2, \ldots, G_k$ embeds straight-line on P (but not necessarily in a label-preserving fashion).*

Proof. Define $\mathcal{P}_n$ by selecting for each of the $t(n, 2)$ distinct labelled order types one representative element, and adding it to $\mathcal{P}_n$. Then $|\mathcal{P}_n| = t(n, 2) \leq n^{(4+o(1))n}$. Now, let any collection $\mathcal{G} = \{G_1, \ldots, G_k\}$ of labelled n-vertex planar graphs be given, and suppose that $\mathcal{G}$ is simultaneously embeddable. Let Q be an n-point set such that each of $G_1, \ldots, G_k$ embeds straight-line on Q. Let $V(G_1) = \{v_1, \ldots, v_n\}$ be the labelling of the vertices of G_1 and let $\pi_1 : V(G_1) \to Q$ be a straight-line embedding of G_1 on Q. Now consider the labelling $Q = \{q_1, \ldots, q_n\}$ of Q, defined by setting $q_i := \pi_1(v_i)$ for $i = 1, \ldots, n$. Then clearly, G_1 has a label-preserving straight-line embedding on $\{q_1, \ldots, q_n\}$. Let $P \in \mathcal{P}_n$ be such that P and $\{q_1, \ldots, q_n\}$ are isomorphic labelled point sets. By Observation 1, (ii) we have that G_1 admits a label-preserving straight-line embedding on P. Since Q and P have isomorphic labellings, they are combinatorially equivalent. Thus, by Observation 1, (i) each of $G_2, \ldots, G_k$ embeds straight-line on P, as desired. This concludes the proof. ☐

3   Proof of Theorem 1

We prepare the proof by introducing a specific family of labelled planar graphs (which are stacked triangulations), that has already been studied in a similar context in [5, 7, 9].

Definition 2 (cf. [9], Definition 2 and [5], Section 3). For every integer $n \geq 4$, we define a set $\mathcal{T}_n$ of labelled planar graphs, each with vertex-set $V_n = \{v_1, \ldots, v_n\}$, inductively as follows:

– $\mathcal{T}_4$ consists only of the complete graph K_4 on vertex-set $V_4 = \{v_1, v_2, v_3, v_4\}$.
– If T is a labelled graph in $\mathcal{T}_{n-1}$ with $n \geq 5$, and $v_i v_j v_k$ is a facial triangle in T, then the labelled planar graph obtained from T by adding the new vertex v_n and connecting it to v_i, v_j, v_k is a member of $\mathcal{T}_n$.

It is crucial to notice that in the above definition and in the following $\mathcal{T}_n$ is to be understood as a set of *labelled planar graphs*, and distinct elements of the set are distinguished even if their underlying planar graphs are isomorphic. Concretely, it may happen that the same planar graph but with different labellings occurs several times in $\mathcal{T}_n$. Regardless, these different labellings are considered as different members of $\mathcal{T}_n$. For our proof of Theorem 1 we will need a few properties of the family $\mathcal{T}_n$, summarized in the following lemma.

Lemma 1 (cf. [9], Lemma 1 and [5], Lemma 1,2).

(i) For every $n \geq 4$, the family $\mathcal{T}_n$ contains exactly $2^{n-4}(n-3)!$ distinct labelled planar graphs.
(ii) Let P be any set of n points in the plane. Then for every bijection $\pi : \{v_1, \ldots, v_n\} \to P$ there exists at most one $T \in \mathcal{T}_n$ with the property that placing the vertex v_i of T at the point $\pi(v_i)$ of P, for every $i = 1, \ldots, n$, defines a straight-line embedding of T.

We need the following simple and direct consequence of the previous lemma.

Corollary 2. *Let $n \geq 4$ and let $\mathbf{G}$ denote a random labelled planar graph drawn uniformly from the set $\mathcal{T}_n$. Then the following hold.*

(i) If $P = \{p_1, \ldots, p_n\}$ is a labelled n-point set in the plane, then

$$\mathbb{P}(\mathbf{G} \text{ has label-preserving straight-line embedding on } P) \leq \frac{1}{2^{n-4}(n-3)!}.$$

(ii) If P is an n-point set in the plane, then

$$\mathbb{P}(\mathbf{G} \text{ embeds straight-line on } P) \leq \frac{16n(n-1)(n-2)}{2^n} = 2^{-(1-o(1))n}.$$

Proof. Let us first prove (i). Note that by definition a graph $T \in \mathcal{T}_n$ has a label-preserving straight-line embedding on $\{p_1, \ldots, p_n\}$ if and only if the mapping $\pi : \{v_1, \ldots, v_n\} \to P$, $\pi(v_i) := p_i$ forms a straight-line embedding of T. By item (ii) of Lemma 1 there is at most one member of $\mathcal{T}_n$ with this property. Since $\mathbf{G}$ follows a uniform distribution on $\mathcal{T}_n$, this implies, using item (i) of Lemma 1,

$$\mathbb{P}(\mathbf{G} \text{ has label-preserving straight-line embedding on } P) \leq \frac{1}{|\mathcal{T}_n|} = \frac{1}{2^{n-4}(n-3)!},$$

as desired. Let us now prove item (ii). To do so, let $p_1, \ldots, p_n$ be some enumeration of the points in P. Now, note that any labelled planar graph G embeds straight-line on P if and only if it has a label-preserving straight-line embedding on the labelled point-set $\{p_{\tau(1)}, \ldots, p_{\tau(n)}\}$ for some permutation τ of $\{1, \ldots, n\}$. We therefore have, using a union bound over the choices of τ,

$$\mathbb{P}(\mathbf{G} \text{ embeds straight-line on } P)$$

$$\leq \sum_{\tau \in S_n} \mathbb{P}(\mathbf{G} \text{ has label-preserving straight-line embedding on } \{p_{\tau(1)}, \ldots, p_{\tau(n)}\})$$

$$\leq \frac{n!}{2^{n-4}(n-3)!} = \frac{16n(n-1)(n-2)}{2^n},$$

as claimed. $\square$

We are now ready for the proof of our main result.

Proof (Proof of Theorem 1). Let $\varepsilon > 0$ be given, and let $n_0 = n_0(\varepsilon)$ be a sufficiently large integer (the conditions for "sufficiently large" are specified further below). Let $n \geq n_0$ be given. Let $k = \lceil (3 + \varepsilon) \log_2(n) \rceil$, and let $\mathbf{G}_1, \ldots, \mathbf{G}_k$ denote a sequence of k random planar graphs sampled independently from the uniform distribution on the set $\mathcal{T}_n$.

Let $\mathcal{G} := \{\mathbf{G}_1, \ldots, \mathbf{G}_k\}$ denote the resulting multi-set of k planar graphs. In the following we work towards arguing that the probability that the randomly generated multi-set $\mathcal{G}$ of planar n-vertex graphs is simultaneously embeddable is strictly less than 1. To bound the probability that $\mathcal{G}$ is simultaneously embeddable, we recall Corollary 1 which gives us a collection $\mathcal{P}_n$ of at most $n^{(4+o(1))n}$ labelled n-point sets with the property that if $\mathcal{G}$ is simultaneously embeddable, then there exists $P \in \mathcal{P}_n$ such that $\mathbf{G}_1$ has a label-preserving straight-line embedding on P, and each of $\mathbf{G}_2, \ldots, \mathbf{G}_k$ embeds straight-line on P.

For any fixed set $P \in \mathcal{P}_n$, we can use the independence of $\mathbf{G}_1, \ldots, \mathbf{G}_k$ and Corollary 2 to compute that

$$\mathbb{P}\left(\{\mathbf{G}_1 \text{ has label-pres. str.-line emb. on } P\} \wedge \bigwedge_{i=2}^{k} \{\mathbf{G}_i \text{ emb. str.-line on } P\} \right)$$

$$= \mathbb{P}(\mathbf{G}_1 \text{ has label-pres. str.-line emb. on } P) \cdot \prod_{i=2}^{k} \mathbb{P}(\mathbf{G}_i \text{ emb. str.-line on } P)$$

$$\leq \frac{1}{2^{n-4}(n-3)!} \cdot \left(2^{-(1-o(1))n} \right)^{k-1} = \frac{1}{n!} \cdot 2^{-(1-o(1))kn}.$$

Using a union-bound over all possible choices of $P \in \mathcal{P}_n$ we conclude:

$$\mathbb{P}(\mathcal{G} \text{ is simultaneously embeddable})$$

$$\leq \sum_{P \in \mathcal{P}_n} \mathbb{P}\left(\{\mathbf{G}_1 \text{ has label-pres. str.-line emb. on } P\} \wedge \bigwedge_{i=2}^{k} \{\mathbf{G}_i \text{ emb. str.-line on } P\} \right)$$

$$\leq |\mathcal{P}_n| \cdot \frac{1}{n!} \cdot 2^{-(1-o(1))kn}$$

$$\leq n^{(4+o(1))n} \cdot \frac{1}{n!} \cdot 2^{-(1-o(1))kn}.$$

Using that $n! = n^{(1-o(1))n}$ and $k \geq (3+\varepsilon)\log_2(n)$, we obtain:

$$\mathbb{P}(\mathcal{G} \text{ is simultaneously embeddable})$$

$$\leq 2^{(4+o(1))n\log_2(n)-(1-o(1))n\log_2(n)-(1-o(1))(3+\varepsilon)n\log_2(n)}$$

$$= 2^{-(\varepsilon-o(1))n\log_2(n)}.$$

As the exponent of the last term tends to $-\infty$ for $n \to \infty$, we have that the last term tends to 0 for $n \to \infty$. In particular, if $n_0 = n_0(\varepsilon)$ was chosen large enough, then we have $\mathbb{P}(\mathcal{G} \text{ is simultaneously embeddable}) < 1$ for every $n \geq n_0$. The probabilistic method now implies that there exists at least one multi-set consisting of $k = \lceil (3+\varepsilon)\log_2(n) \rceil$ planar graphs that is not simultaneously embeddable. The corresponding set $\mathcal{G}_n$, obtained by removing repeated elements, now confirms the statement of the theorem.   □

4   Conclusive Remarks

A careful review of our proof of Theorem 1 shows that it in fact proves an upper bound on $\sigma(n)$ of the form

$$\sigma(n) \leq \left\lfloor \frac{\log_2(t_s(n,2)) - (n-4) - \log_2((n-3)!)}{n - \log_2(16n(n-1)(n-2))} \right\rfloor + 2$$

for every $n \geq 16$. Here, $t_s(n,2)$ denotes the number of labelled order types on n points in the plane in general position. In order to obtain good upper bounds on $t_s(n,2)$, it is actually better to directly estimate $t_s(n,2)$ using Warren's inequality [10], instead of the upper bound of Alon [2] on $t(n,2)$. While the bounds are asymptotically the same, Warren's bound has the better lower-order terms, which matters for small values of n. Using Warren's inequality (cf. Theorem 2 in [10]), one may easily obtain the upper bound

$$t_s(n) \leq 2 \cdot 16^n \cdot \sum_{i=0}^{2n} 2^i \binom{\binom{n}{3}}{i}$$

for every $n \geq 3$. Plugging this estimate into the above bound, we obtain that $\sigma(n) \leq 30$ for all integers $n \in [107, 193]$. In particular, there exists a conflict collection consisting of 30 planar graphs on 107 vertices. This improves upon the previously smallest known conflict collection due to Scheucher et al. [9], which consisted of 49 planar graphs on 11 vertices.

References

1. Aichholzer, O., Aurenhammer, F., Krasser, H.: Enumerating order types for small point sets with applications. Order **19**, 265–281 (2002)
2. Alon, N.: The number of polytopes, configurations and real matroids. Mathematika **33**(1), 62–71 (1986)
3. Bläsius, T., Kobourov, S.G., Rutter, I.: Simultaneous embeddings of planar graphs. In: Tamassia, R. (ed.) Handbook of Graph Drawing and Visualization, Discrete Mathematics and Its Applications, ch. 11, pp. 349–382 (2013)
4. Brass, P., et al.: On simultaneous planar graph embeddings. Comput. Geom.: Theory Appl. **36**(2), 117–130 (2007)
5. Cardinal, J., Hoffmann, M., Kusters, V.: On universal point sets for planar graphs. J. Graph Algorithms Appl. **19**(1), 529–547 (2015)
6. Fary, I.: On straight line representation of planar graphs. Acta Scientiarum Mathematicarum **11**, 229–233 (1948)
7. Goenka, R., Semnani, P., Yip, C.H.: An exponential bound for simultaneous embeddings of planar graphs. Graphs Comb. **39**, 100 (2023)
8. Goaoc, X., Welzl, E.: Convex hulls of random order types. J. ACM **70**(1) (2023). Article No. 8
9. Scheucher, M., Schrezenmaier, H., Steiner, R.: A note on universal point sets for planar graphs. J. Graph Algorithms Appl. **24**(3), 247–267 (2020)
10. Warren, H.E.: Lower bounds for approximation by nonlinear manifolds. Trans. Am. Math. Soc. **133**, 167–178 (1968)

Manipulating Weights to Improve Stress-Graph Drawings of 3-Connected Planar Graphs

Alvin Chiu$^{(\boxtimes)}$, David Eppstein, and Michael T. Goodrich

Department of Computer Science, University of California, Irvine, USA
chiua13@uci.edu

Abstract. We study methods to manipulate weights in stress-graph embeddings to improve convex straight-line planar drawings of 3-connected planar graphs. Stress-graph embeddings are weighted versions of Tutte embeddings, where solving a linear system places vertices at a minimum-energy configuration for a system of springs. A major drawback of the unweighted Tutte embedding is that it often results in drawings with exponential area. We present a number of approaches for choosing better weights. One approach constructs weights (in linear time) that uniformly spread all vertices in a chosen direction, such as parallel to the x- or y-axis. A second approach morphs x- and y-spread drawings to produce a more aesthetically pleasing and uncluttered drawing. We further explore a "kaleidoscope" paradigm for this xy-morph approach, where we rotate the coordinate axes so as to find the best spreads and morphs. A third approach chooses the weight of each edge according to its depth in a spanning tree rooted at the outer vertices, such as a Schnyder wood or BFS tree, in order to pull vertices closer to the boundary.

Keywords: Tutte embedding · convex drawing · vertex spreading

1 Introduction

Sixty years ago, Tutte provided what is arguably one of the first graph drawing algorithms [16] Given a simple, undirected 3-connected planar graph, G, Tutte's algorithm produces a[1]. straight-line, planar drawing of G such that each face is convex. Tutte's algorithm produces such a drawing of G by solving a set of linear equations that determine the x- and y-coordinates of points to which the vertices of G are assigned. Intuitively, the equations are based on "pinning" the vertices of the outer face of G to the vertices of a convex polygon, and then considering all the edges of G to be springs with an idealized length of 0. Solving

---

[1] Proofs of Fáry's Theorem, that any simple, planar graph can be embedded in the plane without crossings so each edge is drawn as a straight line segment, came earlier [7,15,17], but these proofs do not give specific coordinates for the vertices; hence, it is not clear they can be called "graph drawing algorithms.".

© The Author(s), under exclusive license to Springer Nature Switzerland AG 2023
M. A. Bekos and M. Chimani (Eds.): GD 2023, LNCS 14466, pp. 141–149, 2023.
https://doi.org/10.1007/978-3-031-49275-4_10

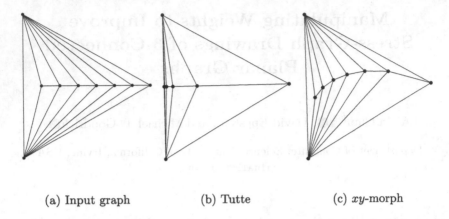

(a) Input graph   (b) Tutte   (c) xy-morph

Fig. 1. Tutte drawings can have exponential area.

the set of equations amounts to finding a minimum-energy configuration for the springs given the pinned vertices of the outer face [4,13].

One unfortunate drawback of Tutte's algorithm is that it can produce drawings with exponential area or exponentially small edge lengths, depending on the normalization of coordinates. Indeed, Eades and Garvan [5] show that this undesirable result occurs even for the planar graphs formed by connecting two outer vertices to each vertex of a simple path and to each other, as shown in Figs. 1a and 1b. Intuitively, the idealized springs representing graph edges have equal stress, which, in turn, "pull" groups of springs into unsightly vertex clusters.

Hopcroft and Kahn [11] generalize Tutte's algorithm to spring systems with different stress weights. In this framework, which we explore in this paper, we assign a stress weight, $w_{u,v}$, to each edge, (u, v), of G.[2] We begin as in the Tutte framework by "pinning" the vertices of an outer face, f, to be the vertices of a convex polygon, and we then formulate two linear equations for each internal vertex, u, of G, as follows:

$$\sum_{(u,v)\in E} w_{u,v}(x_u - x_v) = 0, \quad \text{and} \quad \sum_{(u,v)\in E} w_{u,v}(y_u - y_v) = 0, \qquad (1)$$

where $p_v = (x_v, y_v)$ is the point to which vertex v is assigned. Note that for a vertex, v, on the outer face, we pin $p_v = (x_v^*, y_v^*)$; hence, $x_v = x_v^*$ and $x_v = y_v^*$ are constants in our linear system. As Hopcroft and Kahn [11], as well as Floater [9], show, if the stresses, $w_{u,v}$, are all positive, except possibly for the edges of the outer face, then the resulting drawing is a planar straight-line drawing with each face being convex. In this paper, we experimentally explore the aesthetic improvements to a Tutte embedding that can be achieved by manipulating the stresses in such stress-graph drawings of 3-connected planar graphs.

Related Prior Results. We are not familiar with any prior work on the manipulation of the weights in stress-graph drawings strictly for the purpose of improv-

---

[2] Tutte's approach can be viewed as being for the case when $w_{u,v} = 1$ for each edge.

ing the aesthetic qualities. Nevertheless, the general technique of manipulating stresses in stress-graph drawings is not without precedent. For example, Hopcroft and Kahn [11] and Eades and Garvan [5] give conditions for stresses so that the resulting drawing is the projection of a 3-dimensional convex polyhedron onto the plane. Chrobak, Goodrich, and Tamassia [3] further explore this approach, claiming to produce a 3-dimensional realization of a 3-connected planar graph as the 1-skeleton of a 3-dimensional convex polyhedron with vertex resolution $\Omega(1)$ and with linear volume.[3] Indeed, their approach comes close to ours, in that they first compute weights for a weighted Tutte drawing with good vertex resolution (using a flow-based approach) and then apply the Maxwell–Cremona correspondence to lift this drawing to a convex polyhedron. Their method does not necessarily result in aesthetically pleasing drawings or convex polyhedra, despite the good spacing for the x-coordinates. Researchers have also explored interpolating between stress-graph drawings to morph from one layout to another. For example, Floater and Gotsman [10] use interpolation of the weights for two convex embeddings to morph between them, albeit with vertex movements that are represented implicitly. They also devise a method to obtain weights that will produce a given drawing. Erickson and Lin [6] morph between two convex via unidirectional morphs, where vertices move parallel to the direction of an edge. Kleist et al. [12] turn drawings of planar 3-connected graphs into strictly convex planar drawings with similar morphs.

Our Results. We propose several methods of weight manipulation. In the first, we simplify (and correct) the approach of Chrobak, Goodrich, and Tamassia [3] for finding drawings in which vertices have uniformly distributed coordinates. Instead of using iterated flows, we find suitable weights in linear time by counting certain paths in an st-orientation of the graph. Our implementation fixes the outer face as a regular polygon; in the full version on arXiv we show that an alternative choice allows all vertices, including outer face vertices, to have uniform x-coordinates. We experiment with a modified version of this method that produces two planar straight-line drawings that evenly spread the x-coordinates and the y-coordinates, respectively. We then construct a morph that averages the weights of the x- and y-spread drawings. The idea is that this morph will have fairly even spacing on both directions, e.g., as shown in Fig. 1c and Figs. 2a to 2d. We also explore a "kaleidoscope" version of this approach, where we rotate the coordinate axes to find the best spreads. In another approach, we weight edges based on depth in spanning trees rooted at the outer vertices. Edges closer to the outer vertices will have higher weight and thus more "pull", spreading the internal vertices away from the center of the outer face in a manner that preserves the general structure. We explore two types of spanning trees: BFS and (for fully triangulated graphs) Schnyder woods [1,8,14].

---

[3] However, their proof is only valid for polyhedra that have a triangle face.

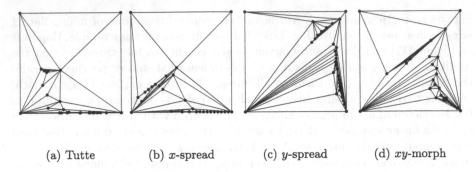

| (a) Tutte | (b) x-spread | (c) y-spread | (d) xy-morph |

Fig. 2. Drawings of a planar graph with 30 vertices and 80 edges, $G(30, 80)$.

2   Algorithms

Weight Manipulation to Spread Vertices Uniformly. To find weights whose stress-graph embedding spreads vertices evenly, we first begin with an unweighted Tutte drawing, rotating it if necessary so no edge is vertical. We sort the vertices by x-coordinates in this drawing, and orient edges from left to right, producing an st-orientation: an acyclic orientation in which each vertex v_i with $1 < i < n$ has both incoming and outgoing edges. Next, we choose new x-coordinates x_i for the interior vertices that are as evenly spaced as possible under the constraint that they respect the sorted x-ordering of all the vertices. (The same constraint is also present in the flow-based method of Chrobak et al. [3]) We can choose new positive edge weights for the Tutte drawing to produce the chosen x-coordinates in linear time. Conceptually, we gradually increase weights along a sequence of paths in the graph, starting with all weights zero. For each edge e, we find a directed path from v_1 to v_n through e, and increase weights on the edges of this path.

Along a single path through consecutive vertices v_i, v_j, v_k, the spacing between the vertex placements should be in the proportion $x_j - x_i : x_k - x_j$, which can be achieved by giving edges $v_i v_j$ and $v_j v_k$ the weights $1/(x_j - x_i)$ and $1/(x_k - x_j)$ respectively. Because these weights do not depend on the other edges of the path, we can use this weight for each edge in all of the paths that it belongs to and preserve the x-equilibrium. In total, the weight of any edge $v_i v_j$ in the whole graph (summing its weights for each path it appears in) will be $n_{ij}/(x_j - x_i)$, where n_{ij} is the number of paths containing edge $v_i v_j$.

To calculate these numbers efficiently, we compute two spanning trees in the oriented graph: tree T_1 directed out of v_1, and tree T_n directed into v_n (shortest-path trees via BFS were used for the implementation). For each edge $v_i v_j$, include a path that follows T_1 from v_1 to v_i, then edge $v_i v_j$, then follow T_n from v_j to v_n. We can count the number of these paths that use $v_i v_j$ as follows:

– There is one path defined in this way from $v_i v_j$.

- Let D_j be the set of descendants of v_j in T_1 (including v_j itself) and $d^+(v_k)$ be the number of outgoing edges from v_k. If $v_i v_j$ belongs to T_1, then $\sum_{v_k \in D_j} d^+(v_k)$ paths pass through $v_i v_j$ in T_1 before crossing to T_n.
- Let A_i be the set of descendants of v_i in T_n and $d^-(v_k)$ be the number of incoming edges at v_k. If $v_i v_j$ belongs to T_n, then symmetrically $\sum_{v_k \in A_i} d^-(v_k)$ paths pass through $v_i v_j$ in T_n after crossing to T_n.

The sums of descendant out-degrees in T_1, and of descendant in-degrees in T_n, can be computed in linear time by a simple bottom-up tree traversal, after which we can calculate the weight $n_{ij}/(x_j - x_i)$ of all edges in linear time. A weighted Tutte drawing with positive weights and convex outer face cannot introduce crossings, so we get a convex drawing with spread out x-coordinates using these new weights. To spread by a different direction, we can rotate the initial unweighted Tutte drawing before doing the spread. Indeed, as we explore experimentally, we consider a number of distinct rotation angles, producing drawings similar to the way a kaleidoscope produces patterns as it is turned.

Moreover, we can produce an "xy-morph" drawing of the input graph. Let a weighted Tutte drawing be represented by $\Gamma = (\Lambda, \mathcal{P})$, where Λ is the coefficient matrix containing the edge weights and $\mathcal{P}$ is the convex polygon chosen to be the outer face. One can morph between the x-coordinate spread drawing $\Gamma_0 = (\Lambda_0, \mathcal{P})$ and y-coordinate spread drawing $\Gamma_1 = (\Lambda_1, \mathcal{P})$ to obtain a more balanced graph drawing $\Gamma_{1/2}$. Intuitively, this is like stopping halfway in Floater and Gotsman's morphing algorithm [10], where we construct $\Gamma_{1/2} = (\Lambda_{1/2}, \mathcal{P})$ where $\Lambda_{1/2} = \frac{1}{2} \cdot \Lambda_0 + \frac{1}{2} \cdot \Lambda_1$. (See Fig. 2.)

Weight Manipulation via Spanning Tree Depth. In our spanning-tree approach, we first do a Tutte drawing, then we find a set of edge-covering spanning trees, T, for the graph rooted at the outer vertices, such as BFS trees or Schnyder woods [1,8,14]. Next, we assign weights to the edges of each tree, T, in a top-down manner according to its depth in the spanning tree. With these new weights, we do another stress-graph drawing.

Let the *depth* of an edge in a tree be the number of edges from the root to the edge plus one (to include the edge itself). Then we assign an edge at depth i with weight a/r^i, where a is some initial constant and r is a scaling parameter. When using BFS to find the shortest-path tree T_v rooted at an outer vertex v, we assign weights to an edge according to its lowest depth from any of the outer vertices. To do this, we create a dummy "super"-vertex, v_s, connected to all the outer vertices and run BFS from v_s, which is akin to running BFS on all the outer vertices simultaneously. For the case when the outer face is a triangle, we also consider Schnyder woods, which form an edge-covering set of three spanning trees that have nice "flow" properties [1,8,14]. (See Fig. 3.)

3   Experiments

Our experimental setup modifies the Open Graph Drawing Framework (OGDF) C++ library [2]. One of our goals is to compare our weight manipulation methods against Tutte's algorithm, which often produces exponentially small edge

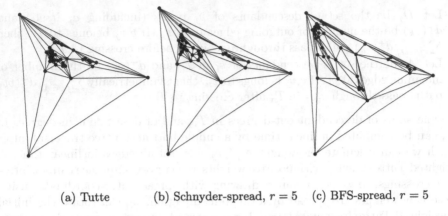

(a) Tutte   (b) Schnyder-spread, $r = 5$ (c) BFS-spread, $r = 5$

Fig. 3. Drawings of a pseudorandom graph, $G(50, 144)$.

lengths. Thus, the main metric we use is the *edge-length ratio* $\rho(\Gamma)$ of drawing Γ, which is the longest edge length divided by the smallest edge length in the drawing. In Table 1, we compare the edge-length ratios of the Tutte embeddings of several pseudorandom planar graphs against the x-spread, the y-spread, the xy-morph between the previous two, and the BFS-spread. For the BFS-spread, we choose the parameter r to be the integer that minimizes the edge-length ratio $p(\Gamma)$. We do not show the results for the Schnyder-spread, as they were almost always worse than the BFS-spread.

Not surprisingly, our testing demonstrates that the x-spread and y-spread drawings achieve edge-length ratio close to the number of vertices, n, because of the uniform vertex spacing that they produce. Nevertheless, optimizing exclusively for edge-length ratio can result in vertices that cluster close to a straight line as can be seen in Table 1. In contrast, the xy-spread drawing often is more aesthetically pleasing, as it tends to have better symmetry visualization than either the x- or y-spread drawings without clustering. However, it usually results in higher edge-length ratio than either of the two drawings it morphs. It may even have a higher edge-length ratio than its corresponding Tutte drawing, as seen by the xy-morph for $G(50, 130)$ in Table 1.

The edge-length ratio of BFS-spread drawings tends to be smaller than Tutte embeddings, while still preserving those drawings' general structure and symmetry visualization.

We also experimented with a "kaleidoscope" drawing paradigm, where we rotate the x- and y-axes by small angular increments and compute an xy-morph for each angle. The edge-length ratios can vary dramatically in such drawings, so the minima offer good choices. We show an example plot of edge-length ratios in Fig. 4, with its worst and best rotations in Table 2. More figures can be found in the full version on arXiv: https://arxiv.org/abs/2307.10527.

Table 1. Drawing Gallery. $\rho(\Gamma)$ is the edge-length ratio, r is the scaling parameter.

	Tutte	x-spread	y-spread	xy-morph	BFS-spread
$G(60, 150)$	$\rho(\Gamma) = 52717$	$\rho(\Gamma) = 86$	$\rho(\Gamma) = 63$	$\rho(\Gamma) = 2074$	$\rho(\Gamma) = 16156, r = 3$
$G(100, 200)$	$\rho(\Gamma) = 3973$	$\rho(\Gamma) = 43$	$\rho(\Gamma) = 71$	$\rho(\Gamma) = 229$	$\rho(\Gamma) = 380, r = 3$
$G(70, 200)$	$\rho(\Gamma) = 1165$	$\rho(\Gamma) = 42$	$\rho(\Gamma) = 69$	$\rho(\Gamma) = 216$	$\rho(\Gamma) = 283, r = 3$
$G(50, 130)$	$\rho(\Gamma) = 766$	$\rho(\Gamma) = 31$	$\rho(\Gamma) = 31$	$\rho(\Gamma) = 1020$	$\rho(\Gamma) = 272, r = 3$
$G(400, 1100)$	$\rho(\Gamma) = 56162$	$\rho(\Gamma) = 538$	$\rho(\Gamma) = 414$	$\rho(\Gamma) = 5054$	$\rho(\Gamma) = 9092, r = 2$

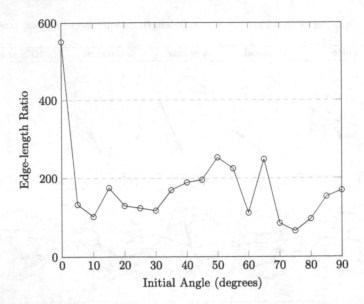

Fig. 4. Edge-length ratios for kaleidoscope xy-morphs for $G(90, 240)$, increments of $5°$C.

Table 2. Worst and best rotations for the graph of Fig. 4.

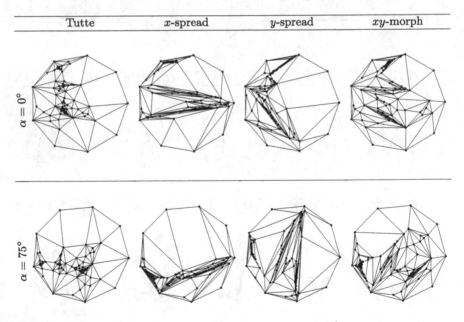

Acknowledgements. This research was supported in part by NSF grant CCF-2212129.

References

1. Bonichon, N., Felsner, S., Mosbah, M.: Convex drawings of 3-connected plane graphs. Algorithmica **47**(4), 399–420 (2007)
2. Chimani, M., Gutwenger, C., Jünger, M., Klau, G., Klein, K., Mutzel, P.: The open graph drawing framework (OGDF). In: Handbook of Graph Drawing and Visualization, pp. 543–569. CRC Press (2013)
3. Chrobak, M., Goodrich, M.T., Tamassia, R.: Convex drawings of graphs in two and three dimensions. In: 12th Symposium on Computational Geometry (SoCG), pp. 319–328. New York, NY, USA (1996). https://doi.org/10.1145/237218.237401
4. Di Battista, G., Eades, P., Tamassia, R., Tollis, I.G.: Graph Drawing: Algorithms for the Visualization of Graphs. Prentice Hall, Hoboken (1999)
5. Eades, P., Garvan, P.: Drawing stressed planar graphs in three dimensions. In: Brandenburg, F.J. (ed.) GD 1995. LNCS, vol. 1027, pp. 212–223. Springer, Heidelberg (1996). https://doi.org/10.1007/BFb0021805
6. Erickson, J., Lin, P.: Planar and toroidal morphs made easier. In: Purchase, H.C., Rutter, I. (eds.) GD 2021. LNCS, vol. 12868, pp. 123–137. Springer, Cham (2021). https://doi.org/10.1007/978-3-030-92931-2_9
7. Fáry, I.: On straight-line representation of planar graphs. Acta Scientiarum Mathematicarum **11**(2), 229–233 (1948)
8. Felsner, S.: Lattice structures from planar graphs. The Electronic Journal of Combinatorics, pp. R15–R15 (2004)
9. Floater, M.S.: Parametric tilings and scattered data approximation. Int. J. Shape Model. **04**(03n04), 165–182 (1998). https://doi.org/10.1142/S021865439800012X
10. Floater, M.S., Gotsman, C.: How to morph tilings injectively. J. Comput. Appl. Math. **101**(1), 117–129 (1999). https://doi.org/10.1016/S0377-0427(98)00202-7, https://www.sciencedirect.com/science/article/pii/S0377042798002027
11. Hopcroft, J.E., Kahn, P.J.: A paradigm for robust geometric algorithms. Algorithmica **7**(1–6), 339–380 (1992)
12. Kleist, L., Klemz, B., Lubiw, A., Schlipf, L., Staals, F., Strash, D.: Convexity-increasing morphs of planar graphs. Comput. Geom. **84**, 69–88 (2019)
13. Kobourov, S.G.: Spring embedders and force directed graph drawing algorithms. arXiv preprint arXiv:1201.3011 (2012)
14. Schnyder, W.: Embedding planar graphs on the grid. In: 1st ACM-SIAM Symposium on Discrete Algorithms (SODA), pp. 138–148 (1990)
15. Stein, S.K.: Convex maps. Proc. Am. Math. Soc. **2**(3), 464–466 (1951)
16. Tutte, W.T.: How to draw a graph. Proc. Lond. Math. Soc. **3**(1), 743–767 (1963)
17. Wagner, K.: Bemerkungen zum Vierfarbenproblem. Jahresber. Deutsch. Math.-Verein. **46**, 26–32 (1936)

Frameworks

Computing Hive Plots: A Combinatorial Framework

Martin Nöllenburg⬚ and Markus Wallinger(⬚)⬚

Algorithms and Complexity Group, TU Wien, Vienna, Austria
{noellenburg,mwallinger}@ac.tuwien.ac.at

Abstract. Hive plots are a graph visualization style placing vertices on a set of radial axes emanating from a common center and drawing edges as smooth curves connecting their respective endpoints. In previous work on hive plots, assignment to an axis and vertex positions on each axis were determined based on selected vertex attributes and the order of axes was prespecified. Here, we present a new framework focusing on combinatorial aspects of these drawings to extend the original hive plot idea and optimize visual properties such as the total edge length and the number of edge crossings in the resulting hive plots. Our framework comprises three steps: (1) partition the vertices into multiple groups, each corresponding to an axis of the hive plot; (2) optimize the cyclic axis order to bring more strongly connected groups near each other; (3) optimize the vertex ordering on each axis to minimize edge crossings. Each of the three steps is related to a well-studied, but NP-complete computational problem. We combine and adapt suitable algorithmic approaches, implement them as an instantiation of our framework and show in a case study how it can be applied in a practical setting. Furthermore, we conduct computational experiments to gain further insights regarding algorithmic choices of the framework. The code of the implementation and a prototype web application can be found on OSF.

Keywords: hive plots · graph clustering · circular arrangement · layered crossing minimization

1 Introduction

Hive plots [16] are a visualization style for network data, where vertices of a graph are mapped to positions on a predefined number of radially emanating axes[1]. Mapping and positioning is usually done based on vertex attributes and not with the intention to optimize layout aesthetics. Due to this strict, rule-based definition, hive plots are a deterministic network visualization style; see Fig. 1 for an example. Similarly to parallel coordinate plots [26], the idea behind hive

---

This research has been funded by the Vienna Science and Technology Fund (WWTF) [10.47379/ICT19035].

[1] https://osf.io/6zqx9/(10.17605/OSF.IO/6ZQX9).

ⓒ The Author(s), under exclusive license to Springer Nature Switzerland AG 2023
M. A. Bekos and M. Chimani (Eds.): GD 2023, LNCS 14466, pp. 153–169, 2023.
https://doi.org/10.1007/978-3-031-49275-4_11

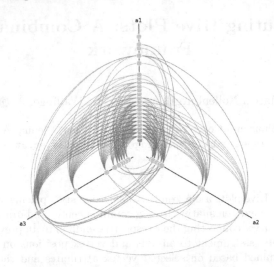

Fig. 1. A hive plot created with jhive [16]. Vertices are mapped to axis according to their degree. The position on each axis is determined by vertex attributes such as degree (axis a_1), vertex betweenness (axis a_2), and reachability (axis a_3).

plots is to quantitatively understand and compare network structures, a task that can quickly get very difficult with force-based layouts due to the 'hairball' effect for large and dense graphs and their often unpredictable behaviour when optimizing for conflicting aesthetic criteria.

Usually, edges are drawn as Bézier curves connecting their respective end points while being restricted to three axes to avoid the problem of routing longer edges around axes; this is considered beneficial for visual clarity. In case of edges between vertices on the same axis, the axis and its associated vertices are cloned and positioned closely such that edges are either drawn twice (symmetrically) or once (asymmetrically). The latter case reduces visual complexity but increases ambiguity as an edge is only explicitly connected to one copy of each vertex; see Fig. 2 for a sketch of the different concepts.

Multiple hive plots can also be arranged in a matrix, called a hive panel [16], where columns and rows represent different axis mapping functions. Differential hive plots visualize networks changing over time [17]. Since their inception a decade ago, hive plots have been utilized in various applications and use-cases, e.g., cyber security [15], machine learning of visual patterns [23], life sciences [22], biological data [28], or sports data [21]. Although various use-cases exist, hive plots have not yet been investigated from a formal graph drawing perspective.

This is a rather unexpected observation, especially, as hive plots have some inherent properties that make them an interesting layout style. For example, by placing vertices on axes the layout is predictable and has usually a good aspect ratio. Similarly, edges can be routed in a predictable manner. Thus, edges overlapping with unrelated vertices is not an issue in hive plot layouts and increases the overall faithfulness of the drawing. Furthermore, it is also relatively straight forward to position labels and avoid label-edge and label-vertex overlaps. Lastly, edges between vertices on the same axis can be hidden or shown on demand, thus, reducing unnecessary information and decreasing the cognitive load.

Contributions and Related Work. In this paper, we present a formal model of hive plots and identify their associated computational optimization problems from a combinatorial point of view. Based on this model, we investigate several unused degrees of freedom that can be utilized for optimizing hive plot layouts for arbitrary undirected graphs.

First, in our investigation we take a new angle on assigning vertices to axes. Basically, the idea is to partition the graph into some number k of densely connected clusters, where each cluster is assigned to exactly one axis. In terms of visual design this allows us to show or hide intra-cluster edges on demand and focus on representing the sparse connectivity between clusters. We find such clusters by applying techniques from the area of community detection in networks [11]. Even though a similar assignment strategy is presented in the original hive plot publication [16], the focus there is on visually clustering vertices according to their community membership and assigning vertex clusters to segments on subdivided axis.

Second, we are free to assign any cyclic order over the k different axes. Here we optimize the total length of inter-axis edges by placing axes with many edges between them close to each other. This is essentially the circular arrangement problem. In the circular arrangement problem vertices are positioned evenly spaced on a circle such that the total weighted length of edges is minimized. Finding the minimum circular arrangement of undirected and directed graphs is NP-complete [12,18]. However, a polynomial-time $O(\log n)$-approximation for undirected graphs exists [18]. Similarly, the problem of minimizing the crossings in a circular arrangement of a graph is NP-complete [19]. The concept of circular arrangements has been applied to circular drawings [13] where a subset of edges is drawn outside of the circle to reduce edge crossings.

Lastly, once the order of axes is fixed we want to minimize the number of inter-axis edge crossings. Here, the problem is similar to multi-layer crossing minimization which has been studied in the context of the Sugiyama framework [24,25] for hierarchical level drawings of directed graphs. In this type of drawing vertices are assigned to horizontal layers with edges either drawn in upward or downward direction. In case of cycles in the graph, some edges need to be reversed in the drawing. Cyclic level drawings have already been mentioned by Sugiyama et al. as an alternative nach ablauf to reversing edges, and they have been thoroughly investigated in more recent years [1–3,13].

Crossing minimization in cyclic level drawings and layered drawings is repeatedly performing a 2-layer crossing minimization step. The 2-layer crossing minimization problem is already NP-hard [8], even if one layer is fixed. Heuristics [9,20] have been proposed and adapted for cyclic level drawings [1]. We adapt the barycenter algorithm [9] to efficiently reduce the number of crossings while adding novel constraints to force long edges to not cross over axes. Adding constraints to 2-layer crossing minimization heuristics has been applied previously, e.g., for fixing the relative positions of a subset of vertex pairs [10].

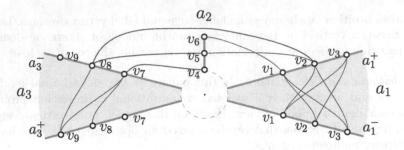

Fig. 2. Schematized hive plot with three axes showing different concepts. Axis a_2 is collapsed. Axis a_1 and a_3 are expanded with edges in a_1 being drawn symmetrically. A long edge between v_2 and v_8 is bypassing a_2.

We combined and implemented the three above mentioned problems into a 3-step framework. Finally, we show in a case study how hive plots generated by our framework can be applied in a practical context of co-authorship networks and conduct a small-scale computational experiment on the crossing minimization aspect of our framework.

2 Formal Model

A *hive plot layout* $H(G) = (A, \alpha, \phi, \Pi)$ of a graph $G = (V, E)$ is a tuple consisting of a set $A = \{a_0, \ldots, a_{k-1}\}$ of k axes, a surjective function $\alpha : V \to A$ mapping vertices to axes, a bijective function $\phi : A \to \{0, ..., |A| - 1\}$ representing a cyclic ordering of axes and a set $\Pi = \{\pi_0, \ldots, \pi_{|A|-1}\}$ of orderings over the vertices assigned to each axis. Each vertex is assigned to exactly one axis $a_i \in A$ imposing a disjoint grouping $V_i := \alpha^{-1}(a_i)$ such that $V_i \cap V_j = \emptyset$ for each $i \neq j$ with $i, j \in \{0, \ldots, |A| - 1\}$. Each π_i is a bijective function $\pi_i : V_i \to \{0, \ldots, |V_i| - 1\}$.

We use the shorthand notation $\phi(u) = \phi(\alpha(u))$ whenever the order of the axis $\alpha(u)$ of a vertex u is needed. The *span* of two axes a_i, a_j or two vertices u, v is defined as $\mathrm{span}(a_i, a_j) = \min\{\phi(a_i) - \phi(a_j) \pmod{k}, \phi(a_j) - \phi(a_i) \pmod{k}\}$ or $\mathrm{span}(u, v) = \mathrm{span}(\alpha(u), \alpha(v))$. Based on the span we can classify edges into three different categories. An edge $e = (u, v)$ is called *proper* if $\mathrm{span}(u, v) = 1$. Otherwise, an edge is considered *long* if $\mathrm{span}(u, v) > 1$ or *intra-axis* if $\mathrm{span}(u, v) = 0$. *Inter-axis* edges are all edges that are either proper or long. A long edge (u, v) can be subdivided and replaced by $\mathrm{span}(u, v) - 1$ dummy vertices assigned to the appropriate axes between $\alpha(u)$ and $\alpha(v)$. A long edge in a hive plot layout needs to *bypass* axes to connect source and target vertices without creating axis-edge overlaps. Combinatorially this can be realized by enforcing dummy vertices to appear at certain positions in each axis order. In our model, a hive plot layout can have up to g gaps per axis, see Fig. 6 for examples. If $g = 1$ then all dummy vertices have to be at the end of each order. If $g = 2$ then dummy vertices have to be at either the beginning or end of each order. In cases where $g > 2$ all dummy vertices form a partition into up to g groups, where they must appear consecutively within each group.

To consider intra-axis edges during optimization an adaption is necessary. Basically, all axes and their associated vertices are duplicated such that for each axis a_i there are two copies a_i^+ and a_i^-, respectively, and vertex sets V_i^+ and V_i^-; see Fig. 2. The vertex order on duplicate axes remains the same, i.e., $\pi_i^+ = \pi_i^- = \pi_i$.

We consider two inter-axis edges (u, v) and (x, y) to be *crossing* if $u, x \in V_i$ and $v, y \in V_j$ such that $\pi_i(u) < \pi_i(x)$ and $\pi_j(y) < \pi_j(v)$. Similarly, if the end points of two long edges (u, v) and (x, y) are on four different axes such that $\phi(u) < \phi(x) < \phi(v) < \phi(y)$ (mod k) or on three different axes such that, w.l.o.g., $\phi(u) = \phi(x) = i$, $\pi_i(u) < \pi_i(x)$, and $\phi(x) < \phi(y) < \phi(v)$ (mod k) a crossing is unavoidable. The *neighborhood* of a vertex u is defined as $N(u) = \{v \mid (u, v) \in E, \text{span}(u, v) = 1\}$.

3 Framework for Computing Hive Plots

Next we present our framework for creating a hive plot layout $H(G) = (A, \alpha, \phi, \Pi)$ of an undirected simple graph $G = (V, E)$. The framework itself is modeled as a pipeline consisting of three stages. In stage (1) we partition the vertices into multiple groups each corresponding to an axis of the hive plot. Next, we (2) optimize the cyclic axis order to bring strongly connected groups near each other. Finally, we (3) optimize the vertex ordering on each axis to minimize edge crossings. Furthermore, edge crossing minimization is performed under the constraint that long edges need to be routed through gaps in the axis.

3.1 Vertex Partitioning

In the first stage we partition the vertex set V into subsets $\{V_0, \ldots, V_{k-1}\}$ such that each subset maps to exactly one axis a_i in the hive plot. The core idea is that the subsets of the partition represent dense induced subgraphs. In our implementation we use three different strategies to compute a partition. First, if we consider the parameter k as an additional input we use the Clauset-Newman-Moore greedy modularity maximization [5] to compute a partition of size k. Second, if k is not specified we apply the Louvain [4] community detection algorithm instead. Here, the size of the partition is determined by how many communities are detected. Lastly, this step of the framework is not necessary if a partition is given in the input. Note, any other algorithm to partition the graph into meaningful groups can be used.

3.2 Axis Ordering

The second stage orders the axes such that the total *span* of edges is minimized. Our approach assumes that edges are always drawn along the shortest path around the circle between endpoints. Basically, we want to maximize the number of proper edges while minimizing the number and length of long edges. We do not consider the individual position of vertices on their respective axes, but rather look at the aggregated edges incident to the subsets of the axis partition.

Let w_{ij} be the number of edges between V_i and V_j for $i < j$. The cost function of an axis order ϕ is defined as follows:

$$\text{cost}(\phi) = \sum_{i=0}^{k-1} \sum_{j=i+1}^{k-1} w_{ij} \; \text{span}(a_i, a_j)$$

We can afford using an exact brute-force approach for instances with $k \leq 8$ to minimize ϕ, as it takes less than $0.5\,\text{s}$ on our reference machine (see Sect. 6); otherwise we use simulated annealing.

3.3 Crossing Minimization

In the third stage of our framework we are concerned with minimizing edge crossings under the assumption that assignment to axes and the cyclic axis order are already fixed. On each axis, the vertices are initially in random order. Here, we employ a two-step approach, where first crossings of long edges and then intra-axis edge crossings are minimized. Additionally, we assume that a global parameter $g \geq 1$, which represents the maximal number of gaps per axis, is given. If $g = 1$ we assume that edges are routed on the outside. In case of $g = 2$ we assume that gaps are on the outside and inside of each axis. If $g > 2$ we evenly distribute the gaps along each axis.

First, we process all long edges to turn them into sequences of proper edges. Each edge $e = (u, v)$ with $\text{span}(u, v) > 1$ is subdivided by inserting $\text{span}(u, v) - 1$ dummy vertices assigned to the appropriate axes between $\alpha(u)$ and $\alpha(v)$. Isolated vertices which are not an endpoint of at least one long edge can be ignored in the first step of the crossing minimization.

Next, we use the barycenter heuristic [25] for crossing minimization. Our approach works by iterating several times in clockwise or counter-clockwise order over all axes while performing a layer-by-layer crossing minimization sweep. At each iteration we process all vertices of an axis by computing a new barycenter position as follows:

$$\text{pos}(u) = \frac{1}{|N(u)|} \sum_{v \in N(u)} \frac{\pi_{\alpha(v)}(v)}{|\pi_{\alpha(v)}|}.$$

As it is necessary to avoid cases where axes are imbalanced, we normalize both axes and consider the neighbourhood $N(u)$ of vertex u in the next and previous axes.

The reason for considering both axes is that when only considering the previous axis crossings might be introduced that are overall worse from the reverse direction.

Once barycenter positions are calculated we sort all vertices $v \in V_i$ of axis a_i by their positions $\text{pos}(v)$. Now, to consider gaps we have to apply a case distinction on g. If $g = 1$ we simply want dummy vertices on the outside to route the long edges around axes. We constrain the sorting algorithm to put all normal vertices before all dummy vertices. For $g > 1$ we perform the following

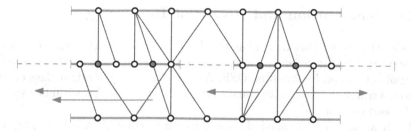

Fig. 3. An axis with $g = 3$ gaps. Gaps are indicated by dashed lines while axis segments are solid lines. Vertices colored red are dummy vertices. After computing new positions we move dummy vertices into either a gap to the left or right. The side is determined by counting the crossings and greedily picking the better option. The order of dummy vertices remains unchanged after the procedure. (Color figure online)

procedure. We create a list of g empty lists representing the gaps and a list of $g-1$ empty lists representing the segments of the axis between gaps. We initialize a counter $j = 0$ that represents the index of the current list. Next, we iteratively process vertices according to the previously computed order and distinguish between normal and dummy vertices. Whenever we encounter a normal vertex we append it to the list of axis segments at position j. If the list contains more than $\frac{|V_i|}{g-1}$ vertices we increase j by one. Here, $|V_i|$ represents the number of normal vertices on axis a_i. The main idea behind this is that normal vertices are evenly assigned to axis segments to increase symmetry. If we encounter a dummy vertex we have to decide if we assign it to the gap to the left or the right of the current axis segment. Here, the decision can be made by looking at all vertices in the same axis segment that are to the left and compute the crossings if we put the dummy vertex in the gap to the left. Similarly, we repeat the process for the right-hand side and choose the gap which induces less crossings. Thus, appending the dummy vertex in the list at index j or $j + 1$. Figure 3 illustrates how the gap assignment works. Finally, we assemble a new list by adding all vertices alternating between gap and axis list. The new position of a vertex is determined by the respective index in the list.

We terminate the overall process after either no change is detected for one cycle or some iteration threshold is reached.

In the second phase of the crossing minimization we aim to further reduce intra-axis crossings by applying the barycenter heuristic. As the focus of the layout is on edges between different axes, we introduce the additional constraint that the relative order of vertices incident to inter-axis edges are not allowed to change in this phase any more. Basically, this is again a classic 2-layer crossing minimization for a vertex subset, however both layers have the same order. Moreover, we apply the same procedure as described above to constrain dummy vertices to gap positions. We process each axis individually and terminate processing of an axis once no change is detected or the iteration threshold is reached.

4   Implementation and Hive Plot Rendering

In this section we will briefly explain the design decisions regarding the visualization; see Fig. 4 or Fig. 8 for examples. The implementation code and a prototype web application can be found on OSF. Axes are drawn as straight lines emanating from a common center with equal angular distribution. Axes can be expanded or collapsed on demand.

When an axis is expanded the background is colored in a light grey color with low opacity. When expanded, the available space is distributed 40 : 60 between intra-axis and inter-axis edges. Vertices are drawn as small circles and their positions on their respective axis a_i are computed based on their position in π_i. Labels are placed in clockwise direction next to an axis horizontally. If a vertex' assigned axis differs by $\pm 25°$ to the horizontal reference direction, the label is rotated by $45°$. Edges are drawn as Bezíer curves. For edges between neighbouring axes control points are set perpendicular to the axes. If long edges are routed around axes, the positions of their dummy vertices are converted to control points. The color of vertices is computed by mapping the angle of the assigned axis to a radial color map [6]. For edges, we assign the color of the first endpoint in counter-clockwise direction. The ideas behind coloring edges are that it becomes easier to follow individual edges and that it is, for half of the vertices, immediately clear to which axis the edge connects.

Interactivity. Figure 4d shows an example how interactivity was realized in our visualization. First, when hovering a vertex, the vertex itself, all neighbours and incident edges are highlighted by a color contrasting the color scheme. Initially, each axis in the visualization is collapsed. By clicking on a single axis it is expanded to show more details on demand. Furthermore, it is also possible to expand all axes with a button click. Naturally, it is also possible to unexpand individual axis. Similarly, by clicking a button in the interface vertices can be scaled to represent their respective degree; see Fig. 4b. Lastly, labeling can be toggled on or off.

5   Case Study

We evaluate our framework by a case study using the *citation* dataset [7] from the creative contest at the 2017 Graph Drawing conference. This dataset contains all papers published at GD from 1994 to 2015. We created co-authorship graphs for different years by extracting researchers from papers and connecting them with edges whenever they co-authored a paper.

In Fig. 4 we show a hive plot of the co-author network of 2015 and three alternative renderings computed with our framework in less than 10 ms. We used the rendering style described in Sect. 4. We did not specify the number of axes k in the input. We specified the number of gaps as $g = 1$. The network has a total of 75 vertices, which are partitioned into 7 groups. A total of 190 edges is split into 172 intra-axis edges, 12 proper edges and 6 long edges.

Authors mapped to individual axes seem to represent mainly clusters of geographic proximity of researchers' institutions or established close collaborations. Inter-axis edges are emphasized in our hive plots and indicate collaborations between clusters in the form of researchers bridging institutions and forming new connections, e.g., via papers originating from research visits or recent changes in affiliation. Another possible interpretation can be seen when vertices are scaled by degree. Researchers with connections to other axes are often also highly connected inside their own cluster. This could mean that they are well connected and prolific and use existing connections to start new collaborations.

In contrast to a force-based layout, see Fig. 5a, several observations can be made. While cliques are very prominent in the force-based layout the macro structure of the graph is less clear. The hive plot layout on the other hand focuses more on the macro structure of the graph with intra-axis structure only shown on demand. However, with two copies per vertex, cliques are harder to identify. Still, the hive plot layout requires no additional cue, such as color in this case, to highlight the community structure. Furthermore, in the hive plot layout individual vertices are easier to identify, labels are more uniform and edges are routed in a predictable manner which is similar to schematic diagrams. Due to the possibility of expanding axes on demand in the hive plot layout, individual communities can be easily explored without being overwhelming. While it is possible to represent communities in the force-based layout by meta vertices, it is not straight-forward to encode the relationship of each single vertex to the rest of the network. Both layouts show some label-edge overlap. Finally, the hive plot layout has a more balanced space utilization.

Furthermore, we also compared against a hierarchical layout; see Fig. 5b. Here, we assumed edge direction from the hive plot layout by directing edges clockwise. Naturally, the hierarchical layout emphasizes the imposed hierarchy while the communities are dispersed over the layout. Still, the communities are visible although it is questionable without the use of coloring. This gives the layer assignment a different meaning than our approach of axis assignment. The orthogonal layout of edges initially simplifies the process of following an edge, but it becomes progressively more challenging with an increase in bends and crossings. In the hive plot layout this is less of an issue, especially for edges between communities. Lastly, the label placement in the hierarchical layout is optimized and avoids label-edge and label-vertex overlaps. However, this optimization comes at the cost of requiring more space. In contrast, the hive plot layout has a few label-edge overlaps but utilizes space more efficiently.

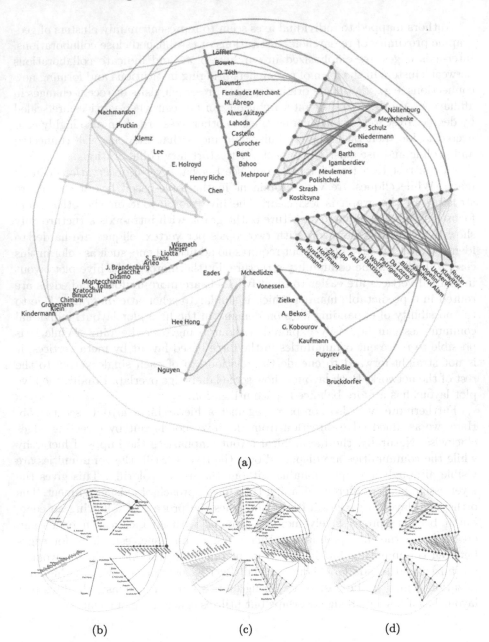

(a)

(b)   (c)   (d)

Fig. 4. Variations of the co-author graph of GD 2015. In (a) some axes of interest are expanded while (c) shows all axes expanded. In (b) vertices are scaled by degree and all axes are collapsed. In (d) interactive highlighting is obtained by hovering the vertex "M. Nöllenburg".

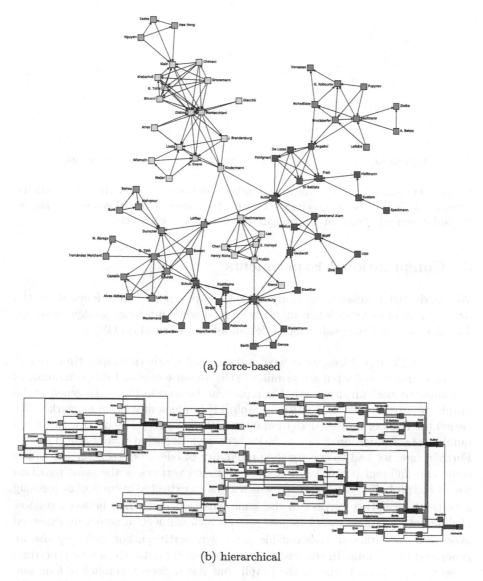

(a) force-based

(b) hierarchical

Fig. 5. Force-based layout (a) of the co-authorship graph of Sect. 5 created with yEd [27]. The smart organic layout functionality was used. The preferred edge length was set to 100 while the minimal vertex distance was set to 20. The option of avoid vertex/edge overlaps was active with a value of 0.8. Also, the labeling was optimized and the graph colored according to the partitions from the case study. (b) shows a hierarchical layout created with yEd. Similar layout optimization steps were applied.

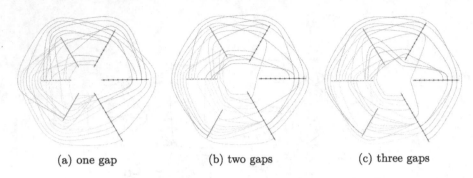

(a) one gap   (b) two gaps   (c) three gaps

Fig. 6. One example graph from the synthetic datasets used in the computational experiment. (a) shows one gap on the outside, (b) shows an additional second gap on the inside and (c) allows for a third gap in the middle of each axis.

6   Computational Experiments

We conducted a small-scale computational experiment which focused on the implications of using different numbers of gaps while minimizing edge crossings. The datasets and the evaluation code can again be found on OSF.

Dataset and Setup. First, we created a dataset of synthetic graphs that tries to capture variations of input sizes similar to the use-case of visualizing communities in a small to medium size graph. Examples can be seen in Fig. 6. To generate the graphs we used a random partition graph [11] implementation of networkX. We varied the number n of vertices from 60 to 510 with a step size of 30 with a fixed number of six partitions $\{V_1, \ldots, V_6\}$ where each partition was of size $|V_i| = \frac{n}{6}$. Furthermore, we had to specify the probability of edges between vertices of the same and different partitions. For edges between vertices of the same partition we set the probability to $p_{in} = \frac{6}{|V_i|}$ which gives an expected average of connecting a vertex to six other vertices in the same partition. For edges between vertices of different partitions we set $p_{out} = \frac{2}{n-|V_i|}$ which connects a vertex on expected average to two other vertices outside of its own partition. For each step size we generated five graphs. In contrast to using real-world data, the above procedure gives us a predictable size of the graph that has a decent number of long and proper edges.

The implementation to compute a hive plot is written in Python 3.11. All experiments where run on a machine with Ubuntu 22.04, 16GiB of RAM and an Intel i7-9700 CPU with 3.00 Hz.

Experiment. In our experiment we evaluated the effect of varying the number of gaps in the input. We computed a hive plot layout for all graphs in the dataset with fixed $k = 6$ but varied $g \in \{1, 2, 3\}$.

We counted the number of crossings of intra-axis edges and the number of crossings of inter-axis edges. We did not consider crossings of inter-axis edges with intra-axis edges as this can only be observed if $g \geq 3$. The resulting plots can be seen in Fig. 7a and Fig. 7b. Interestingly, the number of intra-axis crossings does not substantially differ between the three variants. On the other hand, using two or three gaps drastically decreases the number of inter-axis crossings. While a difference between two and three gaps is visible in the plots, it is not as strong as between one and two, or one and three, gaps. Thus, it is questionable whether sacrificing the zero crossings between intra-axis and long inter-axis edges one gets with only two gaps is worth the slight reduction in inter-axis edge crossings.

The runtime plot can be seen in Fig. 7c. The runtime for one gap stays consistently below 0.4 s while the runtime for two and three gaps increases much steeper. Interestingly, the runtime for three gaps is lower than for two gaps. This can be explained by moving dummy vertices to gap positions. For two gaps we have to inspect all vertices in the axis to the left and right while for three gaps we have to inspect at most half of the vertices as we can stop once we inspect a vertex that is assigned to a different segment of an axis.

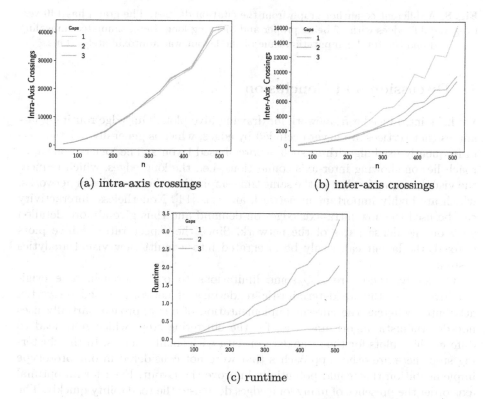

(a) intra-axis crossings

(b) inter-axis crossings

(c) runtime

Fig. 7. Number of crossings and runtime for the synthetic datasets. The x-axis shows an increase in number of vertices which correlates with the number of edges. The y-axis in (a) and (b) show the total number of crossings. In (c) the runtime for our hive plot framework is shown.

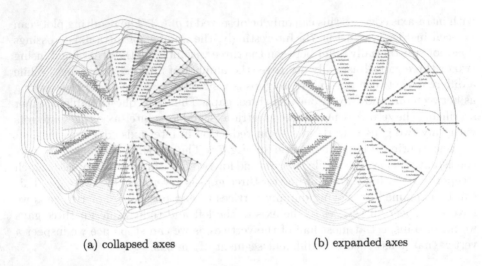

(a) collapsed axes   (b) expanded axes

Fig. 8. A different co-author graph from the citation dataset. The graph has 140 vertices and 397 edges with 47 being proper and 37 being long. Here, simulated annealing was used and the total computation time of the layout was approximately 51 ms.

7   Discussion and Conclusion

We have introduced a framework for drawing hive plots. Our edge routing guarantees that vertices are never occluded by edges, which is generally not the case in frequently used algorithms such as force-based layouts. The focus of our approach lies on showing inter-axis connections, i.e., the long edges, which reduces the visual complexity, but at the same time emphasizes the weak ties in networks, which are highly important in network analysis [14]. Nonetheless, interactivity can be used to show intra-axis edges on demand and thus give a more detailed view on the dense parts of the network. Since the aspect ratio of hive plots is fixed, the layout can easily be integrated into a multi-view visual analytics system.

Obviously, there are also some limitations to our approach. The original hive plots [16] are deterministic renderings of networks based on vertex attributes, whereas the current implementation of our approach partially uses non-deterministic algorithms, e.g., for the clustering step, which may lead to different hive plots for the same data or data with small changes. In the clustering step there are other approaches that were not considered in our prototype implementation that could potentially improve the layout. Even for an optimal axis order the presence of many long edges decreases the readability quickly. The angular resolution for more than 8–12 axes becomes too small to precisely show connectivity details, especially for vertices closer to the origin. Therefore, our framework has limited visual scalability and is recommended mostly for small to medium graphs with less than 500 vertices and no more than 8–12 clusters. However, it is possible to hide some visual complexity by collapsing individual axes. Lastly, the currently implemented crossing minimization heuristic is based on the barycenter algorithm. Here, more sophisticated approaches, such as sifting [20] could potentially be adapted to incorporate gaps.

In terms of future work, several questions arise. While our choice of algorithms in the framework was mostly guided by best-practice from existing literature, a more thorough investigation regarding exact solutions or bounds on typical quality criteria could lead to interesting insights. Also, instead of using a multi-stage framework it could be possible to optimize for multiple criteria simultaneously. Similarly, we only looked at aggregates of axes and the cyclic length when optimizing order, which does not consider the actual Euclidean length of edges. Potentially, we can improve scalability in the number of axes if we arrange them on an ellipse to increase space between axes. From a human-computer interaction perspective it would be interesting to see how our hive plot framework compares to other layouts for visualizing small to medium sized graphs in a formal human-subject study. Finally, adding more interactivity and integrating our framework as an alternative view for data exploration into a visual analytics platform could provide additional insights.

References

1. Bachmaier, C., Brandenburg, F.J., Brunner, W., Hübner, F.: A global k-level crossing reduction algorithm. In: Rahman, M.S., Fujita, S. (eds.) WALCOM 2010. LNCS, vol. 5942, pp. 70–81. Springer, Heidelberg (2010). https://doi.org/10.1007/978-3-642-11440-3_7
2. Bachmaier, C., Brandenburg, F.J., Brunner, W., Lovász, G.: Cyclic leveling of directed graphs. In: Tollis, I.G., Patrignani, M. (eds.) GD 2008. LNCS, vol. 5417, pp. 348–359. Springer, Heidelberg (2009). https://doi.org/10.1007/978-3-642-00219-9_34
3. Bachmaier, C., Brunner, W., König, C.: Cyclic level planarity testing and embedding. In: Hong, S.-H., Nishizeki, T., Quan, W. (eds.) GD 2007. LNCS, vol. 4875, pp. 50–61. Springer, Heidelberg (2008). https://doi.org/10.1007/978-3-540-77537-9_8
4. Blondel, V.D., Guillaume, J.L., Lambiotte, R., Lefebvre, E.: Fast unfolding of communities in large networks. J. Stat. Mech: Theory Exp. **2008**(10), P10008 (2008). https://doi.org/10.1088/1742-5468/2008/10/P10008
5. Clauset, A., Newman, M.E.J., Moore, C.: Finding community structure in very large networks. Phys. Rev. E **70**, 066111 (2004). https://doi.org/10.1103/PhysRevE.70.066111
6. Crameri, F., Shephard, G.E., Heron, P.J.: The misuse of colour in science communication. Nat. Commun. **11**(1), 5444 (2020). https://doi.org/10.1038/s41467-020-19160-7
7. Devanny, W., Kindermann, P., Löffler, M., Rutter, I.: Graph drawing contest report. In: Frati, F., Ma, K.-L. (eds.) GD 2017. LNCS, vol. 10692, pp. 575–582. Springer, Cham (2018). https://doi.org/10.1007/978-3-319-73915-1_44
8. Eades, P., Whitesides, S.: Drawing graphs in two layers. Theoret. Comput. Sci. **131**(2), 361–374 (1994). https://doi.org/10.1016/0304-3975(94)90179-1
9. Eades, P., Wormald, N.C.: Edge crossings in drawings of bipartite graphs. Algorithmica **11**(4), 379–403 (1994). https://doi.org/10.1007/BF01187020

10. Forster, M.: A fast and simple heuristic for constrained two-level crossing reduction. In: Pach, J. (ed.) GD 2004. LNCS, vol. 3383, pp. 206–216. Springer, Heidelberg (2005). https://doi.org/10.1007/978-3-540-31843-9_22

11. Fortunato, S.: Community detection in graphs. Phys. Rep. **486**(3), 75–174 (2010). https://doi.org/10.1016/j.physrep.2009.11.002

12. Ganapathy, M.K., Lodha, S.P.: On minimum circular arrangement. In: Diekert, V., Habib, M. (eds.) STACS 2004. LNCS, vol. 2996, pp. 394–405. Springer, Heidelberg (2004). https://doi.org/10.1007/978-3-540-24749-4_35

13. Gansner, E.R., Koren, Y.: Improved circular layouts. In: Kaufmann, M., Wagner, D. (eds.) GD 2006. LNCS, vol. 4372, pp. 386–398. Springer, Heidelberg (2007). https://doi.org/10.1007/978-3-540-70904-6_37

14. Granovetter, M.S.: The strength of weak ties. Am. J. Sociol. **78**(6), 1360–1380 (1973). https://doi.org/10.1086/225469

15. Guarino, M., Rivas, P., DeCusatis, C.: Towards adversarially robust DDoS-attack classification. In: 2020 11th IEEE Annual Ubiquitous Computing, Electronics Mobile Communication Conference (UEMCON), pp. 0285–0291 (2020). https://doi.org/10.1109/UEMCON51285.2020.9298167

16. Krzywinski, M., Birol, I., Jones, S.J., Marra, M.A.: Hive plots-rational approach to visualizing networks. Brief. Bioinform. **13**(5), 627–644 (2012). https://doi.org/10.1093/bib/bbr069

17. Krzywinski, M., Nip, K.M., Birol, I., Marra, M.: Differential hive plots: seeing networks change. Leonardo **50**(5), 504 (2017). https://doi.org/10.1162/LEON_a_01278

18. Liberatore, V.: Circular arrangements and cyclic broadcast scheduling. J. Algorithms **51**(2), 185–215 (2004). https://doi.org/10.1016/j.jalgor.2003.10.003

19. Masuda, S., Kashiwabara, T., Nakajima, K., Fujisawa, T.: On the NP-completeness of a computer network layout problem. In: Proceedings IEEE International Symposium on Circuits and Systems, pp. 292–295. IEEE Computer Society Press, Los Alamitos (1987)

20. Matuszewski, C., Schönfeld, R., Molitor, P.: Using sifting for k-layer straightline crossing minimization. In: Kratochvíyl, J. (ed.) GD 1999. LNCS, vol. 1731, pp. 217–224. Springer, Heidelberg (1999). https://doi.org/10.1007/3-540-46648-7_22

21. Perin, C., Vuillemot, R., Fekete, J.D.: SoccerStories: a kick-off for visual soccer analysis. IEEE Trans. Visual Comput. Graphics **19**(12), 2506–2515 (2013). https://doi.org/10.1109/TVCG.2013.192

22. Pils, D., et al.: Cyclin E1 (CCNE1) as independent positive prognostic factor in advanced stage serous ovarian cancer patients – a study of the OVCAD consortium. Eur. J. Cancer **50**(1), 99–110 (2014). https://doi.org/10.1016/j.ejca.2013.09.011

23. Rivas, P., et al.: Machine learning for DDoS attack classification using hive plots. In: 2019 IEEE 10th Annual Ubiquitous Computing, Electronics Mobile Communication Conference (UEMCON), pp. 0401–0407 (2019). https://doi.org/10.1109/UEMCON47517.2019.8993021

24. Sugiyama, K., Tagawa, S., Toda, M.: Methods for visual understanding of hierarchical system structures. IEEE Trans. Syst. Man Cybern. **11**(2), 109–125 (1981). https://doi.org/10.1109/TSMC.1981.4308636

25. Tamassia, R.: Handbook of Graph Drawing and Visualization, 1st edn. Chapman & Hall/CRC (2016)

26. Wegman, E.J.: Hyperdimensional data analysis using parallel coordinates. J. Am. Stat. Assoc. **85**(411), 664–675 (1990)
27. yWorks GmbH: yEd. https://www.yworks.com/products/yed
28. Zoppi, J., Guillaume, J.F., Neunlist, M., Chaffron, S.: MiBiOmics: an interactive web application for multi-omics data exploration and integration. BMC Bioinform. **22**(1), 6 (2021). https://doi.org/10.1186/s12859-020-03921-8

A Simple Pipeline for Orthogonal Graph Drawing

Tim Hegemann[✉][iD] and Alexander Wolff[iD]

Universität Würzburg, Würzburg, Germany
hegemann@informatik.uni-wuerzburg.de

Abstract. Orthogonal graph drawing has many applications, e.g., for laying out UML diagrams or cableplans. In this paper, we present a new pipeline that draws multigraphs orthogonally, using few bends, few crossings, and small area. Our pipeline computes an initial graph layout, then removes overlaps between the rectangular nodes, routes the edges, orders the edges, and nudges them, that is, moves edge segments in order to balance the inter-edge distances. Our pipeline is flexible and integrates well with existing approaches. Our main contribution is (i) an effective edge-nudging algorithm that is based on linear programming, (ii) a selection of simple algorithms that together produce competitive results, and (iii) an extensive experimental comparison of our pipeline with existing approaches using standard benchmark sets and metrics.

Keywords: Orthogonal graph drawing · Edge routing · Edge nudging · Experimental evaluation

1 Introduction

Due to its many applications, the orthogonal drawing style has been studied extensively in Graph Drawing. One of the milestones in the development of efficient algorithms for this domain was Tammasia's Topology–Shape–Metric framework [19] which showed that embedded planar graphs with a vertex degree of at most 4 can be laid out efficiently, using the minimum number of bends. The restriction to degree 4 comes from the fact that the framework represents vertices by (grid) points. For practical purposes, however, the restriction to (embedded) planar graphs of constant degree is prohibitive. This triggered many practical approaches to orthogonal graph drawing. For example, the three-phase method of Biedl et al. [1] draws *normalized* graphs (that is, graphs without self-loops and leaves) with vertices in general position (that is, on different grid lines) with "small" vertex boxes on an quadratic size grid using at most one bend per edge. For compaction, Biedl et al. referred to Lengauer's book [10] on VLSI layout, which is also relevant for orthogonal graph drawing. Building on earlier work [3,5,16,20], Kieffer et al. [8] introduced HOLA ("Human-like Orthogonal Network Layout"), a multi-step approach for drawing graphs orthogonally. They partition the input graph and use different layout strategies for different parts: stress minimization for the graph core and specialized code for tree-like subgraphs.

Schulze et al. [17] presented the orthogonal graph drawing library KIELER [15] that took special care of so-called *port constraints*. They allow the user to specify on which side of a vertex box an edge must be attached, which is important, for example, for UML diagrams. Zink et al. [21] presented the PRALINE library, which generalizes port constraints by introducing port groups and port pairings, which are useful for drawing cableplans. Both the approaches of Schulze et al. and Zink et al. arrange nodes on layers and use the framework of Sugiyama et al. [18] for layered graph drawing (although they do not assume input graphs to be directed), among others, in order to reduce edge crossings.

Whereas most graph drawing algorithms place labels into vertex boxes, Binucci et al. [2] also incorporated *edge labels*. Using mixed integer programming, they can draw sparse graphs with vertex and edge labels of up to 100 vertices.

We considered orthogonal graph layout in a third-party-funded project with two industrial partners with different backgrounds. One of them produces network management software; the other produces software for drawing cableplans of complicated, highly configurable machines. Both asked for layouts that work well on mobile devices such as tablet computers used by, for example, technicians who service harvesting machines in the field.

Rather than a monolithic software package, they were interested in a highly configurable, flexible pipeline whose parts can easily be exchanged in order to meet the various needs of their customers. Still, they insisted on a number of basic requirements. The drawings computed by our algorithm must be orthogonal, that is, edges are drawn as sequences of axis-aligned segments and vertices are represented by non-overlapping boxes (i.e., axis-aligned rectangles). Also, the user must be able to specify a minimum object distance $\delta_{\min}$ to be respected by vertex boxes and edge segments.

In terms of quality, we agreed upon standard graph drawing criteria such as few crossings, few bends, small area, good aspect ratio, small total edge length, and small edge length variance. With mobile applications in mind, using small area becomes our key objective. However, rather than an algorithm that excels in one of these metrics (and fails badly in another), our partners were interested in allrounders that are sufficiently fast and generally perform well.

Our Contribution. We set up a layout pipeline with three variants. For the first variant (FORCE), we use a force-directed layout algorithm [6] to place the vertices of a given graph G as points (ignoring their boxes). Then we center the vertex boxes on these points. If some of them overlap, we call an overlap removal algorithm of Nachmanson et al. [11]. Instead of these three steps, for the second variant (HYBRID1), we used the vertex placement computed by PRALINE. In both cases, we apply the following steps that we describe in detail in Sect. 2. *Port assignment:* We assign the endpoints of the edges to the sides of the vertex boxes. *Routing graph construction:* We construct an auxiliary grid-like graph H. *Edge routing* and *path ordering:* We route (and order) the edges of G along the edges of H. Our path ordering is based on existing techniques [7,12,13]. Now each edge of G consists of a path of axis-aligned segments. *Edge nudging:* In this final step, we distribute the path segments so that they partition the available space between the vertex boxes as evenly as possible. As a third variant (HYBRID2), we initialize our pipeline with both the vertex positions and the edge routing computed by PRALINE and apply only the nudging step as a post-processing.

Our main contribution is (i) the edge nudging step, (ii) our simple and flexible pipeline as a whole, and (iii) an experimental comparison with the state-of-the-art orthogonal layout libraries HOLA[1] [8] and PRALINE[2] [21] on two standard benchmark sets (see Sect. 3). PRALINE has already been compared with KIELER, and performed similarly well or even slightly better [21]. HOLA has been compared to the *orthogonal style* automatic layout of yFILES[3] ; HOLA outperformed yFILES almost universally in a user study with 89 participants [8].

As it turns out, our pipeline proves to be a good allrounder that performs well in many of the metrics mentioned above. Due to careful nudging, our pipeline yields very compact layouts, whereas its competitors usually produce fewer crossings and bends. Depending on the variant, our pipeline takes slightly less or about twice as much time than PRALINE. It is much faster than HOLA.

Our source code is available at https://github.com/hegetim/wueortho.

2   Our Pipeline

In order to fine-tune the layout for different requirements (as discussed in Sect. 1), we designed our pipeline as a sequence of mostly independent, interchangeable, and self-contained steps that we describe in detail in this section.

The input for our pipeline is a multigraph G with vertex set $V(G)$ of size n and edge multiset $E(G)$ of size m. We explicitly allow our graphs to have self-loops and handle them in the *port assignment* step. Each vertex comes with a (textual) vertex label or directly with a *vertex box*, that is, an axis-aligned rectangle. Given a text label, we compute a box that fits the label (in some standard font).

Our pipeline consists of several simple algorithms for specific subproblems of orthogonal graph drawing; see Fig. 1 for an overview. The main steps in our pipeline are vertex layout, overlap removal, port assignment, construction of the routing graph, edge routing, path ordering, and edge nudging. Below we detail most of these steps. For the remaining, we use standard algorithms, see "Pipeline Variants", Sect. 3.

Port Assignment. Each edge connects one or two vertices that are represented by rectangular boxes. We call the start point and the end point of an edge its *ports*. Ports lie on the boundary of the vertex boxes that an edge connects. Port positions can either be specified in the input, or they are set by our pipeline as follows (before the edge routes are determined).

---

[1] see https://www.adaptagrams.org/.
[2] see https://github.com/j-zink-wuerzburg/praline.
[3] see https://www.yworks.com/products/yfiles.

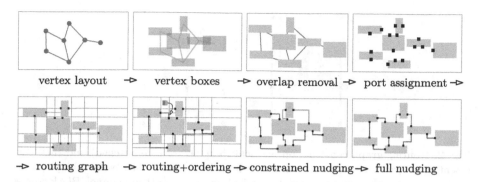

vertex layout –▷ vertex boxes –▷ overlap removal –▷ port assignment –▷

–▷ routing graph –▷ routing+ordering –▷ constrained nudging–▷ full nudging

Fig. 1. Our flexible pipeline of simple algorithms for orthogonal graph drawing.

For each edge uv, we first determine the sides of the boxes of u and v on which we place the ports of uv. Let s_{uv} be the straight-line segment that connects the centers of the boxes of u and v. We usually assign the ports to those box sides that intersect s_{uv}. To avoid situations where an edge uv would have to be drawn as a Z-shape according to the rule above, we adjust the rule as follows. We split each side evenly into four pieces. If s_{uv} intersects a side in the first or last piece of the side, we instead reassign one of the two ports such that uv can be drawn as an L-shape that bends away from the barycenter of the vertex boxes' center points. (We do not actually route uv now, we just use geometry to place its ports.) The order along a side of vertex u is then determined by the circular order of the segments of type s_{uv}, where v is a neighbor of u. Multi-edges require special care regarding their order within a side. They are routed next to each other without crossings. Self-loops get neighboring ports assigned to the least populated side. When all ports have been assigned to a box, we evenly distribute the ports on each side.

Routing Graph Construction. We route each edge of the input graph along an obstacle-avoiding path in a *routing graph* H whose vertices are the ports and the potential bend points of edge routes. This graph forms a partial grid with gaps around the vertex boxes. A precise definition follows below.

The intuition for our routing graph is that edges get routed through horizontal and vertical channels. We describe only *vertical channels*. *Horizontal channels* are defined symmetrically. For a pair of vertex boxes (u, v) where v is entirely to the right of u, we define its *vertical channel* as the largest axis-aligned rectangle whose vertical sides touch the right side of u and the left side of v and that is interior-disjoint from all vertex boxes – if such an empty rectangle exists for (u, v). For each box u we keep only the channel to the right of u that has the smallest width. For this step, we interpret the left and right boundaries of the drawing as vertex boxes of zero width and infinite height. In Fig. 2, the orange boxes depict the vertical channels. Note that some of these channels overlap. We can find all channels in $\mathcal{O}(n \log n)$ time with a sweepline algorithm.

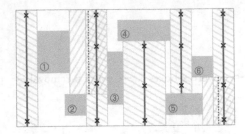

Fig. 2. Routing graph construction: vertical channels (orange hatched) and their representatives (red). Dotted representatives have been sorted out. Note that ports are omitted and all representatives are the vertical center line of their channel. Black crosses mark where horizontal representatives (not shown) intersect. Those will host vertices in the final routing graph. (Color figure online)

For each vertical channel, we define a vertical line segment that spans the channel's entire height as its *representative*. If possible, we choose an appropriate line segment starting in a port. Otherwise, we choose the center line of the channel. As a further optimization, we ignore every vertical channel C that intersects another channel C' such that the projection of C on the y-axis is contained in that of C'. See the dotted red segments in Fig. 2. We define representatives for horizontal channels symmetrically. We add additional representatives for each remaining port. We can find all representatives using the same sweepline algorithm as above in $\mathcal{O}(m \log m)$ time, assuming $n \in \mathcal{O}(m)$.

The routing graph H has a vertex for each port and for each intersection point between a vertical and a horizontal representative. It has an edge between each pair of vertices that are consecutive along a representative. Let M be the number of edges of H. Again assuming $n \in \mathcal{O}(m)$, we have $M \in \mathcal{O}(m^2)$ since there are at most $4n$ channels (one per side for each vertex box) and $2m$ ports.

Edge Routing. In the next step, for each edge we connect its endpoints (which are ports and thus vertices in the routing graph H) by a shortest path in H. Our approach for edge routing is similar to that of Wybrow et al. [20] except that they use the A* search for computing shortest paths whereas we use Dijkstra's algorithm (for simplicity). Recall that H is a partial grid graph. Among all shortest paths, we choose a bend-minimal one by augmenting the state used in Dijktra's algorithm. For each partial path ending in a node v, in addition to the length of the path, we store the number of bends and the direction of entry when entering v. In the following, we refer to routed edges as paths. Each such path can be found in $\mathcal{O}(M \log M)$ time.

The edge routing algorithm of Wybrow et al. does not take crossings into account. Therefore, as a post-processing, we apply an additional *crossing reduction step*. Whenever two paths cross each other more than once, we replace the section between the first and last shared vertex in one path with the corresponding section of the other. This ensures that, eventually, every pair of edges crosses at most once.

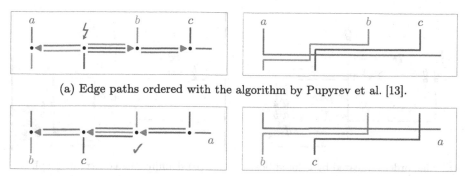

(a) Edge paths ordered with the algorithm by Pupyrev et al. [13].

(b) Edge paths ordered with our modified version of the algorithm.

Fig. 3. Path ordering: The algorithm of Pupyrev et al. assigns each segment an arbitrary direction. This can result in additional bends in the geometric realization (to the right). For orthogonal routing, we therefore preassign directions for horizontal and vertical segments.

Path Ordering. We now know, for each edge of the routing graph H, the set of edge paths routed over this segment. Where several edge paths (forming an *edge bundle*) share a segment, we determine a path order that minimizes crossings.

For non-orthogonal routing graphs, Pupyrev et al. [13] observed that the problem of determining such a path order is computationally equivalent to the metro-line crossing minimization (MLCM) problem: Let $\widetilde{G}$ be a plane graph (such as the routing graph), and let P be a set of simple paths in $\widetilde{G}$. For each edge e in $E(\widetilde{G})$, find an order of all paths that contain e that minimizes the number of crossings among all pairs of paths.

We use the algorithm of Pupyrev et al. [13] for our case where all edges are incident to unique ports (which makes MLCM efficiently solvable). When sorting paths in an edge bundle, for each pair of paths, they consider the directions the paths take after leaving their common subpath at a *fork vertex*. In Fig. 3a, for example, the leftmost vertex is a fork vertex for paths a and b. With a leaving to the top, it will be ordered above of b. For each segment (i.e., edge in H) such an ordering of paths has to be found. They fix an arbitrary direction that determines where to look for either a fork vertex or a segment that has already been processed (see the gray arrowheads in Fig. 3). Crossings are unavoidable if the path orders at the start and end of the common subpath differ. Pupyrev et al. process the segments in arbitrary order. Still, they introduce at most one crossing for each pair of paths in an edge bundle and only if such a crossing is unavoidable. For example, paths a and b in Fig. 3a have an unavoidable crossing.

In our orthogonal setting, if two paths change order between two adjacent edges of H (see the red lightning in Fig. 3a (left)), then we have to introduce two additional bends in a Z-shaped fashion in their geometric realization (see Fig. 3a (right)). In order to avoid such situations, we preassign directions for all segments based on their orientation (left for horizontal segments and down for vertical ones; see Fig. 3b). Crossings now happen only where the orientation of

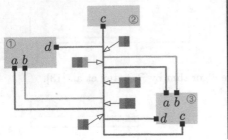 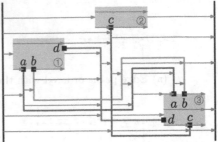

(a) edge order indicated by colored boxes   (b) defining the constraint graph

Fig. 4. Edge nudging: Given an edge order on the vertical edge segments, we define horizontal separation constraints between vertical segments, left and right borders of vertex boxes, and two dummy segments (black bars). In (b) segments are partially nudged for readability. (Color figure online)

segments changes (i.e., at bend points of paths). Such crossings can be realized without additional edge bends; see the crossing at the green checkmark in Fig. 3b. Therefore, before the next step, we join consecutive collinear segments of the same path. Then, in any path, the orientation of the segments alternates.

Applied to the graph H, our modification of the algorithm of Pupyrev et al. runs in $\mathcal{O}(Mk \log m)$ time, where k is the total number of segments in all paths.

Edge Nudging. In the last step of our pipeline, we aim to balance the distances between the path segments within their channels. Our algorithm has two modes called *constrained nudging*, when vertex and port positions must not be altered, and *full nudging*, when a minimum distance between path segments and vertex boxes must be met, but boxes can be moved and, if necessary, enlarged. Both modes use basic linear programming (LP) to optimize segment distances. They process horizontal and vertical distances independently.

We now describe the horizontal pass. The vertical pass works symmetrically. First, we determine the horizontal order χ of all vertical path segments, the left and right borders of all vertex boxes, and the two vertical sides of a (slightly enlarged) bounding box of our instance (see the black bars in Fig. 4b). The order of the objects in χ is determined by their x-coordinate. The two dummy segments are the first and last elements of χ. It remains to define the order of objects with identical x-coordinate. We assume non-intersecting, non-touching vertex boxes. Where path segments overlap, the path order determined in the previous section applies; see the colored boxes in Fig. 4a. Right (left) borders of vertex boxes are inserted into χ before (after) any path segment with the same x-coordinate. The order of non-overlapping path segments with the same x-coordinate is arbitrary.

Given χ, we define the *constraint graph* G_χ, the directed acyclic graph that has a vertex for each object as defined above, and an arc from object u to object v if the vertical dimensions of these objects overlap, u comes before v in χ, and there is no other vertically overlapping object in between. If this is the

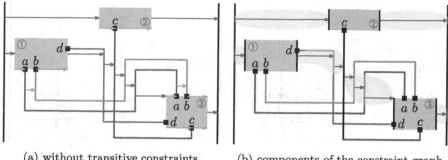

(a) without transitive constraints   (b) components of the constraint graph

Fig. 5. Common steps of edge nudging: (a) After removing transitive arcs, pink arcs remain between unmovable objects. Brown arcs get distance variables. (b) The constraint graph is split at barriers (black bars). All constraints from arcs in the same component share their distance variables. (Color figure online)

case, u will be drawn to the left of v. Edges of this constraint graph will yield *separation constraints* of the form $u_x + \delta \leq v_x$ in the LP, where u_x and v_x are the x-coordinates of u and v, respectively, and δ is either a non-negative variable or the user-defined minimum distance $\delta_{\min}$ between them.

Let $N = 4n+m+b$ be the number of sides of the vertex boxes plus the number of edge segments (i.e., the number of edges plus the number of bends). Then the constraint graph can be constructed in $\mathcal{O}(N \log N)$ time, using a sweepline algorithm of Dwyer et al. [5]. They showed that the number of edges in the constraint graph is linear in N and that the separation constraints derived from the edges of G_χ guarantee a horizontally overlap-free drawing. Next, we decide which constraints share the same *distance variables* of type δ, which we will then maximize. Wider channels with few segments allow for larger gaps than small crowded channels. To obtain a balanced solution, we need to avoid situations where two distance variables work against each other.

To identify preferably small sets of constraints that share the same distance variable, we apply the following operations to the constraint graph. We remove all transitive arcs, i.e., arcs uw where also arcs uv and vw exist in the graph. Constraints from these arcs are redundant. We remove all arcs between objects that do not move in constrained nudging mode, that is, the sides of vertex boxes and edge segments incident to ports; see the purple arrows in Fig. 5a. Then, the graph is split into components (the green areas in Fig. 5b) that are confined by unmovable objects or by dummy segments (the big black bars in Fig. 5b). All constraints derived from arcs of the same component get the same distance variable.

In constrained nudging mode, for each movable or dummy segment, we replace its position by a *position variable* in all related constraints. Finally, our LP minimizes

$$|W|(\omega - \alpha) - \sum_{\delta \in W} \delta,$$

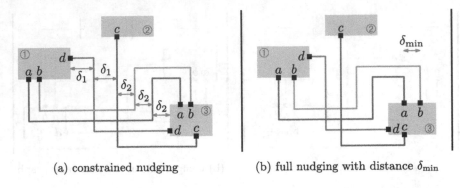

(a) constrained nudging   (b) full nudging with distance $\delta_{\min}$

Fig. 6. Results of a horizontal nudging phase: (a) optimization nudges objects apart, (b) in full nudging mode, objects must maintain a given minimum distance $\delta_{\min}$. Note that vertex ③ has been slightly enlarged to make room for the ports of edges a and b.

where α and ω are the position variables of the left and right dummy segments, respectively, and W is the set of distance variables. The factor $|W|$ is required to prevent the constraints involving distance variables from pushing the dummy segments towards infinity. The result is shown in Fig. 6a. Objects are separated with space between them equivalent to at least the values of the respective distance variables.

In full nudging mode we allow both, the port segments and the borders of the vertex boxes, to be moved by the nudging procedure in order to maintain a minimum object distance $\delta_{\min}$ and to control the total edge length. We allow vertex boxes to grow, if necessary, but not to shrink. Therefore, we use position variables for all sides of vertex boxes and edge segments instead of fixed positions.

In addition to the constraints from the constrained mode (brown arrows in Fig. 5b), we introduce, for each vertex box b of original width w_b (as specified in the input) a separation constraint $b^{\mathrm{R}} - b^{\mathrm{L}} \geq w_b$ where b^{L} and b^{R} are the position variables of the left and right sides of b, respectively. For all arcs that have not been transitively removed (see brown and pink arrows in Fig. 5a), we add separation constraints with a constant distance of $\delta_{\min}$.

In order to establish a hierarchy in optimization, we weight our objective by adding constant factors. Let W be the set of distance variables, let S_{H} be the set of horizontal segments, and let B the set of vertex boxes. We use the term $b^{\mathrm{R}} - b^{\mathrm{L}}$ for the width of box $b \in B$ and $s^{\mathrm{R}} - s^{\mathrm{L}}$ for the length of segment $s \in S_{\mathrm{H}}$. Now we have our LP minimize the sum of the widths of the vertex boxes and the lengths of the segments minus the sum of the distance variables, that is,

$$2(|W| + |S_{\mathrm{H}}|)\left(\omega + \sum_{b \in B}(b^{\mathrm{R}} - b^{\mathrm{L}})\right) + 2\sum_{s \in S_{\mathrm{H}}}(s^{\mathrm{R}} - s^{\mathrm{L}}) - \sum_{\delta \in W}\delta.$$

Figure 6b shows the result for the example depicted in Fig. 4a. Multiple phases of nudging can be repeatedly applied to optimize compactness and edge lengths. To get rid of unnecessary bends, we simply set the separation distance of constraints between segments of the same path to zero.

3   Experiments

We considered three variants (described below) of our pipeline, and compared them to the state-of-the-art orthogonal layout libraries PRALINE and HOLA.

Benchmark Sets. Our pipeline has been implemented as part of a third-party-funded project with two industrial partners that suggested two benchmark sets from their respective domains. The first benchmark set is called *Internet Topology Zoo*[4] [9]. The data set includes textual vertex labels of varying length.

The second dataset is called *Pseudo-cableplans*. The graphs have been part of a benchmark set for orthogonal graph drawing by Zink et al.[5] [21]. We removed some domain-specific peculiarities such as special vertex pairing and port grouping constraints, and we replaced each hyperedge e by a new dummy vertex v_e that we connected to every vertex in e. The labels in this dataset are fixed-length or empty (the dummy vertices).

In both benchmark sets, we kept only the largest connected component of each graph. Although preliminary tests showed some good results, HOLA officially does not support multigraphs. Therefore, we removed all but the first occurrence of each multi-edge and all self-loops. Furthermore, we removed all graphs where HOLA crashed or took more than 10 min. Note that our pipeline correctly draws every connected instance in the original datasets, including multi-edges and self-loops.

So for our experiments we used 260 simple graphs derived from the original 261 multigraphs of the Internet Topology Zoo and 1,026 graphs derived from the original 1,139 Pseudo-cableplans. Figure 7 shows the edge density distribution of the graphs in the two benchmark sets. We set the default dimensions of the vertex boxes to 12×38 (pixels) and widened them if necessary to accommodate the label text and to fit all incident edges (with gaps of 18 pixels).

Pipeline Variants. We set up three variants of our pipeline. In the first variant, FORCE, we use a simple force-directed layout algorithm [6] to place the vertices as points (ignoring their boxes). Then, we apply the GTREE Algorithm by Nachmanson et al. [11] to remove overlaps. In the second variant, HYBRID1, we use vertex positions computed by PRALINE [21]. These variants both go through the steps port distribution, routing graph construction, edge routing, edge ordering, and full edge nudging as described in the previous sections. Full nudging is applied horizontally, then vertically, and then once more horizontally. As a third variant HYBRID2, we initialize our pipeline with both the vertex positions and the edge routing computed by Praline and apply only the nudging step as a post-processing. All pipeline steps are implemented in Scala and dynamically configurable for various setups. We use the GLOP[6] optimizer for LP-solving.

---

[4] see http://www.topology-zoo.org/index.html.
[5] see https://github.com/j-zink-wuerzburg/pseudo-praline-plan-generation.
[6] see https://developers.google.com/optimization.

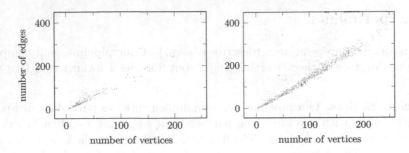

Fig. 7. Number of vertices and edges for each graph in the two datasets. Semi-transparent markers represent the original multigraphs.

Metrics. To assess the quality of graph drawings many metrics have been proposed. In our experiments we use edge crossings, edge bends, total edge length, variance in edge length, area, aspect ratio, and minimum object distance (δ_{min}). These will be discussed below.

It has been shown (e.g., in a study by Purchase [14]) that drawings with fewer edge crossings and bends simplify several tasks related to graph understanding and navigation. A study by Dwyer et al. [4] suggests that users benefit from graph drawings with low variance in edge length. When drawing graphs with more than 30 vertices, scaling becomes an issue as text labels tend to become unreadably small and overly long edges become hard to follow. Therefore, we include metrics assessing the compactness of drawings in our comparisons, namely total edge length and the area of the bounding box. For a drawing with a bounding box of width w and height h, we define aspect ratio as $\max(w, h)/\min(w, h)$. This yields a value in the range $[1, \infty)$. We consider lower aspect ratios better and squares (with aspect ratio 1) optimal. In order to ensure a fair comparison with metrics sensitive to scaling, we also include the minimum object distance δ_{min}. We configure a minimum value (or target value, if no minimum is supported) of 12 pixels and report deviations.

Comparison. We compared our pipeline to the following two libraries. PRALINE [21] is based on the well-known Sugiyama framework [18] for layered graph drawing. PRALINE differs from the original framework especially in terms of edge routing and port placement. Layering-based algorithms tend to produce few crossings and balanced results. HOLA [8] is a multi-stage algorithm that decomposes the input into trees and a connected core that is drawn using stress-minimization and overlap removal. The trees are drawn using a specialized layout algorithm, and the tree drawings are then inserted into the drawing of the whole graph. For our comparison, we used the default settings regarding vertex distances and ideal edge length in HOLA. We conducted small-scale experiments

Table 1. Experimental results on two datasets. The mean μ is relative to PRALINE (abbreviated PRAL); β measures the percentage of cases where an algorithm achieved the best result. Sums over 100 % are possible due to ties.

(a) The Internet Topology Zoo benchmark set.

	FORCE		HYBRID1		HYBRID2		HOLA		PRAL.	
	μ	β	μ	β	μ	β	μ	β	μ	β
crossings	3.90	27	1.30	54	1.00	70	.85	**77**	1	70
edge bends	.92	4	.76	6	.49	**51**	.50	47	1	2
edge length variance	.47	**39**	.39	21	.37	38	11.43	3	1	1
total edge length	.73	20	.61	6	.49	**73**	1.83	0	1	2
bounding box area	.68	30	.56	32	.56	**38**	3.94	0	1	0
aspect ratio	.93	**35**	1.03	14	1.07	11	1.35	22	1	18
$\delta_{\min}$	1.11	**88**	1.10	87	1.10	87	1.01	45	1	69

(b) The Pseudo-cableplans benchmark set.

	FORCE		HYBRID1		HYBRID2		HOLA		PRAL.	
	μ	β	μ	β	μ	β	μ	β	μ	β
crossings	1.58	9	1.03	19	1.00	25	.73	**89**	1	25
edge bends	.96	1	.76	2	.66	12	.30	**88**	1	1
edge length variance	.34	**52**	.35	31	.40	19	1.79	0	1	0
total edge length	.59	29	.52	29	.54	**43**	1.20	0	1	0
bounding box area	.55	25	.50	24	.50	**51**	2.21	0	1	0
aspect ratio	.99	**37**	.97	12	.96	17	.97	24	1	10
$\delta_{\min}$	1.42	**93**	1.42	**93**	1.42	**93**	1.24	60	1	33

that confirm that the defaults yield a good compromise between compactness and readability (i.e., sufficiently large $\delta_{\min}$).

Results. See Table 1 and Fig. 8 for the results of our experiments. Concerning the number of crossings, we see a weakness of the simplistic approach of our pipeline. On average, FORCE produced over twice as many crossings as the best results on the Pseudo-cableplans and nearly five times as many on the Internet Topology Zoo graphs. HYBRID1, combining PRALINE vertex positions and our pipeline, on the other hand, produced only 41 % more crossings on the Pseudo-cableplans and 54 % more crossings on the Internet Topology Zoo graphs compared to the best results. HYBRID2 per construction produces the same number of crossings as PRALINE. Bad vertex placement also hurts down the pipeline. As we can see, the HYBRID variants are almost consistently better than FORCE with two exceptions: edge length variance and aspect ratio. However, in terms of aspect ratio, the only outlier is HOLA, performing nearly 30 % worse than the others on the Internet Topology Zoo.

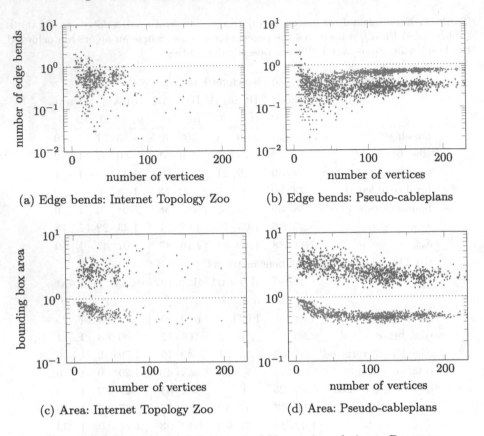

(a) Edge bends: Internet Topology Zoo   (b) Edge bends: Pseudo-cableplans

(c) Area: Internet Topology Zoo   (d) Area: Pseudo-cableplans

Fig. 8. Selected metrics of HOLA ♦, and HYBRID2 ▲ relative to PRALINE.

HOLA shows an impressive performance in terms of crossings and creates by far the fewest bends. But this comes at a cost of a very large drawing area and overly long edges. HOLA considers δ_{min} an optimization goal, not a strict requirement. In the majority of cases the target value of 12 px could be maintained. PRALINE, however, surprised us, too. Not maintaining δ_{min} was confirmed to us being a bug in the current PRALINE implementation by the authors.

Overall, the HYBRID variants of our pipeline show good and very consistent results with HYBRID2, surpassing PRALINE in all quality metrics. It produces leading results with respect to area, total edge length, and edge length variance while reliably maintaining the given minimum object distance.

Running Time. We evaluated the running times of our pipeline using 300 random multigraphs with 5 to 150 vertices and average vertex degree 4. To ensure that every graph is connected, we first created a tree with all vertices and then added the remaining edges at random. Our experiments ran on an Intel®

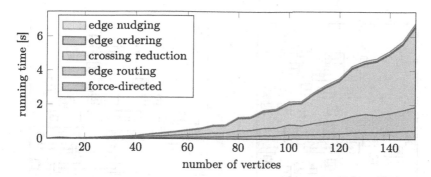

Fig. 9. Running times of the FORCE pipeline on random multigraphs with an average vertex degree of 4. Stages with less than 10 ms average running time are omitted.

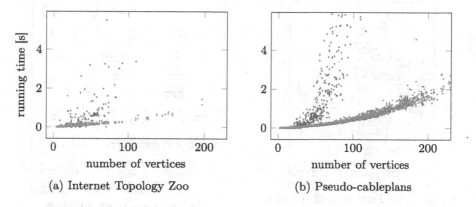

(a) Internet Topology Zoo   (b) Pseudo-cableplans

Fig. 10. Running times of FORCE ▲, HOLA ♦, and PRALINE ● for our benchmark sets.

Core^{TM} i7-8565U. We measured runtimes using Java's `nanoTime` function (for our pipeline and PRALINE) and the GNU `time` command (for HOLA).

See Fig. 9 for different steps of our pipeline. Steps that consistently require less than 10 ms to complete are omitted. The overall runtime is clearly dominated by edge routing and crossing reduction (that is, finding pairs of edges that cross more than once and then joining their common subpaths), followed by force-directed vertex layout. The time spent on nudging, edge ordering, and on creating the routing graph was insignificant.

The time for drawing the graphs from the two benchmark sets is shown in Fig. 10. The HYBRID setups are omitted. In HYBRID1 just like with FORCE the edge routing dominates the runtime, in HYBRID2 the runtime is dominated by performing the PRALINE layout. Note that PRALINE by default does ten repetitions with different initial vertex positions of which it keeps the best. Depicted is the sum of all repetitions. For PRALINE and FORCE only the bare layouting time was measured whereas for HOLA, for technical reasons, the measurements include file handling. However, this increases the runtime by less than 50 ms.

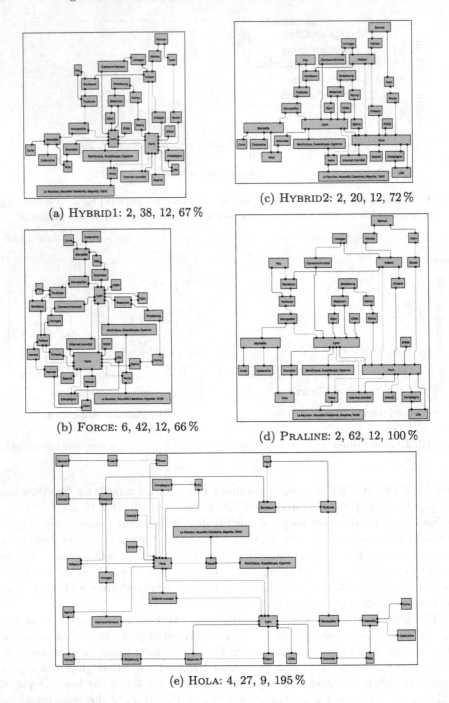

(a) HYBRID1: 2, 38, 12, 67%

(c) HYBRID2: 2, 20, 12, 72%

(b) FORCE: 6, 42, 12, 66%

(d) PRALINE: 2, 62, 12, 100%

(e) HOLA: 4, 27, 9, 195%

Fig. 11. An example from the Internet Topology Zoo drawn by the five layout algorithms. The figures are scaled proportionally. The numbers refer to: crossings, bends, $\delta_{\min}$ (in pixels), and area (in percent w.r.t. PRALINE).

4   Conclusions

Our experiments show that HYBRID1 is a good allrounder. The edge nudging step performs particularly well and leads to compact drawings. On the other hand, due to our current rather simple edge routing, the drawings tend to have more bends and crossings than those of its competitors. When combined with a more sophisticated layouting (the HYBRID2 setup), we can significantly improve compactness (almost half the bounding box area) and number of edge bends with the same number of crossings as PRALINE.

Nonetheless, we intend to improve edge routing, especially in terms of crossings. To this end, Wybrow et al. [20] suggested to take into account edges that have already been routed. Also it may help to reorder the ports around the boundary of the vertex boxes. To reduce the number of bends, we want to add a postprocessing that straightens Z-shaped edges whose middle piece is short. Currently, such unnecessary double bends tend to occur quite frequently; see Fig. 11b.

Acknowledgments. We thank Steve Kieffer, Micheal Wybrow, and Tobias Czauderna who helped us with HOLA in our experiments, Johannes Zink who helped us with PRALINE, and our very supportive reviewers. This work was supported by BMBF grant 01IS22012C.

References

1. Biedl, T.C., Madden, B., Tollis, I.G.: The three-phase method: a unified approach to orthogonal graph drawing. Int. J. Comput. Geom. Appl. **10**(6), 553–580 (2000). https://doi.org/10.1142/S0218195900000310
2. Binucci, C., Didimo, W., Liotta, G., Nonato, M.: Orthogonal drawings of graphs with vertex and edge labels. Comput. Geom. Theory Appl. **32**(2), 71–114 (2005). https://doi.org/10.1016/j.comgeo.2005.02.001
3. Dwyer, T., Koren, Y., Marriott, K.: IPSep-CoLa: an incremental procedure for separation constraint layout of graphs. IEEE Trans. Visual Comput. Gr. **12**(5), 821–828 (2006). https://doi.org/10.1109/TVCG.2006.156
4. Dwyer, T., et al.: A comparison of user-generated and automatic graph layouts. IEEE Trans. Visual Comput. Gr. **15**(6), 961–968 (2009). https://doi.org/10.1109/TVCG.2009.109
5. Dwyer, T., Marriott, K., Stuckey, P.J.: Fast node overlap removal. In: Healy, P., Nikolov, N.S. (eds.) GD 2005. LNCS, vol. 3843, pp. 153–164. Springer, Heidelberg (2006). https://doi.org/10.1007/11618058_15
6. Fruchterman, T.M.J., Reingold, E.M.: Graph drawing by force-directed placement. Softw. Pract. Exper. **21**(11), 1129–1164 (1991). https://doi.org/10.1002/spe.4380211102
7. Groeneveld, P.: Wire ordering for detailed routing. IEEE Design Test Comput. **6**(6), 6–17 (1989). https://doi.org/10.1109/54.41670
8. Kieffer, S., Dwyer, T., Marriott, K., Wybrow, M.: HOLA: human-like orthogonal network layout. IEEE Trans. Visual Comput. Gr. **22**(1), 349–358 (2016). https://doi.org/10.1109/TVCG.2015.2467451

9. Knight, S., Nguyen, H.X., Falkner, N., Bowden, R., Roughan, M.: The internet topology zoo. IEEE J. Sel. Areas Comm. **29**(9), 1765–1775 (2011). https://doi.org/10.1109/JSAC.2011.111002

10. Lengauer, T.: Combinatorial Algorithms for Integrated Circuit Layout. Vieweg+Teubner, Wiesbaden (1990). https://doi.org/10.1007/978-3-322-92106-2

11. Nachmanson, L., Nocaj, A., Bereg, S., Zhang, L., Holroyd, A.: Node overlap removal by growing a tree. J. Graph Alg. Appl. **21**(5), 857–872 (2017). https://doi.org/10.7155/jgaa.00442

12. Nöllenburg, M.: An improved algorithm for the metro-line crossing minimization problem. In: Eppstein, D., Gansner, E.R. (eds.) GD 2009. LNCS, vol. 5849, pp. 381–392. Springer, Heidelberg (2010). https://doi.org/10.1007/978-3-642-11805-0_36

13. Pupyrev, S., Nachmanson, L., Bereg, S., Holroyd, A.E.: Edge routing with ordered bundles. Comput. Geom. Theory Appl. **52**, 18–33 (2016). https://doi.org/10.1016/j.comgeo.2015.10.005

14. Purchase, H.: Effective information visualisation: a study of graph drawing aesthetics and algorithms. Interact. Comput. **13**(2), 147–162 (2000). https://doi.org/10.1016/S0953-5438(00)00032-1

15. Real-Time and Embedded Systems group. Kiel Integrated Environment for Layout Eclipse Rich Client (KIELER) (2020). https://rtsys.informatik.uni-kiel.de/confluence/display/KIELER/Overview

16. Rüegg, U., Kieffer, S., Dwyer, T., Marriott, K., Wybrow, M.: Stress-minimizing orthogonal layout of data flow diagrams with ports. In: Duncan, C., Symvonis, A. (eds.) GD 2014. LNCS, vol. 8871, pp. 319–330. Springer, Heidelberg (2014). https://doi.org/10.1007/978-3-662-45803-7_27

17. Schulze, C.D., Spönemann, M., von Hanxleden, R.: Drawing layered graphs with port constraints. J. Vis. Lang. Comput. **25**(2), 89–106 (2014). https://doi.org/10.1016/j.jvlc.2013.11.005

18. Sugiyama, K., Tagawa, S., Toda, M.: Methods for visual understanding of hierarchical system structures. IEEE Trans. Syst. Man Cybern. **11**(2), 109–125 (1981). https://doi.org/10.1109/TSMC.1981.4308636

19. Tamassia, R.: On embedding a graph in the grid with the minimum number of bends. SIAM J. Comput. **16**(3), 421–444 (1987). https://doi.org/10.1137/0216030

20. Wybrow, M., Marriott, K., Stuckey, P.J.: Orthogonal connector routing. In: Eppstein, D., Gansner, E.R. (eds.) GD 2009. LNCS, vol. 5849, pp. 219–231. Springer, Heidelberg (2010). https://doi.org/10.1007/978-3-642-11805-0_22

21. Zink, J., Walter, J., Baumeister, J., Wolff, A.: Layered drawing of undirected graphs with generalized port constraints. Comput. Geom. Theory Appl. **105–106**(101886), 1–29 (2022). https://doi.org/10.1016/j.comgeo.2022.101886

Algorithmics

Parameterized and Approximation Algorithms for the Maximum Bimodal Subgraph Problem

Walter Didimo[1][ID], Fedor V. Fomin[2][ID], Petr A. Golovach[2][ID],
Tanmay Inamdar[2][ID], Stephen Kobourov[3][ID], and Marie Diana Sieper[4(✉)][ID]

[1] Department of Engineering, University of Perugia, Perugia, Italy
`walter.didimo@unipg.it`
[2] Department of Computer Science, University of Bergen, Bergen, Norway
`{fedor.fomin,petr.golovach,tanmay.inamdar}@uib.no`
[3] Department of Computer Science, University of Arizona, Tucson, USA
`kobourov@cs.arizona.edu`
[4] Department of Computer Science, University of Würzburg, Würzburg, Germany
`marie.sieper@uni-wuerzburg.de`

Abstract. A vertex of a plane digraph is *bimodal* if all its incoming edges (and hence all its outgoing edges) are consecutive in the cyclic order around it. A plane digraph is bimodal if all its vertices are bimodal. Bimodality is at the heart of many types of graph layouts, such as upward drawings, level-planar drawings, and L-drawings. If the graph is not bimodal, the *Maximum Bimodal Subgraph (MBS)* problem asks for an embedding-preserving bimodal subgraph with the maximum number of edges. We initiate the study of the MBS problem from the parameterized complexity perspective with two main results: (i) we describe an FPT algorithm parameterized by the branchwidth (and hence by the treewidth) of the graph; (ii) we establish that MBS parameterized by the number of non-bimodal vertices admits a polynomial kernel. As the byproduct of these results, we obtain a subexponential FPT algorithm and an efficient polynomial-time approximation scheme for MBS.

Keywords: bimodal graphs · maximum bimodal subgraph · parameterized complexity · FPT algorithms · polynomial kernel · approximation scheme

Research started at the Dagstuhl Seminar 23162: New Frontiers of Parameterized Complexity in Graph Drawing, April 2023, and partially supported by: (*i*) University of Perugia, Ricerca Base 2021, Proj. "AIDMIX - Artificial Intelligence for Decision Making: Methods for Interpretability and eXplainability"; (*ii*) MUR PRIN Proj. 2022TS4Y3N - "EXPAND: scalable algorithms for EXPloratory Analyses of heterogeneous and dynamic Networked Data", (*iii*) MUR PRIN Proj. 2022ME9Z78 - "NextGRAAL: Next-generation algorithms for constrained GRAph visuALization", (*iv*) the Research Council of Norway project BWCA 314528, (*v*) the European Research Council (ERC) grant LOPPRE 819416, and (*vi*) NSF-CCF 2212130.

M. A. Bekos and M. Chimani (Eds.): GD 2023, LNCS 14466, pp. 189–202, 2023.
https://doi.org/10.1007/978-3-031-49275-4_13

1   Introduction

Let G be a plane digraph, that is, a planar directed graph with a given planar embedding. A vertex v of G is *bimodal* if all its incoming edges (and hence all its outgoing edges) are consecutive in the cyclic order around v. In other words, v is bimodal if the circular list of edges incident at v can be split into at most two linear lists, where all edges in the same list are either all incoming or all outgoing v. Graph G is *bimodal* if all its vertices are bimodal. Bimodality is a key property at heart of many graph drawing styles. In particular, it is a necessary condition for the existence of *level-planar* and, more generally, *upward planar* drawings, where the edges are represented as curves monotonically increasing in the upward direction according to their orientations [12–14, 24]; see Fig. 1a. Bimodality is also a sufficient condition for *quasi-upward planar* drawings, in which edges are allowed to violate the upward monotonicity a finite number of times at points called *bends* [5–7]; see Fig. 1b. It has been shown that bimodality is also a sufficient condition for the existence of *planar L-drawings* of digraphs, in which distinct L-shaped edges may overlap but not cross [1–3]; see Fig. 1c. A generalization of bimodality is *k-modality*. Given a positive even integer k, a plane digraph is *k-modal* if the edges at each vertex can be grouped into at most k sets of consecutive edges with the same orientation [26]. In particular, it is known that 4-modality is necessary for planar L-drawings [10].

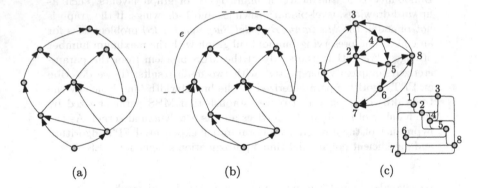

(a)   (b)   (c)

Fig. 1. (a) An upward planar drawing. (b) A quasi-upward planar drawing, where edge e makes two bends (the two horizontal tangent points). (c) A bimodal digraph (above) and a corresponding planar L-drawing (below).

While testing if a digraph G admits a bimodal planar embedding can be done in linear time [5], a natural problem that arises when G does not have such an embedding is to extract from G a subgraph of maximum size (i.e., with the maximum number of edges) that fulfills this property. This problem is NP-hard, even if G has a given planar embedding and we look for an embedding-preserving maximum bimodal subgraph [8]. We address exactly this fixed-embedding version of the problem, and call it the *Maximum Bimodal Subgraph* (MBS) problem.

Contribution. While a heuristic and a branch-and-bound algorithm are given in [8] to solve MBS (and also to find a maximum upward-planar digraph), here we study this problem from the parameterized complexity and approximability perspectives (refer to [11,18] for an introduction to parameterized complexity). More precisely, we consider the following more general version of the problem with weighted edges; it coincides with MBS when we restrict to unit edge weights.

MWBS(G, w) (*Maximum Weighted Bimodal Subgraph*). *Given a plane digraph G and an edge-weight function $w : E(G) \to \mathbb{Q}^+$, compute a bimodal subgraph of G of maximum weight, i.e., whose sum of the edge weights is maximum over all bimodal subgraphs of G.*
Our contribution can be summarized as follows.

− *Structural Parameterization.* We show that MWBS is FPT when parameterized by the *branchwidth* of the input digraph G or, equivalently, by the *treewidth* of G (Sect. 3). Our algorithm deviates from a standard dynamic approach for graphs of bounded treewidth. The main difficulty here is that we have to incorporate the "topological" information about the given embedding in the dynamic program. We accomplish this via the sphere-cut decomposition of Dorn et al. [16].

− *Kernelization.* Let b be the number of non-bimodal vertices in an input digraph G. We construct a polynomial kernel for the decision version of MWBS parameterized by b (Sect. 4). Our kernelization algorithm performs in several steps. First we show how to reduce the instance to an equivalent instance whose branchwidth is $\mathcal{O}(\sqrt{b})$. Second, by using specific gadgets, we compress the problem to an instance of another problem whose size is bounded by a polynomial of b. In other words, we provide a polynomial compression for MWBS. Finally, by the standard arguments, [18, Theorem 1.6], based on a polynomial reduction between any NP-complete problems, we obtain a polynomial kernel for MWBS.

By pipelining the crucial step of the kernelization algorithm with the branchwidth algorithm, we obtain a parameterized subexponential algorithm for MWBS of running time $2^{\mathcal{O}(\sqrt{b})} \cdot n^{\mathcal{O}(1)}$. Since $b \leq n$, this also implies an algorithm of running time $2^{\mathcal{O}(\sqrt{n})}$. Note that our algorithms are asymptotically optimal up to the *Exponential Time Hypothesis* (ETH) [21,22]. The NP-hardness result of MBS (and hence of MWBS) given in [8] exploits a reduction from PLANAR-3SAT. The number of non-bimodal vertices in the resulting instance of MBS is linear in the size of the PLANAR-3SAT instance. Using the standard techniques for computational lower bounds for problems on planar graphs [11], we obtain that the existence of an $2^{o(\sqrt{b})} \cdot n^{\mathcal{O}(1)}$-time algorithm for MBWS would contradict ETH.

− *Approximability.* We provide an Efficient Polynomial-Time Approximation Scheme (EPTAS) for MWBS, based on Baker's (or shifting) technique [4]. Namely, using our algorithm for graphs of bounded branchwidth, we give an $(1 + \epsilon)$-approximation algorithm that runs in $2^{\mathcal{O}(1/\epsilon)} \cdot n^{\mathcal{O}(1)}$ time.

Full proofs of the results marked with an asterisk (*), as well as additional definitions and technical details, are given in [15].

2   Definitions and Terminology

Let G be a digraph. We denote by $V(G)$ and $E(G)$ the set of vertices and the set of edges of G. Throughout the paper we assume that G is planar and that it comes with a planar embedding; such an embedding fixes, for each vertex $v \in V(G)$, the clockwise order of the edges incident to v. We say that G is a *planar embedded digraph* or simply that G is a *plane digraph*.

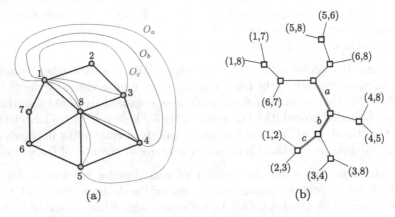

Fig. 2. A plane graph G and a sphere-cut decomposition of G; three nooses are highlighted on G for the arcs a, b, and c of the decomposition tree.

Branch Decomposition and Sphere-Cut Decomposition. A *branch decomposition* of a graph G defines a hierarchical clustering of the edges of G, represented by an unrooted proper binary tree, that is a tree with non-leaf nodes of degree three, whose leaves are in one-to-one correspondence with the edges of G. More precisely, a branch decomposition of G consists of a pair $\langle T, \xi \rangle$, where T is an unrooted proper binary tree and $\xi : \mathcal{L}(T) \leftrightarrow E(G)$ is a bijection between the set $\mathcal{L}(T)$ of the leaves of T and the set $E(G)$ of the edges of G. For each arc a of T, denote by T_1^a and T_2^a the two connected components of $T \setminus \{a\}$, and, for $i = 1, 2$, let G_i^a be the subgraph of G that consists of the edges corresponding to the leaves of T_i^a. The *middle set* $\mathrm{mid}(a) \subseteq V(G)$ is the intersection of the vertex sets of G_1^a and G_2^a, i.e., $\mathrm{mid}(a) := V(G_1^a) \cap V(G_2^a)$. The *width* $\beta(\langle T, \xi \rangle)$ of $\langle T, \xi \rangle$ is the maximum size of the middle sets over all arcs of T, i.e., $\beta(\langle T, \xi \rangle) = \max\{|\mathrm{mid}(a)| : a \in E(T)\}$. An *optimal branch decomposition* of G is a branch decomposition with minimum width; this width is called the *branchwidth* of G and is denoted by $\mathrm{bw}(G)$.

A sphere-cut decomposition is a special type of branch decomposition (see Fig. 2). Let G be a connected planar graph, topologically drawn on a sphere Σ. A *noose* O of G is a closed simple curve on Σ that intersects G only at vertices and that traverses each face of G at most once. The *length* of O is the number of vertices that O intersects. Note that, O bounds two closed discs Δ_O^1

and Δ_O^2 in Σ; we have $\Delta_O^1 \cap \Delta_O^2 = O$ and $\Delta_O^1 \cup \Delta_O^2 = \Sigma$. Let $\langle T, \xi \rangle$ be a branch decomposition of G. Suppose that for each arc a of T there exists a noose O_a that traverses exactly the vertices of $\mathrm{mid}(a)$ and whose closed discs $\Delta_{O_a}^1$ and $\Delta_{O_a}^2$ enclose the drawings of G_1^a and of G_2^a, respectively. Denote by π_a the circular clockwise order of the vertices in $\mathrm{mid}(a)$ along O_a and let $\Pi = \{\pi_a : a \in E(T)\}$ the set of all circular orders π_a. The triple $\langle T, \xi, \Pi \rangle$ is a *sphere-cut decomposition* of G. We assume that the vertices of $\mathrm{mid}(a) = V(G_1^a) \cap V(G_2^a)$ are enumerated according to π_a. Since a noose O_a traverses each face of G at most once, both graphs G_1^a and G_2^a are connected. Also, the nooses are pairwise non-crossing, i.e., for any pair of nooses O_a and O_b, we have that O_b lies entirely inside $\Delta_{O_a}^1$ or entirely inside $\Delta_{O_a}^2$. For a noose O_a, we define $\mathrm{mid}(O_a) = \mathrm{mid}(a)$, or in general, we define $\mathrm{mid}(\phi)$ to be the vertices cut by ϕ. We rely on the following result on the existence and computation of a sphere-cut decomposition [23] (see also [16]).

Proposition 1 ([23]). *Let G be a connected graph embedded in the sphere with n vertices and branchwidth $\ell \geq 2$. Then there exists a sphere-cut decomposition of G with width ℓ, and it can be computed in $\mathcal{O}(n^3)$ time.*

We remark that the branchwidth $\mathrm{bw}(G)$ and the treewidth $\mathrm{tw}(G)$ of a graph G are within a constant factor: $\mathrm{bw}(G) - 1 \leq \mathrm{tw}(G) \leq \lfloor \frac{3}{2} \mathrm{bw}(G) \rfloor - 1$ (see [25]).

3   FPT Algorithms for MWBS by Branchwidth

In this section we describe an FPT algorithm parameterized by branchwidth. We first introduce configurations, which encode on which side of a closed curve and in what order in a bimodal subgraph for a vertex v the switches between incoming to outgoing edges happen.

Definition 1 *(Configuration).* *Let $C = \{(i), (o), (i, o), (o, i), (o, i, o), (i, o, i)\}$. Let G be a graph embedded in the sphere Σ, ϕ be a noose in Σ with a prescribed inside, $v \in \mathrm{mid}(\phi)$, and $X \in C$. Let $E^{v,\phi}$ be the set of edges incident to v in ϕ. We say v has configuration X in ϕ, if $E^{v,\phi}$ can be partitioned into sets such that:*

1. *For every $x \in X$, there is a (possibly empty) set E_x associated with it.*
2. *Every set associated with an i (o) contains only in- (/out-) edges of v.*
3. *For every set, the edges contained in it are successive around v.*
4. *The sets E_x appear clockwise (seen from v) in the same order in G inside ϕ as the x appear in X.*

For every $v \in \mathrm{mid}(\phi)$, let X_v be a configuration of v in ϕ. We say $X_\phi = \{X_v \mid v \in \mathrm{mid}(\phi)\}$ is a configuration set of ϕ.

If G is bimodal, then for every noose ϕ and every vertex $v \in \mathrm{mid}(\phi)$, v must have at least one configuration $X \in C$ in ϕ. Note that configurations and configuration sets are not unique, as seen in Fig. 3(a). A vertex can even have all configurations if it has no incident edges in ϕ. The next definition is needed to encode when configurations can be combined in order to obtain bimodal vertices.

Definition 2 *(Compatible configurations).* *Let* $X, X', X^* \in C$ *be configurations. We say* X, X' *are* compatible configurations *or short* compatible, *if by concatenating* X, X' *and deleting consecutive equal letters, the result is a substring of* (o, i, o) *or* (i, o, i). *Note that it is not important in which order we concatenate* X, X'. *See Fig. 3(b). We say* X *and* X' *are* compatible with respect to X^* *if by concatenating* X, X' *(in this order) and deleting consecutive equal letters, the result is a substring of* X^*.

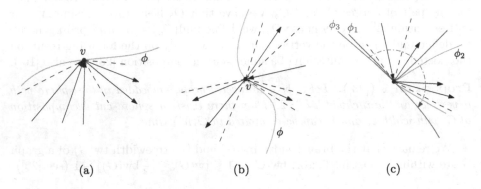

(a)   (b)   (c)

Fig. 3. (a) A vertex with configurations $(o, i), (o, i, o)$ and (i, o, i) in ϕ. The most restricted and thus minimal configuration is (o, i). (b) A vertex with configuration (o, i, o) in ϕ and (o) outside of ϕ. Concatenating (o, i, o) with (o) and deleting consecutive equal letters results in (o, i, o), the result is a substring of (o, i, o), thus (o, i, o) and (o) are compatible. (c) Note that ϕ_3 is composed of ϕ_1 and ϕ_2; the inside of ϕ_1, the inside of ϕ_2 and the outside of ϕ_3 are clockwise in this order around v with configuration (i, o) in ϕ_1 and (o) in ϕ_2. They can be concatenated to configuration (i, o) in ϕ_3, while (i, o) and (o) are compatible w.r.t. (i, o), but not (o, i).

A configuration X can have several compatible configurations, for example $(i, o) \in C$ is compatible with $(o), (i)$ and (o, i). From these (o, i) is in some sense maximal, meaning that configurations (o) and (i) are substrings of (o, i). Given a configuration X, a *maximal compatible configuration* X' of X is a configuration that is compatible with X, and all other compatible configurations of X are substrings of X'. Observe that every configuration has a unique maximal compatible configuration, they are pairwise: $(i) - (i, o, i)$, $(o) - (o, i, o)$ and $(o, i) - (i, o)$.

We say a noose ϕ_3 is *composed* of the nooses ϕ_1 and ϕ_2, if the edges of G in ϕ_3 are partitioned by ϕ_1 and ϕ_2. If a noose ϕ_3 is composed of nooses ϕ_1 and ϕ_2, and there exists a vertex $v \in \text{mid}(\phi_1) \cap \text{mid}(\phi_2) \cap \text{mid}(\phi_3)$, such that in ϕ_3 around v, all adjacent edges of v in ϕ_1 are clockwise before all adjacent edges of v in ϕ_2. If X, X' and X^* are nooses and X and X' are compatible with respect to X^*, and v has configuration X in ϕ_1 and configuration X' in ϕ_2, then it has configuration X^* in ϕ_3. See Fig. 3(c).

If a curve ϕ contains only one edge on its inside, finding maximal subgraphs for a configuration inside ϕ is easy.

Lemma 1 (*). *Let G be a graph embedded in the sphere Σ, let $e = \{u, v\}$ be an edge and let ϕ be a noose that cuts G only in u and v, such that e is in ϕ and all other edges are on the outside of ϕ. Let X_u, X_v be prescribed configurations. Then we can compute in $\mathcal{O}(1)$ time the maximum subgraph G' of G such that u, v have configuration X_u respectively X_v in ϕ in G'.*

We will now see how we can compute optimal subgraphs bottom-up.

Lemma 2 (*). *Let G be a graph embedded in the sphere Σ, let ϕ_1, ϕ_2, ϕ_3 be nooses with length at most ℓ each, and let $E_{\phi_1}, E_{\phi_2}, E_{\phi_3}$ be the sets of edges contained inside the respective noose with E_{ϕ_1}, E_{ϕ_2} being a partition of E_{ϕ_3}. Let X_{ϕ_3} be a configuration set for ϕ_3. Let further for every configuration set X_{ϕ_1} (X_{ϕ_2}) of ϕ_1 (ϕ_2), the maximum subgraph that has configuration set X_{ϕ_1} (X_{ϕ_2}) and is bimodal in ϕ_1 (ϕ_2) be known. Then a maximum subgraph G' of G that has configuration set X_{ϕ_3} and is bimodal in ϕ_3 can be computed in $\mathcal{O}(6^{2\ell}) \cdot n^{\mathcal{O}(1)}$ time.*

If a noose ϕ contains only $e \in E$, we have only two options in ϕ: delete e or do not. Testing which is optimal can be done in constant time, this leads to Lemma 1. Now let ϕ_3 be a noose that contains more than one edge, let ϕ_1, ϕ_2 be two nooses that partition the inside of ϕ_3, and let X_{ϕ_3} be a given configuration set. If we already know optimal solutions for any given configuration set in ϕ_1 (ϕ_2) (which we already computed when traversing the sphere-cut decomposition bottom up), we can guess for some optimal solution for ϕ_3 for every $v \in \text{mid}(\phi_1) \cap \text{mid}(\phi_2)$ the configuration it has in ϕ_1 and in ϕ_2. This gives us configuration sets X_{ϕ_1} and X_{ϕ_2} for ϕ_1 and ϕ_2, respectively (for every $v \in \text{mid}(\phi_1) \setminus \text{mid}(\phi_2)$ we take its configuration in X_{ϕ_3}). We obtain the corresponding solution G' that coincides with the optimal solution for ϕ_1 (ϕ_2) in ϕ_1 (ϕ_2) respecting X_{ϕ_1} (X_{ϕ_2}) and that coincides with G outside of ϕ_3. Since $|\text{mid}(\phi_1) \cap \text{mid}(\phi_2)| \leq \ell$, we achieve the same by enumerating all possible configurations for $\text{mid}(\phi_1) \cap \text{mid}(\phi_2)$, compute the corresponding solutions and take the maximum in $\mathcal{O}(6^{2\ell}) \cdot n^{\mathcal{O}(1)}$ time, leading to Lemma 2. We now obtain the following theorem.

Theorem 1 (*). *There is an algorithm that solves MWBS(G, w) in $2^{\mathcal{O}(\text{bw}(G))} \cdot n^{\mathcal{O}(1)}$ time. In particular, MWBS is FPT when parameterized by branchwidth.*

Proof (Sketch). Assume that G is connected (otherwise process every connected component independently). If $\text{bw}(G) = 1$, G is a star and we can compute an optimal solution in polynomial time. Otherwise, according to Proposition 1 we can compute a sphere-cut decomposition $\langle T, \xi, \Pi \rangle$ for G with optimal width ℓ. We pick any leaf of T to be the root r of T. For every noose O corresponding to an arc of T let X_O be a configuration set for O. Then we define $E_{(O, X_O)}$ to be edge set of minimum weight, such that $G \setminus E_{(O, X_O)}$ is bimodal inside of O and has configuration set X_O in O. We now compute the $E_{(O, X_O)}$ bottom-up. For a noose O corresponding to a leaf-arc in T, Lemma 1 shows that we can compute all possible values of $E_{(O, X_O)}$ in linear time. For a noose O corresponding to a non-leaf arc in T, Lemma 2 shows that we can compute E_{O, X_O} for a given X_O in $\mathcal{O}(6^{2\ell}) \cdot n^{\mathcal{O}(1)}$ time, and thus all entries for O in $\mathcal{O}(6^{3\ell}) \cdot n^{\mathcal{O}(1)}$ time. Let $e \in E$

be the edge associated with r. We have only two options left, delete e or do not. In both cases we obtain the optimal solution for the rest of G from the values $E_{(O,X_O)}$. The overall running time is $2^{\mathcal{O}(\ell)} \cdot n^{\mathcal{O}(1)}$. $\qquad\square$

Since our input graphs are planar, we immediately obtain a subexponential algorithm for MWBS because for a planar graph G, $\mathrm{bw}(G) = \mathcal{O}(\sqrt{n})$ [19].

Theorem 2. $MWBS(G = (V, E), w)$ *can be solved in* $2^{\mathcal{O}(\sqrt{n})}$ *time.*

4   Compression for MWBS by b

Throughout this section we assume that (i) the weights are rational, that is, for (G, w), $w \colon V(G) \to \mathbb{Q}^+$ and (ii) we consider the decision version of MWBS, that is, additionally to (G, w), we are given a target value $W \in \mathbb{Q}^+$ and the task is to decide whether G has a bimodal subgraph G^* with $w(E(G^*)) \geq W$.

Further Definitions. For simplicity, we say that a bimodal vertex of G is a *good* vertex, and that a non-bimodal vertex is a *bad* vertex. We denote by $\mathcal{G}(G)$ and $\mathcal{B}(G)$ the sets of good and bad vertices of G, respectively. Given a vertex $v \in V(G)$, an *in-wedge* (resp. *out-wedge*) of v is a maximal circular sequence of consecutive incoming (resp. outgoing) edges of v. Clearly, if v is bimodal it has at most one in-wedge and at most one out-wedge. Given a vertex $v \in \mathcal{B}(G)$, a *good edge-section* of v is a maximal consecutive sequence of in- and out- wedges of v, such that no edge is incident to another bad vertex.

Observation 1. Let (G, w) be an instance of MWBS with b bad vertices, and let $v \in \mathcal{B}(G)$. Then v can have at most $b - 1$ good edge-sections.

We introduce a generalization of MWBS called CUT-MWBS$(G, w, \mathcal{E})$ (*maximum weighted bimodal subgraph with prescribed cuts*). Given a plane digraph G, an edge-weight function $w \colon E(G) \to \mathbb{Q}^+$, and a partition $\mathcal{E}$ of $E(G)$, compute a bimodal subgraph G' of G of maximum weight, i.e., whose sum of the edge weights is maximum over all bimodal subgraphs of G, under the condition that for every set $E_i \in \mathcal{E}$, either all $e \in E_i$ are still present in G' or none of them are. We can see that every instance (G, w) of MWBS is equivalent to the instance $(G, w, \{\{e\} \mid e \in E(G)\})$ of Cut-MWBS, and thus Cut-MWBS is NP-hard. Also, the decision variant of the problem is NP-complete.

We now give reduction rules for the MWBS to Cut-MWBS compression, and prove that each of them is *sound*, i.e., it can be performed in polynomial time and the reduced instance is solvable if and only if the starting instance is solvable.

Reduction Rule 1. Let (G, w) be an instance of MWBS, and $v \in V(G)$ be an isolated vertex. Then, let (G', w) be the new instance, where $V(G') = V(G) \backslash \{v\}$.

Reduction Rule 2. Let (G, w) be an instance of MWBS with the target value W, and $u, v \in \mathcal{G}(G)$ be such that (u, v) is an edge. Then, the resulting instance is (G', w), where $G' = G - (u, v)$, and the new target value is $W' = W - w(u, v)$.

Reduction Rule 3. Let (G, w) be an instance of MWBS and $v \in \mathcal{G}(G)$ of degree ≥ 2. Let (G', w) be the new instance, where in G' we replace each edge $e = (u, v)$ (resp. $e = (v, u)$) where $u \in \mathcal{G}(G)$ with another edge $e' = (u, x_{uv})$ (resp. $e' = (x_{uv}, u)$), where x_{uv}'s are distinct vertices created for each such edge, and each e' is embedded within the embedding of e, where $w(e') = w(e)$ (see Fig. 4).

Claim 1 (*). Reduction rules 1, 2 and 3 are sound.

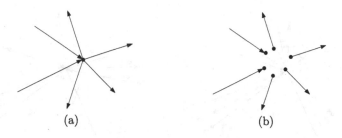

(a)   (b)

Fig. 4. A bimodal vertex (a) before and (b) after Reduction rule 3 is applied.

By applying Reductions 1, 2 and 3 exhaustively, we get Lemma 3, which is already enough to give a subexponential FPT algorithm by b (Theorem 3).

Lemma 3 (*). *Given an instance (G, w) of MWBS, there exists a polynomial-time algorithm to obtain an equivalent instance (G', w) with G' being a subgraph of G, such that (i) $|\mathcal{B}(G')| \leq |\mathcal{B}(G)|$, (ii) $\mathcal{G}(G')$ is an independent set in G', and (iii) for all $v \in \mathcal{G}(G')$, $\deg(v) = 1$ in the underlying graph of G'.*

Theorem 3. *There exists an algorithm that solves MWBS(G, w) with b bad vertices in $2^{\mathcal{O}(\sqrt{b})} \cdot n^{\mathcal{O}(1)}$ time.*

Proof. By Lemma 3, (G, w) is equivalent to (G', w) with at most b vertices of degree > 1, which we can compute in polynomial time. This implies bw$(G') = \mathcal{O}(\text{tw}(G')) = \mathcal{O}(\sqrt{b})$, and we can apply Theorem 1 to obtain an algorithm that computes a solution for (G', w) in $2^{\mathcal{O}(\text{bw}(G'))}|V(G')|^{\mathcal{O}(1)}$ time.   □

We now describe how we can partition, for a given input, all good-edge sections into edge sets in such a way that there exists an optimal solution in which every set is either contained or deleted completely, and the total number of sets is bounded in a function of b. We will then show how we can replace the sets with edge sets of size at most two. The main difficulty will be to ensure that sets that exclude each other continue to do so in the reduced instance.

Lemma 4 (*). *Let (G, w) be an instance of MWBS with n vertices and b bad vertices, such that $\mathcal{G}(G)$ is an independent set in G and $\deg(v) = 1$ for all $v \in \mathcal{G}(G)$. Let further $v \in \mathcal{B}(G)$, and let S be a good edge-section of v. Then S can be partitioned into at most 26 sets $S_1, \ldots, S_{26}$, such that for every optimal* solution $G' \subseteq G$ of $MWBS(G, w)$, there exists an optimal solution $G^* \subseteq G$ of $MWBS(G, w)$, such that G' and G^* coincides on $G \setminus S$, and for every i, S_i is either contained or removed completely in G^*.

Further, there exists a partition $P_1, \ldots, P_j$ of $\{S_1, \ldots, S_{26}\}$, such that for all P_i: (1) $|P_i| \leq 2$, (2) the edges in P_i are consecutive in S and (3) if $P_i = \{S_1, S_2\}$, then S_1 consists of outgoing edges of v iff S_1 consists of incoming edges of v, and at least one of S_1, S_2 does not form a set of consecutive edges in S.

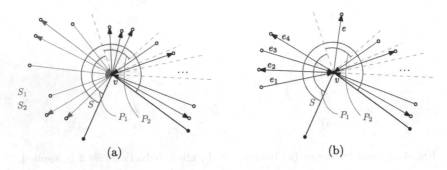

(a)   (b)

Fig. 5. (a) Illustration for Lemmas 4 and 5. The gray dashed lines correspond to a set of switches between the optimal solution we will choose; they impose the partition $P_1, \ldots, P_{13}$. S_1 (S_2) are the incoming (outgoing) edges of P_1, respectively. (b) The same vertex after transition to CUT-MWBS by Lemma 5, and after Reduction Rule 4 (5) got applied to P_2 (P_1), respectively.

To show this, we enclose S in a curve ϕ, and then compute for every given configuration X the maximal subgraph G' such that v has configuration X in ϕ. This yields a set of at most 12 possible locations for switches between incoming and outgoing edges in S, which gives a partition of S into at most 13 sets (corresponding to $P_1, \ldots, P_j$) that do not contain a switch, and thus at most 26 sets that will not be separated by an optimal solution, corresponding to $S_1, \ldots, S_{26}$. We now describe a parameter-preserving reduction from MWBS to Cut-MWBS.

Lemma 5 (*). Given an instance (G, w) of MWBS with b bad vertices, we can find in polynomial time an instance $(G', w, \mathcal{E})$ of Cut-MWBS, so that: (i) For every $\mathcal{E}_i \subseteq \mathcal{E}$ with $|\mathcal{E}_i| \geq 2$, there exists a bad vertex $v \in G'$ and a good edge-section S of v, so that $\mathcal{E}_i$ is a subset of S and $\mathcal{E}_i$ contains only outgoing or only incoming edges of v. (ii) $|\mathcal{B}(G')| \leq b$, (iii) $|\mathcal{E}| = \mathcal{O}(b^2)$, (iv) (G, w) and $(G', w, \mathcal{E})$ have the same optimal cost, (v) there exists a partition $P_1, \ldots, P_j$ of $\mathcal{E}$, such that $|P_i| \leq 2$ for all P_i, (vi) if $|P_i| = 1$, then the edges-set contained in P_i is either an edge between two bad vertices, or there exists a bad vertex $v \in G'$ and good edge-section S of v, such that the edges contained in P_i are all consecutive in S, and, (vii) if $|P_i| = 2$ with $P_i = \{\mathcal{E}_1, \mathcal{E}_2\}$, there exists some $v \in \mathcal{B}(G')$ and a good edge-section S of v, such that the edges in P_i are all consecutive in S; and $\mathcal{E}_1$ consists of outgoing edges of v if and only if $\mathcal{E}_1$ consists of incoming edges of v, and at least one of $\mathcal{E}_1, \mathcal{E}_2$ does not form a set of consecutive edges in S.

See Fig. 5(a) for a visualization. We obtain this transformation by applying Lemma 3 in order to get a simplified equivalent instance G'. Let E_{rest} be all edges incident to two bad vertices. For every bad vertex v and every good edges section S of v, let $\mathcal{S}_{v,S}$ be the partition of S obtained from Lemma 4. We define $\mathcal{E} = \{e \mid e \in E_{\text{rest}}\} \cup \bigcup_{v,S} \mathcal{S}_{v,S}$. This defines the instance $(G', w, \mathcal{E})$ of Cut-MWBS. We will now further reduce the size of $(G, w, \mathcal{E})$.

Reduction Rule 4. Let $(G, w, \mathcal{E})$ be an instance of Cut-MWBS with properties (i) to (vii) of Lemma 5. Let $v \in \mathcal{B}(G)$, let S be a good edge-section of v, and let $\mathcal{E}_i \in \mathcal{E}$ such that $\mathcal{E}_i \subseteq S$ is a *consecutive* set of edges in S. Then let $(G', w', \mathcal{E}')$ be the new instance that is obtained from $(G, w, \mathcal{E})$ by deleting all edges (and their incident good vertices) but one edge e out of $\mathcal{E}_i$, and assigning $w'(e) = w(\mathcal{E}_i)$.

Reduction Rule 5. Let $(G, w, \mathcal{E})$ be an instance of Cut-MWBS with the properties (i) to (vii) of Lemma 5. Let further $v \in \mathcal{B}(G)$, let S be a good edge-section of v, and let $\mathcal{E}_{\text{in}}, \mathcal{E}_{\text{out}} \in \mathcal{E}$ such that $\mathcal{E}_{\text{in}}, \mathcal{E}_{\text{out}} \subseteq S$, $\mathcal{E}_{\text{in}}$ are all incoming to v, $\mathcal{E}_{\text{out}}$ are all outgoing of v, $\mathcal{E}_{\text{in}} \cup \mathcal{E}_{\text{out}}$ is a consecutive set of edges in S, and at least one of $\mathcal{E}_{\text{in}}$ or $\mathcal{E}_{\text{out}}$ does not form a consecutive set of edges in S. We construct a new edge-set e_1, e_2, e_3, e_4 as follows: e_1, e_3 are incoming for v, e_2, e_4 are outgoing of v, and all of e_1, e_2, e_3, e_4 are incident to a newly inserted (good) vertex v_{e_k} for $k \in \{1, \ldots, 4\}$. We set $w'(e_1) = w'(e_4) = 0$, $w'(e_2) = w(\mathcal{E}_{\text{out}})$ and $w'(e_3) = w(\mathcal{E}_{\text{in}})$. Further we assign $e_1, e_3 \in \mathcal{E}_{\text{in}}$ and $e_2, e_4 \in \mathcal{E}_{\text{out}}$. Let $(G', w', \mathcal{E})$ be the new instance that is obtained from $(G, w, \mathcal{E})$ by replacing the edges in $\mathcal{E}_i \cup \mathcal{E}_j$ with the consecutive sequence e_1, e_2, e_3, e_4.

Claim 2 (*). Reductions 4 and 5 are sound.

Lemma 6 (*). *Let $(G, w, \mathcal{E})$ be an instance of Cut-MWBS with b bad vertices and properties (i) till (vii) of Lemma 5. Then we can compute in polynomial time an equivalent instance $(G', w', \mathcal{E}')$ such that $V(G') = \mathcal{O}(b^2)$.*

See Fig. 5(b) for an illustration. We compute $(G', w', \mathcal{E}')$ by applying Reductions 4 and 5 exhaustively. To bound the size of the weights w, we use the approach of Etscheid et al. [17] and the well-known Theorem 4. This yields the compression of MWBS (Theorem 5) and a kernel for MWBS (Theorem 6).

Theorem 4 ([20]). *There is an algorithm that, given a vector $\omega \in \mathbb{Q}^r$ and an integer N, in polynomial time finds a vector $\bar{\omega}$ such that $\|\bar{\omega}\|_\infty = 2^{\mathcal{O}(r^3)}$ and $\text{sign}(\omega \cdot b) = \text{sign}(\bar{\omega} \cdot b)$ for all vectors $b \in \mathbb{Z}^r$ with $\|b\|_1 \leq N - 1$.*

Theorem 5 (*). *There exists a polynomial-time algorithm that, given an instance (G, w) of MWBS with b bad vertices and a target value W, computes an instance $(G', w', \mathcal{E})$ of Cut-MWBS with size $\mathcal{O}(b^8)$, and a new target value W' with size $\mathcal{O}(b^6)$, such that there exists a solution for (G, w) of cost W if and only if there exists a solution for $(G', w', \mathcal{E})$ of cost W'.*

Theorem 6 (*). *The decision version of MWBS parameterized by the number of bad vertices b admits a polynomial kernel.*

5   Efficient PTAS for MWBS and Final Remarks

We sketch our Efficient Polynomial-Time Approximation Scheme (EPTAS) for MWBS, i.e., a $(1 - \epsilon)$-approximation that runs in $2^{\mathcal{O}(1/\epsilon)} \cdot n^{\mathcal{O}(1)}$ time. We use Baker's technique [4] to design our EPTAS. Our goal is to reduce the problem to (multiple instances of) the problem, where the treewidth (hence, branchwidth) of the graph is bounded by $\mathcal{O}(1/\epsilon)$, at the expense of an ϵ-factor loss in cost. Then, we can use our single-exponential algorithm in the branchwidth to solve each such instance exactly, which implies a $(1 - \epsilon)$-approximation.

We sketch the details of this reduction. W.l.o.g. assume that the graph is connected. We perform a breadth-first search starting from an arbitrary vertex $v \in V(G)$, and partition the vertex-set into layers $L_0, L_1, \ldots$, where L_i is the set of vertices at distance *exactly* i from v in the *undirected* version of G. It is known that the treewidth of the subgraph induced by any d consecutive layers is upper bounded by $\mathcal{O}(d)$ – this follows from a result of Bodlaender [9], which states that the treewidth of a planar graph with diameter D is $\mathcal{O}(D)$. Let $t = 1/\epsilon$, and for each $0 \leq i \leq t$, let $E^{(i,i+1)}$ denote edges uv such that $u \in L_j$, $v \in L_{j+1}$ with j mod $t = i$. By an averaging argument, there exists an index $0 \leq i \leq t$, such that the total contribution of all the edges from an optimal solution (i.e., the set of edges inducing a maximum-weight bimodal subgraph) that belong to $E^{(i,i+1)}$, is at most $1/t = \epsilon$ times the weight of the optimal solution. Since we do not know this index i, we consider all values of i, and consider the subproblems obtained by deleting the edges. Then, the graph breaks down into multiple connected components, and the treewidth of each component is $\mathcal{O}(1/\epsilon)$. We solve each such subproblem optimally in time $2^{\mathcal{O}(1/\epsilon)} \cdot n^{\mathcal{O}(1)}$ using Theorem 1, and combine the solutions for the subproblems to obtain a solution for the original instance. Note that the graph obtained by combining the optimal solutions for the subproblems is bimodal, and for the correct value of i, the weight of the graph is at least $1 - \epsilon$ times the optimal cost. That is, the combined solution is a $(1-\epsilon)$-approximation.

Theorem 7 (*). *There exists an algorithm that runs in time $2^{\mathcal{O}(1/\epsilon)} \cdot n^{\mathcal{O}(1)}$ and returns a $(1-\epsilon)$-approximate solution for the given instance of MWBS. That is, MWBS admits an EPTAS.*

We note that Baker's technique can also be used to obtain an EPTAS with the similar running for the *minimization* variant of MWBS. Although the high level idea is similar, the details are more cumbersome.

Final Remarks. We conclude by suggesting some open questions. One natural problem is to ask for a maximum k-modal subgraph for any given even integer $k \geq 2$; we believe that our ideas can be extended to this more general setting. Another natural variant of MBS is to limit the number of edges that we can delete to get a bimodal subgraph by an integer h; in this setting, h becomes another parameter in addition to those we have considered. Finally, studying MBS in the variable embedding setting is an interesting future direction.

References

1. Angelini, P., Chaplick, S., Cornelsen, S., Da Lozzo, G.: Planar L-drawings of bimodal graphs. J. Graph Algorithms Appl. **26**(3), 307–334 (2022). https://doi.org/10.7155/jgaa.00596

2. Angelini, P., Chaplick, S., Cornelsen, S., Lozzo, G.D.: On upward-planar L-drawings of graphs. In: Szeider, S., Ganian, R., Silva, A. (eds.) 47th International Symposium on Mathematical Foundations of Computer Science. MFCS 2022, 22–26 August 2022, Vienna, Austria. LIPIcs, vol. 241, pp. 10:1–10:15. Schloss Dagstuhl - Leibniz-Zentrum für Informatik (2022). https://doi.org/10.4230/LIPIcs.MFCS.2022.10

3. Angelini, P., et al.: Algorithms and bounds for L-drawings of directed graphs. Int. J. Found. Comput. Sci. **29**(4), 461–480 (2018). https://doi.org/10.1142/S0129054118410010

4. Baker, B.S.: Approximation algorithms for NP-complete problems on planar graphs. J. Assoc. Comput. Mach. **41**(1), 153–180 (1994)

5. Bertolazzi, P., Di Battista, G., Didimo, W.: Quasi-upward planarity. Algorithmica **32**(3), 474–506 (2002). https://doi.org/10.1007/s00453-001-0083-x

6. Binucci, C., Di Giacomo, E., Liotta, G., Tappini, A.: Quasi-upward planar drawings with minimum curve complexity. In: Purchase, H.C., Rutter, I. (eds.) GD 2021. LNCS, vol. 12868, pp. 195–209. Springer, Cham (2021). https://doi.org/10.1007/978-3-030-92931-2_14

7. Binucci, C., Didimo, W.: Computing quasi-upward planar drawings of mixed graphs. Comput. J. **59**(1), 133–150 (2016). https://doi.org/10.1093/comjnl/bxv082

8. Binucci, C., Didimo, W., Giordano, F.: Maximum upward planar subgraphs of embedded planar digraphs. Comput. Geom. **41**(3), 230–246 (2008). https://doi.org/10.1016/j.comgeo.2008.02.001

9. Bodlaender, H.L.: A partial k-arboretum of graphs with bounded treewidth. Theor. Comput. Sci. **209**(1–2), 1–45 (1998). https://doi.org/10.1016/S0304-3975(97)00228-4

10. Chaplick, S., et al.: Planar L-drawings of directed graphs. In: Frati, F., Ma, K.-L. (eds.) GD 2017. LNCS, vol. 10692, pp. 465–478. Springer, Cham (2018). https://doi.org/10.1007/978-3-319-73915-1_36

11. Cygan, M., et al.: Parameterized Algorithms (2015). https://doi.org/10.1007/978-3-319-21275-3

12. Di Battista, G., Eades, P., Tamassia, R., Tollis, I.G.: Graph Drawing: Algorithms for the Visualization of Graphs. Prentice-Hall, Upper Saddle River (1999)

13. Di Battista, G., Nardelli, E.: Hierarchies and planarity theory. IEEE Trans. Syst. Man Cybern. **18**(6), 1035–1046 (1988). https://doi.org/10.1109/21.23105

14. Didimo, W.: Upward graph drawing. In: Encyclopedia of Algorithms, pp. 2308–2312 (2016). https://doi.org/10.1007/978-1-4939-2864-4_653

15. Didimo, W., Fomin, F.V., Golovach, P.A., Inamdar, T., Kobourov, S., Sieper, M.D.: Parameterized and approximation algorithms for the maximum bimodal subgraph problem (2023). https://doi.org/10.48550/arXiv.2308.15635

16. Dorn, F., Penninkx, E., Bodlaender, H.L., Fomin, F.V.: Efficient exact algorithms on planar graphs: exploiting sphere cut decompositions. Algorithmica **58**(3), 790–810 (2010). https://doi.org/10.1007/s00453-009-9296-1

17. Etscheid, M., Kratsch, S., Mnich, M., Röglin, H.: Polynomial kernels for weighted problems. J. Comput. Syst. Sci. **84**, 1–10 (2017). https://doi.org/10.1016/j.jcss.2016.06.004

18. Fomin, F.V., Lokshtanov, D., Saurabh, S., Zehavi, M.: Kernelization. Theory of Parameterized Preprocessing. Cambridge University Press, Cambridge (2019)
19. Fomin, F.V., Thilikos, D.M.: New upper bounds on the decomposability of planar graphs. J. Graph Theory 51(1), 53–81 (2006). https://doi.org/10.1002/jgt.20121
20. Frank, A., Tardos, É.: An application of simultaneous diophantine approximation in combinatorial optimization. Combinatorica 7, 49–65 (1987). https://doi.org/10.1007/BF02579200
21. Impagliazzo, R., Paturi, R.: Complexity of k-sat. In: Proceedings of the 14th Annual IEEE Conference on Computational Complexity, Atlanta, Georgia, USA, 4–6 May 1999, pp. 237–240. IEEE Computer Society (1999). https://doi.org/10.1109/CCC.1999.766282
22. Impagliazzo, R., Paturi, R., Zane, F.: Which problems have strongly exponential complexity? J. Comput. Syst. Sci. 63(4), 512–530 (2001). https://doi.org/10.1006/jcss.2001.1774
23. Jacob, H., Pilipczuk, M.: Bounding twin-width for bounded-treewidth graphs, planar graphs, and bipartite graphs. In: Bekos, M.A., Kaufmann, M. (eds.) WG 2022. LNCS, vol. 13453, pp. 287–299. Springer, Cham (2022). https://doi.org/10.1007/978-3-031-15914-5_21
24. Jünger, M., Leipert, S., Mutzel, P.: Level planarity testing in linear time. In: Whitesides, S.H. (ed.) GD 1998. LNCS, vol. 1547, pp. 224–237. Springer, Heidelberg (1998). https://doi.org/10.1007/3-540-37623-2_17
25. Robertson, N., Seymour, P.D.: Graph minors. x. obstructions to tree-decomposition. J. Comb. Theory, Ser. B 52(2), 153–190 (1991). https://doi.org/10.1016/0095-8956(91)90061-N
26. Vial, J.J.B., Da Lozzo, G., Goodrich, M.T.: Computing k-modal embeddings of planar digraphs. In: Bender, M.A., Svensson, O., Herman, G. (eds.) 27th Annual European Symposium on Algorithms, ESA 2019, 9–11 September 2019, Munich/Garching, Germany. LIPIcs, vol. 144, pp. 19:1–19:16. Schloss Dagstuhl - Leibniz-Zentrum für Informatik (2019). https://doi.org/10.4230/LIPIcs.ESA.2019.19

Upward and Orthogonal Planarity are W[1]-Hard Parameterized by Treewidth

Bart M. P. Jansen[1] , Liana Khazaliya[2]([✉]) , Philipp Kindermann[3] ,
Giuseppe Liotta[4] , Fabrizio Montecchiani[4] , and Kirill Simonov[5]

[1] Eindhoven University of Technology, Eindhoven, The Netherlands
b.m.p.jansen@tue.nl
[2] Technische Universität Wien, Vienna, Austria
lkhazaliya@ac.tuwien.ac.at
[3] Universität Trier, Trier, Germany
kindermann@uni-trier.de
[4] University of Perugia, Perugia, Italy
{giuseppe.liotta,fabrizio.montecchiani}@unipg.it
[5] Hasso Plattner Institute, University of Potsdam, Potsdam, Germany

Abstract. UPWARD PLANARITY TESTING and RECTILINEAR PLANARITY TESTING are central problems in graph drawing. It is known that they are both NP-complete, but XP when parameterized by treewidth. In this paper we show that these two problems are W[1]-hard parameterized by treewidth, which answers open problems posed in two earlier papers. The key step in our proof is an analysis of the ALL-OR-NOTHING FLOW problem, a generalization of which was used as an intermediate step in the NP-completeness proof for both planarity testing problems. We prove that the flow problem is W[1]-hard parameterized by treewidth on planar graphs, and that the existing chain of reductions to the planarity testing problems can be adapted without blowing up the treewidth. Our reductions also show that the known $n^{\mathcal{O}(\mathsf{tw})}$-time algorithms cannot be improved to run in time $n^{o(\mathsf{tw})}$ unless ETH fails.

Keywords: Upward Planarity · Rectilinear Planarity · Parameterized Complexity · Treewidth

Bart M. P. Jansen has received funding from the European Research Council (ERC) under the European Union's Horizon 2020 research and innovation programme (grant agreement No 803421, ReduceSearch).

Liana Khazaliya is supported by Vienna Science and Technology Fund (WWTF) [10.47379/ICT22029]; Austrian Science Fund (FWF) [Y1329]; European Union's Horizon 2020 COFUND programme [LogiCS@TUWien, grant agreement No. 101034440].

G. Liotta and F. Montecchiani—This work was supported, in part, by MUR of Italy, under PRIN Project n. 2022ME9Z78 - NextGRAAL: Next-generation algorithms for constrained GRAph visuALization, and under PRIN Project n. 2022TS4Y3N - EXPAND: scalable algorithms for EXPloratory Analyses of heterogeneous and dynamic Networked Data.

Kirill Simonov acknowledges support by DFG Research Group ADYN via grant DFG 411362735.

M. A. Bekos and M. Chimani (Eds.): GD 2023, LNCS 14466, pp. 203–217, 2023.
https://doi.org/10.1007/978-3-031-49275-4_14

1   Introduction

A graph is *planar* if it admits a drawing in the plane where no two edges cross each other. Testing graph planarity is among the most fundamental problems in graph algorithms and graph drawing. While several papers proposed efficient algorithms for this problem (including the celebrated linear-time algorithm of Hopcroft and Tarjan [25]), notable variants and restrictions have also been investigated, including clustered planarity (see, e.g. [6,22]), constrained planarity (see, e.g. [7,29]), and k-planarity (see, e.g. [24,28,34]); refer to [31] for a survey.

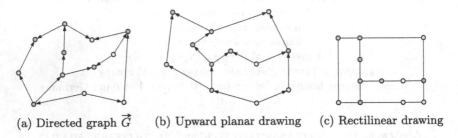

(a) Directed graph $\vec{G}$ (b) Upward planar drawing   (c) Rectilinear drawing

Fig. 1. A directed graph $\vec{G}$, an upward planar drawing of $\vec{G}$, and a rectilinear planar drawing of its underlying undirected graph G.

This paper investigates two of such classical variants, namely *upward planarity testing* and *rectilinear planarity testing*. Given a directed acyclic graph G, upward planarity testing asks whether G admits a crossing-free drawing where all edges are monotonically increasing in a common direction, which is conventionally called the upward direction; see Fig. 1b. For an undirected graph G, rectilinear planarity testing asks whether G admits a crossing-free drawing such that each edge is either a horizontal or a vertical segment; see Fig. 1c. Both upward planarity and rectilinear planarity testing are classical and extensively investigated topics in graph drawing (see, e.g. [2,27,30,33]).

While apparently different, the two problems have a lot in common. Namely, both an upward planar drawing of a digraph and a rectilinear planar drawing of a graph exists provided that the graph has a planar embedding where for every face f there is some "balancing" of the angles that the edges along the boundary of the face form in the interior of f. Consider, for simplicity, biconnected graphs. In a rectilinear planar drawing of a biconnected graph the boundary of every internal face f is an orthogonal polygon and hence the number of $\frac{\pi}{2}$ angles minus the number of $\frac{3\pi}{2}$ angles must be 4. Similarly, if a biconnected graph is upward planar it admits an upward planar drawing where for every internal face the number of angles that are smaller than $\frac{\pi}{2}$ always exceeds by two units the number of angles that are larger than π.

Hence, it is quite natural to see that in the *fixed embedding setting* (i.e. when the combinatorial structure of the faces of the graph is given as part of the

input) classical results solve both upward planarity testing and rectilinear planarity testing in polynomial time by looking for the existence of a feasible flow in
a network where vertices supply angles to faces and faces have a limited capacity
which depends on structure of the graph [3,32]. On the other hand, both problems are NP-complete in the so-called *variable embedding setting*, that is when
the testing algorithm must verify whether the input graph has a combinatorial
structure of its faces which allows the balancing of angles described above. Again
unsurprisingly, both proofs of NP-completeness follow the same logic based on a
reduction from a common flow problem on planar graphs [23].

These NP-completeness results have motivated a flourishing literature
describing both polynomial-time solutions for special classes of graphs and
parameterized solutions for general graphs. For example, polynomial-time solutions are known for both problems when the input graph has treewidth at most
two [15–18,21]; also, rectilinear planarity testing can be solved in linear time
if the maximum degree of the input graph is at most three [19], and upward
planarity testing can be solved in linear time if the digraph has only one source
vertex [4,11,26]. Concerning parameterized solutions, upward planarity testing
is fixed-parameter tractable when parameterized by the number of triconnected
components [12], by the treedepth [13], and by the number of sources [13].

The research in this paper is motivated by the fact that both upward planarity and rectilinear planarity testing are known to lie in XP when parameterized by treewidth [13,16]. Determining whether these two parameterized problems are in FPT are mentioned as open problems in both [13,16]. The main
contribution of this paper is as follows.

Theorem 1. *Upward planarity testing and rectilinear planarity testing parameterized by treewidth are both* W[1]*-hard. Moreover, assuming the Exponential
Time Hypothesis, neither problem can be solved in time* $f(k) \cdot n^{o(k)}$ *for any computable function* f, *where* k *is the treewidth of the input graph.*

Theorem 1 implies that, under the standard hypothesis FPT $\neq$ W[1] in
parameterized complexity, there exists no fixed-parameter tractable algorithm
for either problem parameterized by treewidth, hence answering the above mentioned open problems. To obtain our results we analyze the auxiliary flow problem used as a common starting point in the NP-completeness proof of both
planarity problems. It closely resembles the ALL-OR-NOTHING FLOW problem
(AoNF), which asks for an st-flow of prescribed value in an edge-capacitated
flow network such that each edge is either used fully, or not at all. The AoNF
problem parameterized by treewidth was recently shown to be W[1]-hard (in
fact, even XNLP-complete) on general graphs by Bodlaender et al. [9]. By a
significant adaptation of their construction, we can prove that AoNF parameterized by treewidth remains W[1]-hard on *planar* graphs. By revisiting the
chain of reductions to the planarity testing problems, passing through the CIR-
CULATING ORIENTATION problem in between, we show they can be carried out
without blowing up the treewidth of the graph and thereby obtain Theorem 1.

The rest of the paper is organized as follows. In Sect. 2 we formally define the
problems involved in our chain of reductions. Due to space limitations, all formal

proofs have been deferred to the Full Version[1]. In Sect. 3 we therefore provide high-level sketches of our proofs. We conclude in Sect. 4. For space reasons, results marked with a "$\star$" are proved in the Full Version.

2   Preliminaries

We assume familiarity with the basic notions of graph drawing [2] and of parameterized complexity [14], including the notions of treewidth and pathwidth which are commonly used parameters to capture the complexity of a graph G. The formal definitions, including those for upwards and rectilinear planarity, can be found in the Full Version. Below we define the parameterized problems which are used in the chain of reductions of our lower bounds. Throughout the paper we utilize both undirected and directed graphs, which may have parallel edges but no loops. A graph without parallel arcs or edges is a *simple* graph. We use uv to denote a (directed) arc from u to v and $\{u, v\}$ to denote an (undirected) edge between u and v. The vertex set of a graph G is denoted by $V(G)$ and the (multi)set of edges by $E(G)$. Depending on whether the graph is directed or not, $E(G)$ either contains its undirected edges or its directed arcs. For a vertex in a directed graph G, we denote by $E_G^-(v)$ the (multi)set of arcs leading into v, and by $E_G^+(v)$ the (multi)set of arcs leading out of v.

The starting point of our reductions is the following parameterized version of the CLIQUE problem, which is well-known to be W[1]-complete [14, Thm. 13.25]. Assuming ETH, it cannot be solved in time $f(k)n^{o(k)}$ for any computable function f [14, Cor. 14.32].

MULTICOLORED CLIQUE
Input: An undirected simple graph G and a partition of its vertex set into k sets $V_1, \ldots, V_k$, each consisting of N vertices.
Parameter: k.
Question: Does G contain a clique $C \subseteq V(G)$ such that $|C \cap V_i| = 1$ for each $i \in [k]$?

Note that the assumption that all sets V_i have the same size is without loss of generality, since we may pad the input with isolated vertices if needed.

The next problem in our chain of reductions is a variation of MAXIMUM FLOW; we therefore need some terminology regarding flows. A *flow network* (G, c, s, t) consists of a directed graph G with a capacity function $c \colon E(G) \to \mathbb{Z}_+$ on the arcs, together with two distinct vertices s, t called the *source* and *sink*. We allow a flow network to have parallel arcs. An *st-flow* in the flow network is a function $f \colon E(G) \to \mathbb{Z}_{\geq 0}$ such that for each arc $e \in E(G)$ we have $0 \leq f(e) \leq c(e)$ (capacity constraints), and for each vertex $v \in V(G) \setminus \{s, t\}$, we have $\sum_{e \in E_G^-(v)} f(e) = \sum_{e \in E_G^+(v)} f(e)$ (flow conservation). The *value* of the flow is defined as $\sum_{e \in E_G^+(s)} f(e) - \sum_{e \in E_G^-(s)} f(e)$. For

---

[1] https://arxiv.org/abs/2309.01264.

a vertex v in a flow network (G, c, s, t), we denote the total capacity of arcs leaving v by $d_G^+(v) := \sum_{e \in E_G^+(v)} c(e)$ and use $d_G^-(v) := \sum_{e \in E_G^+(v)} c(e)$ for the total capacity of arcs entering v. While a maximum flow can be found in polynomial time, the following variation is hard.

ALL OR NOTHING FLOW
Input: A flow network (G, c, s, t) and a positive integer $\mathcal{F}$.
Question: Does there exist an st-flow of value exactly $\mathcal{F}$, such that the flow through any arc $uv \in E(G)$ is either 0 or equal to $c(uv)$?

We then reduce to the following problem on undirected graphs that models the combinatorial difficulty encountered in testing upwards or rectilinear planarity, because it captures the problem of deciding orientations of edges to balance certain contributions around a vertex (i.e. a face of the dual graph).

CIRCULATING ORIENTATION
Input: An undirected graph G with an edge-capacity function $c: E(G) \to \mathbb{Z}_{\geq 0}$.
Question: Is it possible to orient the edges of G, such that for each vertex $v \in V(G)$ the total capacity of edges oriented into v is equal to the total capacity of edges oriented out of v? (Such an orientation is called a *circulating orientation*.)

Edges are allowed to have capacity 0 in this problem, which allows us to construct *triconnected* instances in the hardness reduction by inserting capacity-0 edges that do not violate planarity and do not blow up the pathwidth. For an undirected multigraph G, we use $E_G(v)$ to denote the (multi)set of edges incident on a vertex $v \in V(G)$. In the context of an edge-capacity function c, we denote the total capacity of edges incident on v by $d_G(v) := \sum_{e \in E_G(v)} c(e)$.

In the non-planar setting, ALL OR NOTHING FLOW easily reduces to CIRCULATING ORIENTATION by a polynomial-time transformation that increases the pathwidth by only a constant: it suffices to add a super-source and super-sink with properly chosen capacities on their incident edges [9]. Since the addition of a super-source and super-sink typically violates planarity of the graph, in our hardness construction for the flow problem we take special care to produce instances that can later be reduced to CIRCULATING ORIENTATION without violating planarity. We point out that Didimo et al. [20] recently proved the NP-completeness of CIRCULATING ORIENTATION on planar graphs. However, their reduction does not have any consequences for the complexity of the problem parameterized by treewidth.

3 Overview

Since space requirements prohibit us from presenting our reductions in detail, we give an outline that discusses the main technical ideas behind our result and defer the formal proofs to the Full Version.

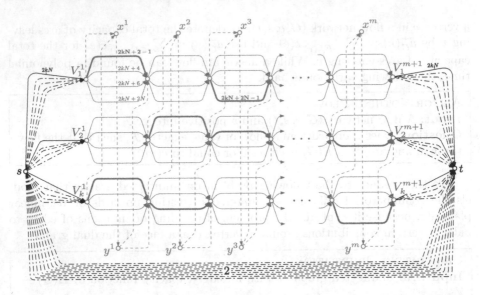

Fig. 2. Illustration for the FPT-reduction from MULTICOLORED CLIQUE with $k = 3, N = 4$ to non-planar ALL OR NOTHING FLOW on a graph of pathwidth $\mathcal{O}(k)$. Dashed edges have capacity 1, dash-dotted edges ($k(N-1)$ of them on the bottom) have capacity 2.

3.1 Hardness of ALL-OR-NOTHING-FLOW

The Non-planar Case. To aid the intuition, we first sketch an FPT-reduction from MULTICOLORED CLIQUE to non-planar ALL OR NOTHING FLOW on a graph of pathwidth $\mathcal{O}(k)$, which is inspired by an XNLP-completeness proof due to Bodlaender et al. [9]. Given an input $(G, V_1, \ldots, V_k, k)$, which asks whether G has a clique containing exactly one vertex from each of the size-N sets $V_1, \ldots, V_k$, we construct a flow network $\mathcal{G}$ as follows. Number the vertices in each set V_i as $v_{i,1}, \ldots, v_{i,N}$. Let $m = |\overline{E}(G)|$, where $\overline{E}(G)$ is the set of unordered vertex pairs which do not form an edge of G. The graph $\mathcal{G}$ contains k *rows* $R_1, \ldots, R_k$. Each row R_i consists of $m + 1$ vertices V_i^j for $j \in [m + 1]$. For each $j \in [m]$, there are N parallel arcs from V_i^j to V_i^{j+1} whose capacities are $2kN + 2q$ for each $q \in [N]$. (Below, we will decrease the capacities of some of these arcs by 1, to model the non-edges of G.) Intuitively, sending flow over an arc with capacity $2kN + 2q$ on row R_i corresponds to selecting the q-th vertex of V_i into the clique. Flow conservation will ensure that the same choice is made for all arcs on the same row R_i. To feed each row R_i, there is an arc of capacity $2kN$ from the source s to the first vertex V_i^1, along with N parallel arcs of capacity 2. This is sufficient to saturate any single edge on the row, but insufficient to saturate two edges. The analogous arcs leave from the last vertex V_i^{m+1} to the sink t.

We will set the target value $\mathcal{F}$ of the flow problem to $k(2kN + 2N)$, which is effectively the amount that we obtain when selecting the largest index on each row. To compensate for the fact that some rows may want to select a smaller index, corresponding to sending less than $2kN + 2N$ flow, we additionally add $k(N-1)$ parallel capacity-2 arcs directly from s to t. This number is chosen so that any feasible flow which sends at least value $2kN + 2$ through each row can be augmented to a flow of value exactly $k(2kN+2N)$, while making it impossible to reach the desired flow value without choosing one index on each row.

The final part of the construction ensures that the choices encoded by the flow correspond to a multicolored k-clique in G, using a small gadget that crucially exploits the all-or-nothing property of the flow; see Fig. 2. For each pair $v_{i,a}, v_{\ell,b}$ of non-adjacent vertices of G with $i \neq \ell$, we incorporate a gadget to ensure that we cannot simultaneously choose flow value $2kN + 2a$ on row R_i and $2kN + 2b$ on row R_ℓ. We pick a unique index z from $[m]$ for this non-edge of G, and adapt the construction as follows:

- Among the multiple arcs from V_i^z to V_i^{z+1}, we consider the arc whose capacity was set to $2kN + 2a$, which corresponds to choosing vertex $v_{i,a}$. We decrease the capacity of this arc by 1.
- Among the multiple arcs from V_ℓ^z to V_ℓ^{z+1}, we consider the arc whose capacity was set to $2kN + 2b$, which correspond to choosing vertex $v_{\ell,b}$. We decrease the capacity of this arc by 1.
- We introduce two new vertices x^z, y^z and an arc $x^z y^z$ of capacity 1. Vertices V_i^z, V_ℓ^z both get an arc to x^z of capacity 1. Vertices V_i^{z+1}, V_ℓ^{z+1} both get an arc from y^z of capacity 1.

The idea behind the gadget is as follows. It is possible to send a flow of value $2kN + 2a$ from V_i^z to V_i^{z+1} (by utilizing the arc of capacity $2kN + 2a - 1$ to V_i^{z+1} along with a flow of value 1 along the path via x^z, y^z). Analogously, it is possible to send a flow of value $2kN + 2b$ from V_ℓ^z to V_ℓ^{z+1}. But we cannot do both simultaneously, due to the capacity-1 bottleneck between x^z and y^z and the fact that the capacity of the other arcs on the row is either too large or too small to be used in an all-or-nothing fashion. Hence we ensure the values a and b cannot be simultaneously selected on rows i and ℓ.

The construction is completed by inserting such a gadget for each non-edge of G. The resulting flow network $\mathcal{G}$ can be shown to have pathwidth $\mathcal{O}(k)$ since it effectively consists of k paths whose interconnections are confined to vertices whose index differ by at most one along the path.

Planarizing the Instance. It is conceptually not difficult to extend the construction to prove hardness also for planar instances of the flow problem, since the all-or-nothing nature of the flow facilitates a simple method to eliminate edge crossings. Suppose we have a crossing between two arcs uv and xy whose capacities are different. Then we may simply replace the crossing arcs uv, xy by a new vertex d of degree four along with arcs ud, dv of capacity $c(uv)$ and arcs xd, dy of capacity $c(xy)$. This transformation preserves the answer to the ALL OR NOTH- ING FLOW problem: in one direction, any flow in the original network trivially yields a flow of the same value in the transformed network. The interesting step is the converse direction. An all-or-nothing flow in the reduced network either sends flow over both halves ud, dv (respectively xd, dy) of an arc, or over neither half of the arc: flow-conservation ensures that all flow entering d must also exit d, while the all-or-nothing property of the flow together with the fact that $c(uv) \neq c(xy)$ means that flow entering on arc ud (xd) cannot leave on arc dy (dv). (The same approach for eliminating edge crossings was used previously by Didimo et al. [20].)

Since the non-planar drawings coming out of the construction above (Fig. 2) have the property that all crossings involve pairs of arcs of different capacities, we can simply planarize the drawing by inserting degree-four vertices where needed. It is not difficult to show that the pathwidth increases by only a constant factor, resulting in the following lemma. Its proof can be found in the Full Version.

Lemma 1 ($\star$). *There is a polynomial-time algorithm that, given an instance of* MULTICOLORED CLIQUE *with parameter k, outputs an equivalent instance of* ALL OR NOTHING FLOW *on a planar graph of pathwidth $\mathcal{O}(k)$ whose edge capacities are bounded by a polynomial in $|V(G)|$.*

The bound on the edge capacities of the instance will later govern the size of *tendril* gadgets that will be created for the planarity testing problems.

3.2   Hardness of CIRCULATING ORIENTATION

We continue by describing the relation between ALL OR NOTHING FLOW and CIRCULATING ORIENTATION, starting with a special case that will be insightful to establish some intuition. Suppose we have a flow network (G, s, t, c) in which we ask for an all-or-nothing flow of value $\mathcal{F}$, satisfying the following conditions: for each vertex $v \in V(G) \setminus \{s, t\}$ we have $d_G^+(v) = d_G^-(v)$, the source has no incoming arcs, the sink has no outgoing arcs, and $\mathcal{F} = d_G^+(s)/2 = d_G^-(t)/2$.

We argue that this flow instance is equivalent to the instance of CIRCULATING ORIENTATION on the edge-capacitated undirected graph G' that is simply obtained from G by dropping the orientation of the edges. An all-or-nothing flow f of value $\mathcal{F}$ in G leads to a circulating orientation of the undirected graph G', as follows: simply start from the orientation of arcs as given by G, but reverse the orientation for each arc $uv \in E(G)$ with $f(uv) = 0$. For each vertex $v \in V(G) \setminus \{s, t\}$, flow conservation ensures that the total capacity of the incoming edges that carry flow is equal to the capacity of outgoing edges that carry flow. By the assumption that $d_G^-(v) = d_G^+(v)$, the total capacity of out-arcs of v which is reversed by the process equals that of the capacity of in-arcs of v which is reversed. Hence the total capacity of arcs which are oriented outwards remains unaffected by the reversals and equals $d_G^-(v) = d_G^+(v) = d_{G'}(v)/2$. For the source s, since the flow has value $\mathcal{F} = d^+(s)/2$ by assumption, reversing the orientation of edges not carrying flow leaves $\mathcal{F} = d_G^+(s)/2$ of capacity oriented out of s, which is exactly half of $d_{G'}(s)$ since there are no incoming arcs. The situation for the sink is analogous. This shows that a solution to the flow problem yields a solution to CIRCULATING ORIENTATION, and it is not difficult to show the converse also holds.

Under the simplifying assumptions above, it is therefore trivial to reduce from ALL OR NOTHING FLOW to CIRCULATING ORIENTATION. In the flow network G

we construct above, most vertices $v \in V(G) \setminus \{s, t\}$ satisfy $d_G^-(v) = d_G^+(v)$. The only exceptions are the x^j-vertices, along with the last vertex of each row (their in-capacity exceeds their out-capacity), and the y^j vertices along with the first vertex of each row (their out-capacity exceeds their in-capacity). In general graphs, the imbalance can be resolved by adding a super-source S and super-sink T. Here we can do something similar while preserving planarity, utilizing the fact that there is a face in the embedding that contains all x^j vertices along with the source s and sink t, and another face that contains all y^j-vertices and $\{s, t\}$. We insert a super-source and super-sink into these faces and use them to resolve the imbalance of the x^j and y^j vertices. Based on a delicate argument, we show that the imbalance of the first and last vertices of each row can be resolved via the standard source and sink, which are already adjacent to them.

For the super-source S and super-sink T to work as desired, we need an edge between them. This cannot be drawn in a planar fashion. However, we argue that the effect of the edge $\{S, T\}$ can be simulated by having a four-cycle (s, S, t, T) involving the standard source and sink with appropriate capacities, which can be added in a planar fashion. By carefully setting the capacities, this four-cycle also resolves the issue caused by the fact that $\mathcal{F} \neq d_G^+(s)/2$.

To transform the resulting instances of CIRCULATING ORIENTATION into the two planarity testing problems, it will be useful for the constructed planar graph to be triangulated (which implies it is triconnected). We can achieve this property by a post-processing step based on a result by Biedl. She proved [5] that any simple planar graph G of pathwidth k can be transformed into a triangulated planar supergraph G' on the same vertex set having pathwidth $\mathcal{O}(k)$, and such a triangulation can be computed efficiently. To make our graph simple, we can start by subdiving all edges which is known to increase the pathwidth by at most a constant. In the context of the CIRCULATING ORIENTATION problem, in which we are allowed to have edges of capacity 0 which do not affect the answer to the problem, we may then compute a triangulation of the simple graph and assign all newly introduced edges capacity 0, thereby leading to the following lemma.

Lemma 2 ($\star$). *There is a polynomial-time algorithm that, given an instance of* MULTICOLORED CLIQUE *with parameter k, outputs an equivalent instance of* CIRCULATING ORIENTATION *on a simple, triconnected, triangulated planar graph of pathwidth $\mathcal{O}(k)$ whose edge capacities are bounded by a polynomial in $|V(G)|$.*

3.3 From CIRCULATING ORIENTATION to Planarity Testing Problems

The main idea behind the reduction from CIRCULATING ORIENTATION to upward/rectilinear planarity testing is largely the same as in the original NP-hardness proof of Garg and Tamassia [23]. In what follows we explain the approach and highlight the differences. Since the reductions for both problems are fairly similar, we mostly focus on UPWARD PLANARITY TESTING. See also Fig. 3 for illustration.

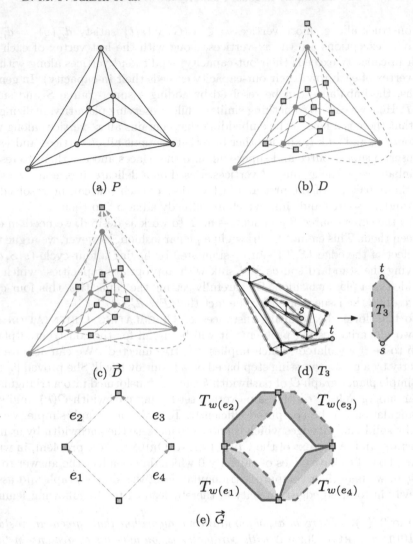

(a) P

(b) D

(c) $\vec{D}$

(d) T_3

(e) $\vec{G}$

Fig. 3. The reduction to UPWARD PLANARITY TESTING: (a) A triangulated planar graph P. (b) The dual D of P (based on the depicted planar embedding). (c) An orientation $\vec{D}$ of D which is an st-planar graph. (d) A tendril T_3 and its schematization; the red boundary has negative contribution, while the blue (dashed) boundary has positive contribution. (e) Construction of $\vec{G}$: replacing the edges of a face of $\vec{D}$ with the corresponding tendrils. (Color figure online)

As per the previous subsection, we start our reduction with an instance (P, c) of CIRCULATING ORIENTATION, where P is a planar triconnected graph of pathwidth $\mathcal{O}(k)$. The first step is to consider the dual graph D of P. By known results [1] about the pathwidth of planar graphs, the pathwidth of D is also $\mathcal{O}(k)$. We also consider the graph D to be weighted by c: the weight $w(e)$

of an edge e in D is set to be the capacity of its dual edge in P. We obtain the final digraph $\vec{G}$ of the reduction as follows: every edge $e \in E(D)$ is replaced by a *tendril* $T_{w(e)}$. The tendrils T_ℓ are special gadget graphs designed by Garg and Tamassia [23]; their properties are: (i) the upward planar embedding is unique; and (ii) one of the boundary walks has contribution 2ℓ to the adjacent face, and the other boundary walk has contribution -2ℓ. Here, contribution refers to the angle assignment characterization of upward planarity: roughly speaking, the graph admits an upward planar embedding if and only if the angles in a planar embedding of the graph could be assigned numbers in $\{-1, 0, 1\}$ according to certain rules so that the sum of angles on the boundary of every inner face is -2, and on the boundary of the outer face is 2.

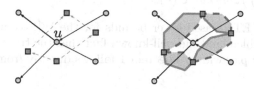

Fig. 4. Constructing a circulating orientation of (P, c).

Now, since the "skeleton" graph D is triconnected and the inserted tendrils are also triconnected, the planar embeddings of G are essentially defined by the flip of each tendril. Picking the flip of a tendril then directly corresponds to picking the orientation of the respective edge in P. Specifically, the property of the target orientation of P is that the sum of the weights of outgoing and incoming edges of a vertex is zero; this translates to the property that the edges of D are to be oriented so that for every face, the sum of the weights of the clockwise edges is equal to the sum of the weights of the counter-clockwise edges. Finally, the latter translates to the upward planarity condition on the face of $\vec{G}$ that originates from a face of $\vec{D}$: the weights of clockwise and counter-clockwise edges are balanced if and only if the total contribution of all tendrils to the face is zero, since the contribution of a tendril is proportional to the weight of the respective edge, and the sign is picked exactly in accordance with the orientation of the edge. See Fig. 4 that illustrates the correspondence.

As opposed to the original NP-hardness proof [23], our starting point is the more special CIRCULATING ORIENTATION problem, so we only require the fixed-contribution tendril gadgets, and not the wiggle gadgets that have variable contribution. We also start the reduction with an arbitrary planar triconnected graph, instead of the special instance originating from the satisfiability problem; all in all this leads to an arguably clearer and more direct NP-hardness proof. We additionally observe that the reduction performed in this way does not blow up the pathwidth, which is crucial for our main result.

Finally, there are a few differences in the case of RECTILINEAR PLANARITY TESTING. First, it is important that we start with a triangulated graph P and so the graph D is of maximum degree 3. The edges of the graph D are then subdivided to obtain the graph F that admits a rectilinear embedding. Finally, to obtain the target graph G, one edge in every subdivision is replaced by a *rectilinear tendril* [23], which play an analogous role to that of tendrils above. In the same way, faces of G originating from faces of D correspond to vertices in P, with the contribution of tendrils to the face being proportional to the capacities of the respective edges.

Combining these transformations with Lemma 2, we obtain the following.

Lemma 3 ($\star$). *There is a polynomial-time algorithm that, given an instance of* Multicolored Clique *with parameter k, outputs an equivalent instance of* Upward planarity testing *(respectively* Rectilinear planarity testing*) on a graph of pathwidth $\mathcal{O}(k)$.*

Due to known ETH-based lower bounds for the W[1]-complete Multicolored Clique problem and the well-known fact that the treewidth of a graph is not larger than its pathwidth, Theorem 1 follows directly from Lemma 3.

4   Conclusion

We proved that Upward planarity and Rectilinear planarity are both W[1]-hard parameterized by treewidth and that the $n^{\mathcal{O}(\mathsf{tw})}$ running times of the existing algorithms for them are tight assuming ETH. Our reduction also provides an alternative NP-completeness proof for these problems, which avoids the use of the so-called *wiggle* gadgets [23].

The All-or-Nothing Flow problem on general graphs was recently shown to be XNLP-complete [9] parameterized by treewidth. This complexity class (which contains $W[1]$) was recently introduced [10] and captures parameterized problems solvable in nondeterministic FPT-time and logarithmic space. It would be interesting to see whether the two planarity testing problems parameterized by treewidth are also XNLP-complete.

Our results show that the parameter treewidth is too general to allow for FPT algorithms for the considered problems. An investigation of more restrictive parameterizations that yield fixed-parameter tractability is left for future work. Based on preliminary investigations, we believe both Upward planarity and Rectilinear planarity may be FPT parameterized by the cutwidth of the dual multigraph. We remark that, since the instances produced by our hardness reduction have pathwidth $\mathcal{O}(k)$ and maximum degree $\mathcal{O}(1)$, the cutwidth of the primal graph is also $\mathcal{O}(k)$ ([8, Thm. 49]). Hence our hardness results extend to the parameterization by the cutwidth of the primal graph.

Acknowledgements. We acknowledge the fruitful working atmosphere of Dagstuhl Seminar 23162 "New Frontiers of Parameterized Complexity in Graph Drawing", where this work was started.

European Research Council
Established by the European Commission

References

1. Amini, O., Huc, F., Pérennes, S.: On the path-width of planar graphs. SIAM J. Discret. Math. **23**(3), 1311–1316 (2009). https://doi.org/10.1137/060670146
2. Battista, G.D., Eades, P., Tamassia, R., Tollis, I.G.: Graph Drawing: Algorithms for the Visualization of Graphs, 1st edn. Prentice Hall PTR, Upper Saddle River (1998)
3. Bertolazzi, P., Di Battista, G., Liotta, G., Mannino, C.: Upward drawings of tri-connected digraphs. Algorithmica **12**(6), 476–497 (1994). https://doi.org/10.1007/BF01188716
4. Bertolazzi, P., Di Battista, G., Mannino, C., Tamassia, R.: Optimal upward planarity testing of single-source digraphs. SIAM J. Comput. **27**(1), 132–169 (1998). https://doi.org/10.1137/S0097539794279626
5. Biedl, T.: Triangulating planar graphs while keeping the pathwidth small. In: Mayr, E.W. (ed.) WG 2015. LNCS, vol. 9224, pp. 425–439. Springer, Heidelberg (2016). https://doi.org/10.1007/978-3-662-53174-7_30
6. Bläsius, T., Fink, S.D., Rutter, I.: Synchronized planarity with applications to constrained planarity problems. In: Mutzel, P., Pagh, R., Herman, G. (eds.) 29th Annual European Symposium on Algorithms, ESA 2021, 6–8 September 2021, Lisbon, Portugal (Virtual Conference). LIPIcs, vol. 204, pp. 19:1–19:14. Schloss Dagstuhl - Leibniz-Zentrum für Informatik (2021). https://doi.org/10.4230/LIPIcs.ESA.2021.19
7. Bläsius, T., Rutter, I.: Simultaneous PQ-ordering with applications to constrained embedding problems. ACM Trans. Algorithms **12**(2), 16:1–16:46 (2016). https://doi.org/10.1145/2738054
8. Bodlaender, H.L.: A partial k-arboretum of graphs with bounded treewidth. Theor. Comput. Sci. **209**(1–2), 1–45 (1998). https://doi.org/10.1016/S0304-3975(97)00228-4
9. Bodlaender, H.L., Cornelissen, G., van der Wegen, M.: Problems hard for treewidth but easy for stable gonality. In: Bekos, M.A., Kaufmann, M. (eds.) WG 2022. LNCS, vol. 13453, pp. 84–97. Springer, Cham (2022). https://doi.org/10.1007/978-3-031-15914-5_7
10. Bodlaender, H.L., Groenland, C., Nederlof, J., Swennenhuis, C.M.F.: Parameterized problems complete for nondeterministic FPT time and logarithmic space. In: 62nd IEEE Annual Symposium on Foundations of Computer Science, FOCS 2021, Denver, CO, USA, 7–10 February 2022, pp. 193–204. IEEE (2021). https://doi.org/10.1109/FOCS52979.2021.00027
11. Brückner, G., Himmel, M., Rutter, I.: An SPQR-tree-like embedding representation for upward planarity. In: Archambault, D., Tóth, C.D. (eds.) GD 2019. LNCS, vol. 11904, pp. 517–531. Springer, Cham (2019). https://doi.org/10.1007/978-3-030-35802-0_39
12. Chan, H.: A parameterized algorithm for upward planarity testing. In: Albers, S., Radzik, T. (eds.) ESA 2004. LNCS, vol. 3221, pp. 157–168. Springer, Heidelberg (2004). https://doi.org/10.1007/978-3-540-30140-0_16

13. Chaplick, S., Di Giacomo, E., Frati, F., Ganian, R., Raftopoulou, C.N., Simonov, K.: Parameterized algorithms for upward planarity. In: Goaoc, X., Kerber, M. (eds.) 38th International Symposium on Computational Geometry, SoCG 2022, 7–10 June 2022, Berlin, Germany. LIPIcs, vol. 224, pp. 26:1–26:16. Schloss Dagstuhl - Leibniz-Zentrum für Informatik (2022). https://doi.org/10.4230/LIPIcs.SoCG.2022.26

14. Cygan, M., et al.: Parameterized Algorithms. Springer, Cham (2015). https://doi.org/10.1007/978-3-319-21275-3

15. Di Battista, G., Liotta, G., Vargiu, F.: Spirality and optimal orthogonal drawings. SIAM J. Comput. **27**(6), 1764–1811 (1998). https://doi.org/10.1137/S0097539794262847

16. Di Giacomo, E., Liotta, G., Montecchiani, F.: Orthogonal planarity testing of bounded treewidth graphs. J. Comput. Syst. Sci. **125**, 129–148 (2022). https://doi.org/10.1016/j.jcss.2021.11.004

17. Didimo, W., Giordano, F., Liotta, G.: Upward spirality and upward planarity testing. SIAM J. Discret. Math. **23**(4), 1842–1899 (2009). https://doi.org/10.1137/070696854

18. Didimo, W., Kaufmann, M., Liotta, G., Ortali, G.: Rectilinear planarity of partial 2-trees. In: Angelini, P., von Hanxleden, R. (eds.) GD 2022. LNCS, vol. 13764, pp. 157–172. Springer, Cham (2022). https://doi.org/10.1007/978-3-031-22203-0_12

19. Didimo, W., Liotta, G., Ortali, G., Patrignani, M.: Optimal orthogonal drawings of planar 3-graphs in linear time. In: Chawla, S. (ed.) Proceedings of the 2020 ACM-SIAM Symposium on Discrete Algorithms, SODA 2020, Salt Lake City, UT, USA, 5–8 January 2020, pp. 806–825. SIAM (2020). https://doi.org/10.1137/1.9781611975994.49

20. Didimo, W., Liotta, G., Patrignani, M.: Hv-planarity: algorithms and complexity. J. Comput. Syst. Sci. **99**, 72–90 (2019). https://doi.org/10.1016/j.jcss.2018.08.003

21. Frati, F.: Planar rectilinear drawings of outerplanar graphs in linear time. Comput. Geom. **103**, 101854 (2022). https://doi.org/10.1016/j.comgeo.2021.101854

22. Fulek, R., Tóth, C.D.: Atomic embeddability, clustered planarity, and thickenability. J. ACM **69**(2), 13:1–13:34 (2022). https://doi.org/10.1145/3502264

23. Garg, A., Tamassia, R.: On the computational complexity of upward and rectilinear planarity testing. SIAM J. Comput. **31**(2), 601–625 (2001). https://doi.org/10.1137/S0097539794277123

24. Grigoriev, A., Bodlaender, H.L.: Algorithms for graphs embeddable with few crossings per edge. Algorithmica **49**(1), 1–11 (2007). https://doi.org/10.1007/s00453-007-0010-x

25. Hopcroft, J.E., Tarjan, R.E.: Efficient planarity testing. J. ACM **21**(4), 549–568 (1974). https://doi.org/10.1145/321850.321852

26. Hutton, M.D., Lubiw, A.: Upward planning of single-source acyclic digraphs. SIAM J. Comput. **25**(2), 291–311 (1996). https://doi.org/10.1137/S0097539792235906

27. Kaufmann, M., Wagner, D. (eds.): Drawing Graphs, Methods and Models (The Book Grow Out of a Dagstuhl Seminar, April 1999). Lecture Notes in Computer Science, vol. 2025. Springer, Heidelberg (2001). https://doi.org/10.1007/3-540-44969-8

28. Korzhik, V.P., Mohar, B.: Minimal obstructions for 1-immersions and hardness of 1-planarity testing. J. Graph Theory **72**(1), 30–71 (2013). https://doi.org/10.1002/jgt.21630

29. Liotta, G., Rutter, I., Tappini, A.: Parameterized complexity of graph planarity with restricted cyclic orders. J. Comput. Syst. Sci. **135**, 125–144 (2023). https://doi.org/10.1016/j.jcss.2023.02.007

30. Nishizeki, T., Rahman, M.S.: Planar Graph Drawing, LNSC, vol. 12. World Scientific, Singapore (2004). https://doi.org/10.1142/5648
31. Patrignani, M.: Planarity testing and embedding. In: Tamassia, R. (ed.) Handbook on Graph Drawing and Visualization, pp. 1–42. Chapman and Hall/CRC, Boca Raton (2013). https://cs.brown.edu/people/rtamassi/gdhandbook/chapters/planarity.pdf
32. Tamassia, R.: On embedding a graph in the grid with the minimum number of bends. SIAM J. Comput. **16**(3), 421–444 (1987). https://doi.org/10.1137/0216030
33. Tamassia, R. (ed.): Handbook on Graph Drawing and Visualization. Chapman and Hall/CRC, Boca Raton (2013). https://www.crcpress.com/Handbook-of-Graph-Drawing-and-Visualization/Tamassia/9781584884125
34. Urschel, J.C., Wellens, J.: Testing gap k-planarity is np-complete. Inf. Process. Lett. **169**, 106083 (2021). https://doi.org/10.1016/j.ipl.2020.106083

On Minimizing the Energy of a Spherical Graph Representation

Matt DeVos, Danielle Rogers, and Alexandra Wesolek[(✉)][iD]

Simon Fraser University, Burnaby, BC V5A 1S6, Canada
{mdevos,danielle_rogers,agwesole}@sfu.ca

Abstract. Graph representations are the generalization of geometric graph drawings from the plane to higher dimensions. A method introduced by Tutte to optimize properties of graph drawings is to minimize their energy. We explore this minimization for spherical graph representations, where the vertices lie on a unit sphere such that the origin is their barycentre. We present a primal and dual semidefinite program which can be used to find such a spherical graph representation minimizing the energy. We denote the optimal value of this program by $\rho(G)$ for a given graph G. The value turns out to be related to the second largest eigenvalue of the adjacency matrix of G, which we denote by λ_2. We show that for G regular, $\rho(G) \leq \frac{\lambda_2}{2} \cdot v(G)$, and that equality holds if and only if the λ_2 eigenspace contains a spherical 1-design. Moreover, if G is a random d-regular graph, $\rho(G) = \left(\sqrt{(d-1)} + o(1)\right) \cdot v(G)$, asymptotically almost surely.

Keywords: Graph representation · Energy · Semidefinite program

1 Introduction

A *representation* of a graph G in $\mathbb{R}^d$ is a function $\mathbf{r} : V(G) \to \mathbb{R}^d$. If e is an edge of G with ends u, v then we associate the straight line segment between $\mathbf{r}(u)$ and $\mathbf{r}(v)$ with the edge e. A representation is an *embedding* if the function r is one-to-one, and the interior of every edge is disjoint from the rest of the graph. The *energy* of a representation $\mathbf{r} : V(G) \to \mathbb{R}^d$ is defined to be the sum of the squares of the lengths of the line segments associated with the edges, $energy(r) = \sum_{uv \in E(G)} ||\mathbf{r}(u) - \mathbf{r}(v)||^2$ (see [6, pg. 285]).

Tutte's seminal paper "How to draw a graph" introduces a natural method to find an embedding of a planar graph: The vertices of a face are associated with the vertices of a convex polygon (in the natural manner) and then all other vertices of the graph follow the rule that they lie in the barycentre of their neighbours (which we call the *barycentre property*). Tutte proves that for a 3-connected graph with no $K_{3,3}$ or K_5 minor this always results in an embedding in the plane (thus reproving the Kuratowski-Wagner theorem characterizing planar graphs). This setup has a natural physical interpretation in which each edge is

M. A. Bekos and M. Chimani (Eds.): GD 2023, LNCS 14466, pp. 218–231, 2023.
https://doi.org/10.1007/978-3-031-49275-4_15

treated as an elastic that wants to have smallest possible length, and then the barycentre property for vertices not in the special face is a physical consequence. This paper is the foundation for a broad area of research on representing graphs, which frequently employ similar methods to draw arbitrary graphs in a manner that makes them easy to understand. Tutte's barycentre property yields

$$\mathbf{r}(v) = \frac{1}{\deg(v)} \sum_{uv \in E(G)} \mathbf{r}(u).$$

For a fixed face C, all vertices in $V(G) \setminus V(C)$ satisfy the barycentre property if and only if the energy of the geometric drawing of G given by its representation $\mathbf{r}$ is minimized.

Another famous work of interest here is the Goemans-Williamson Algorithm. If G is a graph and $X \subseteq V(G)$, then the set of edges with one end in X and one in $V(G) \setminus X$ is called an *edge-cut*. A famous NP-hard problem called MAXCUT, is to determine the maximum size of an edge-cut in G. In [7], Goemans and Williamson introduce a semidefinite programming relaxation of MAXCUT. This (polynomially-solvable) problem is that of representing an n-vertex graph G in $\mathbb{R}^d$ with each vertex on the unit sphere (such an embedding is called a *unit embedding*) in such a way as to maximize the energy. Remarkably, Goemans and Williamson show that choosing a random hyperplane through the origin then gives an edge-cut with expected size at least .868 times the size of the true maximum cut.

Our interest here combines some ideas from Tutte and Goemans-Williamson in a natural manner to consider another kind of graph representation. We are interested in representations of a graph G in $\mathbb{R}^{v(G)}$ in which every vertex lies on the unit sphere, the common barycentre of all vertices is the origin, and subject to this the energy is minimized. We call those representations *unit barycentre* **0** *representations*. Note that the condition that the common barycentre is **0** is required for non-degeneracy. As we will demonstrate in the following section, this problem is naturally encoded by a semidefinite program, and therefore can be computed in polynomial time. This program can be used to find drawings of graphs by projecting the given representation into $\mathbb{R}^2$, and indeed this seems to give nice drawings of some small graphs, but it is limited for use in large graphs thanks to the difficulty in operating with large semidefinite programs.

From a theoretical standpoint, there is a natural graph parameter given by our problem. Throughout we use $\langle \cdot, \cdot \rangle$ to denote the standard inner product. For a unit representation $\mathbf{r} : V(G) \to \mathbb{R}^d$ and $u, v \in V(G)$ we have

$$\begin{aligned}
\|\mathbf{r}(u) - \mathbf{r}(v)\|^2 &= \langle \mathbf{r}(u) - \mathbf{r}(v), \mathbf{r}(u) - \mathbf{r}(v) \rangle \\
&= \|\mathbf{r}(u)\|^2 + \|\mathbf{r}(v)\|^2 - 2\langle \mathbf{r}(u), \mathbf{r}(v) \rangle \qquad (1) \\
&= 2 - 2\langle \mathbf{r}(u), \mathbf{r}(v) \rangle.
\end{aligned}$$

So the energy of this representation is equal to $2|E(G)| - \sum_{uv \in E(G)} \langle \mathbf{r}(u), \mathbf{r}(v) \rangle$. The latter term in this expression is more natural for us to work with, and gives rise to our parameter of interest.

Definition 1. *If* $\mathbf{r} : V(G) \to \mathbb{R}^d$ *is a unit barycentre* $\mathbf{0}$ *representation of* G, *we define*

$$\rho(G, \mathbf{r}) = \sum_{uv \in E(G)} \langle \mathbf{r}(u), \mathbf{r}(v) \rangle.$$

We define $\rho(G)$ *to be the maximum of* $\rho(G, \mathbf{r})$ *over all unit barycentre* $\mathbf{0}$ *representations* $\mathbf{r}$ *of* G *(which must exist by compactness).*

Note that a representation maximizing ρ also minimizes energy. Our main interest here is in understanding this parameter, and approximating it for certain classes of graphs. For regular graphs, there is a straightforward upper bound on ρ as follows. Here $v(G)$ denotes the number of vertices of G and $e(G)$ the number of edges.

Corollary 1. *If* G *is a connected regular graph and* λ_2 *is the second largest eigenvalue of the adjacency matrix, then every unit barycentre* $\mathbf{0}$ *representation of* G *must have energy at least* $2e(G) - \frac{1}{2}\lambda_2 v(G)$, *i.e.* $\rho(G) \le \frac{\lambda_2}{2} v(G)$.

For certain classes of graphs such as vertex-transitive graphs and distance regular graphs we will prove that this upper bound is achieved. We will also show that random regular graphs and regular graphs with high girth come close to achieving this bound.

2   Upper Bound

In this section we introduce the basic notation and definitions we require, and then we establish a key upper bound on $\rho(G)$ for regular graphs. We use λ_2 to denote the second largest eigenvalue of the adjacency matrix of a graph G and we let $Eig(\lambda_2)$ denote the corresponding eigenspace. In the following we will often give the mapping of a representation $\mathbf{r}$ by a $d \times v(G)$ *representation matrix* R in which the column vector v is the representation $\mathbf{r}(v)$.

The *barycentre* of a graph representation $\mathbf{r} : V(G) \to \mathbb{R}^d$ is the point given by $\frac{1}{v(G)} \sum_{v \in V(G)} \mathbf{r}(v)$ (the barycentre of the points). A representation that has the origin as barycentre is called a *barycentre* $\mathbf{0}$ representation. We say that the representation is a *unit* representation if $\|\mathbf{r}(v)\| = 1$ for $v \in V(G)$ (so each point lies on the unit sphere). Our interest here is exclusively in unit barycentre $\mathbf{0}$ representations $\mathbf{r}$, whose point sets $(r(v))_{v \in V(G)}$ are also known as spherical 1-designs.

If $uv \in E(G)$, then we treat this edge as a straight line segment between the corresponding points $\mathbf{r}(u)$ and $\mathbf{r}(v)$. Accordingly, the *length* of the edge uv is defined to be $\|\mathbf{r}(u) - \mathbf{r}(v)\|$ (using the standard Euclidean norm).

There is an alternate way to formulate the parameter ρ that focuses on the rows instead of the columns of the representation matrix. This viewpoint will be especially helpful for us in our investigations.

Lemma 1. *Let* $\mathbf{r}$ *be a unit barycentre* $\mathbf{0}$ *representation of a graph* G *and let* $r_1, \ldots, r_d$ *be the row vectors of its representation matrix* R. *If* A *is the adjacency matrix of* G, *then we have*

$$\rho(G, \mathbf{r}) = \frac{1}{2} \sum_{k=1}^{d} r_k A r_k^\top.$$

Proof. We may denote the entry in the i, u position of the representation matrix R by either $(\mathbf{r}(u))_i$ or $(r_i)_u$. Now the result follows from

$$\rho(G, \mathbf{r}) = \sum_{uv \in E(G)} \langle \mathbf{r}(u), \mathbf{r}(v) \rangle = \sum_{k=1}^{d} \sum_{uv \in E(G)} (\mathbf{r}(u))_k (\mathbf{r}(v))_k$$

$$= \sum_{k=1}^{d} \sum_{uv \in E(G)} (r_k)_u (r_k)_v = \frac{1}{2} \sum_{k=1}^{d} r_k A r_k^\top.$$

□

The above lemma gives rise to a natural upper bound on $\rho(G)$ that we prove next. Our argument relies upon some standard concepts from algebraic graph theory. In particular, if A is the adjacency matrix of a connected d-regular graph, then the largest eigenvalue of A is d and the corresponding eigenspace is spanned by the vector $\mathbf{1}$ (with all entries 1). Further, the min-max theorem for Rayleigh quotients implies that the maximum of $\frac{x^\top A x}{x^\top x}$ over all nonzero vectors orthogonal to $\mathbf{1}$(the all ones vector) is λ_2.

Proposition 1. *If G is a regular graph, then $\rho(G) \leq \frac{\lambda_2}{2} v(G)$. Furthermore $\rho(G, \mathbf{r}) = \frac{\lambda_2}{2} v(G)$ if and only if every row of the representation matrix R is a λ_2-eigenvector of the adjacency matrix of G.*

Proof. Let A be the adjacency matrix of G, let R be a representation matrix of a unit barycentre $\mathbf{0}$ representation $\mathbf{r}$ of G with $\rho(G, \mathbf{r}) = \rho(G)$. Let $r_1, \ldots, r_d$ be the rows of R. Since $\mathbf{r}$ is a unit embedding the sum of the squares of the entries in each column of R is 1. It follows from this that $\sum_{k=1}^{d} \langle r_k, r_k \rangle = v(G)$. Since we have a barycentre $\mathbf{0}$ representation, each row r_k is orthogonal to $\mathbf{1}$. Now we have

$$\rho(G) = \rho(G, \mathbf{r}) = \frac{1}{2} \sum_{k=1}^{d} r_k A r_k^\top \leq \frac{1}{2} \sum_{i=1}^{d} \lambda_2 \langle r_k, r_k \rangle = \frac{\lambda_2}{2} v(G)$$

which gives the desired bound. If this bound is tight we must have $r_k A r_k^\top = \lambda_2 r_k r_k^\top$ but this implies that r_k is a λ_2-eigenvector as desired. □

Our main theoretical results demonstrate that the above bound can be achieved or nearly achieved for certain well-behaved classes of regular graphs. However, we will first demonstrate that ρ can be effectively computed.

3 The Semidefinite Program

In this section we will demonstrate a natural semidefinite program that computes $\rho(G)$ and constructs a minimum energy unit barycentre $\mathbf{0}$ representation of a graph. We will show that for regular graphs this program is strongly dual. We will call upon standard properties of positive semidefinite matrices and semidefinite programming. In particular we define the *dot product* of two $n \times n$ matrices A and B to be $A \bullet B = \text{trace}(A^\top B)$.

We begin with the definition of our semidefinite program, alongside its dual. Here we assume that G is a graph with adjacency matrix A, and for every $v \in V(G)$ the matrix C_v is a $v(G) \times v(G)$ matrix with 1 in position (v, v) and 0 everywhere else. We use J to denote a $v(G) \times v(G)$ matrix with all entries 1.

Primal Program (Primal):	*Dual Program (Dual):*
Maximize: $\frac{1}{2} A \bullet X$ Subject To: $C_v \bullet X = 1$ for $v \in V(G)$ $J \bullet X = 0$ $X \succcurlyeq 0$	Minimize $\sum_{v \in V(G)} y_v$ Subject To: $-\sum_{v \in V(G)} y_v \cdot C_v - y_0 \cdot J \preccurlyeq -\frac{1}{2} A$

Theorem 1. *$X = U^T U$ is feasible for Primal if and only if U is the representation matrix of a unit barycentre $\mathbf{0}$ representation of G. Furthermore X maximizes Primal if and only if U is a minimum energy barycentre $\mathbf{0}$ unit representation of G. The optimum value of Primal is $\rho(G)$.*

Proof. Let R be a $d \times v(G)$ representation matrix of a representation $\mathbf{r}$ and let $X = R^\top R$. The u, u coordinate of X is $\langle \mathbf{r}(u), \mathbf{r}(u) \rangle$ so the first condition in the Primal Program is equivalent to $\mathbf{r}$ being a unit representation. Next observe that $J \bullet X = \langle \sum_{u \in V(G)} \mathbf{r}(u), \sum_{v \in V(G)} \mathbf{r}(v) \rangle = \| \sum_{u \in V(G)} \mathbf{r}(\mathbf{u}) \|^2$ so $J \bullet X = 0$ if and only if $\mathbf{r}$ is a barycentre $\mathbf{0}$ representation. It follows that X is a feasible matrix for the program if and only if $\mathbf{r}$ is a unit barycentre $\mathbf{0}$ representation of G. Now we have

$$\tfrac{1}{2} A \bullet X = \sum_{uv \in E(G)} \langle \mathbf{r}(u), \mathbf{r}(v) \rangle = \rho(G, \mathbf{r}),$$

so the value of this program on a feasible matrix $X = R^\top R$ is precisely $\rho(G, \mathbf{r})$. $\square$

From this we can set up a graph representation algorithm.

Algorithm 1: Semidefinite Graph Representation Algorithm

Input: Adjacency matrix A of a graph G
Result: A geometric representation of G in one of $\mathbb{R}^1, \ldots, \mathbb{R}^{v(G)}$

1. For a graph G, solve *Primal* to obtain an optimal matrix X_G.
2. Use the Cholesky decomposition to obtain a representation matrix R_G with $X_G = R_G^T R_G$.

If we want to get a representation in a particular dimension, for example a graph drawing in dimension $k = 2$, we can perform the following additional steps.

3. If $rank(R_G) \leq k$ then let R'_G be an orthogonal transformation of R_G where all but the first k rows are zero rows. Take for the representations $\mathbf{r}(v)$ the column vector v of R'_G.

4. If $rank(R_G) > k$ then take a random orthogonal transformation of R_G to obtain R'_G. Take for the representations $\mathbf{r}(v)$ the first k entries of the column vector v of R'_G.

5. If vertices u and j are adjacent in G then draw a straight line between their placements.

Naturally the projected representation is a barycentre $\mathbf{0}$ representation. (Orthogonal transformations preserve the origin and are linear, so in particular taking an orthogonal transformation of a matrix with column sum 0 results in a matrix with column sum 0. Further, restricting to the first k rows does not affect the condition that the columns sum up to 0.) In the Appendix of the full version of the paper [1] we show that for $k = 2$ the projection preserves in expectation the energy of an edge (up to a scaling factor which depends on $rank(R_G)$).

An approximate solution to the given semidefinite program can be computed in polynomial time since the Frobenius norm of the solution space is polynomially bounded in n [4]. Standard algorithms for the Cholesky decomposition run in $O(n^3)$ time [4]. Therefore, the above gives us a way to approximately compute $\rho(G)$ and to construct a minimum energy unit barycentre $\mathbf{0}$ representation of a graph in polynomial time. Figure 1 depicts drawings of the 5-dimensional hypercube and the Peterson graph which were made using the semidefinite graph representation algorithm.

For the class of regular graphs, the primal and dual are well-behaved. A semidefinite program is *strongly dual* if both the Primal and Dual programs achieve the same optimum.

Theorem 2. *If G is regular then Primal and Dual are strongly dual.*

Proof. Suppose G is a regular graph on n vertices. We show strong duality by showing that there exists a feasible point $\mathbf{y}$ for *Dual* such that

$$M = -\frac{1}{2}A + \sum_{v \in V(G)} y_v C_i + y_0 J \succ 0,$$

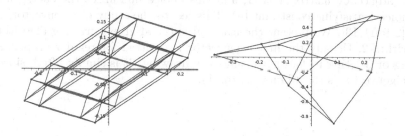

Fig. 1. Projection of the 5-dimensional hypercube and the Peterson graph

which is enough by [4, Theorem 4.1.1]. Set $y_v = \frac{\lambda_2}{2} + 1$ for $v \in V(G)$ and $y_0 = \frac{\lambda_1 - \lambda_2}{2n}$, then

$$M = -\frac{1}{2}A + \left(\frac{\lambda_2}{2} + 1\right)I + \frac{\lambda_1 - \lambda_2}{2n}J.$$

All that is left to show is that M is positive definite. The eigenvalues of $-\frac{1}{2}A + (\frac{\lambda_2}{2} + 1)I$ are

$$\left\{ \frac{-\lambda_1 + \lambda_2}{2} + 1, 1, \ldots, \frac{-\lambda_n + \lambda_2}{2} + 1 \right\}$$

and all eigenvalues are positive except for $\frac{-\lambda_1 + \lambda_2}{2} + 1$ which has eigenvector $\mathbf{1}$. The negative eigenvalue gets is increased to 1 by adding $\frac{\lambda_1 - \lambda_2}{2n}J$, as $\mathbf{1}$ is an eigenvector of $\frac{\lambda_1 - \lambda_2}{2n}J$ with eigenvalue $\frac{\lambda_1 - \lambda_2}{2}$ and noting that the other eigenvalues of $\frac{\lambda_1 - \lambda_2}{2n}J$ are 0, so each eigenvector of $-\frac{1}{2}A + (\frac{\lambda_2}{2} + 1)I$ is an eigenvector of M. $\square$

4   Semidefinite Representations and Eigenvector Representations

Restating our main objective, we want to minimize the energy of a unit barycentre $\mathbf{0}$ graph representation $\mathbf{r}$ (or equivalently maximize ρ) with respect to

$$\|\mathbf{r}(v)\| = 1 \text{ for each } v \in V(G). \tag{2}$$

$$\text{The origin is the barycentre of } (\mathbf{r}(\mathbf{v}))_{v \in V(G)}. \tag{3}$$

Historically, energy minimization has been studied with the following classical constraint (instead of Constraints (2) and (3)). Let r_k be the k-th row of the representation matrix of $\mathbf{r}$. The classical constraint is

$$r_k \cdot r_j = \delta_{ij}. \tag{4}$$

Both constraints $r_k \cdot r_k = 1$ and $r_k \cdot r_j = 0$ for $i \neq j$ are necessary to avoid degeneracy, more precisely, to ensure that the vertices are not mapped to the same point. For regular graphs, these constraints are met by an orthogonal basis of eigenvectors $r_1, \ldots, r_k$ associated with the k largest eigenvalues $\lambda_1 \leq \cdots \leq \lambda_k$ of the adjacency matrix A of G, and this choice minimizes the energy among all choices satisfying constraint (4). This led to the study of eigenvector drawings [2,9–11,13]. We present the classical spectral graph drawing algorithm in Algorithm 2. Usually the first eigenvector is omitted for regular graphs, since adding or deleting it does not change the drawing (it is simply lifted, since the first eigenvector is the all ones vector $\mathbf{1}$).

Algorithm 2: Spectral Graph Drawing Algorithm.

Input: An adjacency matrix A of a regular graph G
Result: A geometric representation of G in $\mathbb{R}^k$

1. Compute the eigenvectors $r_2, \ldots, r_{k+1}$ to the eigenvalues $\lambda_2, \ldots, \lambda_{k+1}$.
2. Let R be the representation matrix formed by the rows $r_2, \ldots, r_{k+1}$.
3. Let the column v be the representation $\mathbf{r}(v)$.

Eigenvector drawings are often chosen, because they are computable in polynomial time and can be used in applications [8,10]. Classic spectral drawings are those of (2-skeletons of) Platonic solids, see Fig. 2.

Fig. 2. Platonic solids.

1-skeletons of platonic solids are distance regular graphs. *Distance regular graphs* are graphs for which the number of vertices which are simultaneously at distance j from a vertex v and at distance k from a vertex w depends only on j, k, and the distance between v and w. For distance regular graphs, optimising the energy of a representation with respect to the constraint (4) or with respect to the constraints (2) and (3) yields a common solution. By omitting the $\mathbf{1}$ vector as an eigenvector for constraint (4), $r_1, \ldots, r_k$ can be taken as vectors from the eigenspace of λ_2. By a result of Godsil [5, Lemma 1.2, Corollary 6.2] there exists an orthonormal basis for the eigenspace of λ_2 in a distance regular graph such that the representation formed by rows of basis vectors is a unit barycentre $\mathbf{0}$ representation.

To showcase that our set of chosen constraints is natural and a suitable extension of the spectral algorithm, we consider vertex-transitive graphs. *Vertex-transitive* graphs are graphs whose automorphism group $Aut(G)$ acts transitively on $V(G)$. An *automorphism* of a graph G is a permutation σ of $V(G)$, such that (u, v) is an edge if and only if $(\sigma(u), \sigma(v))$ is an edge.

Lemma 2. *For each connected vertex-transitive graph G there exists a unit barycentre $\mathbf{0}$ representation $\mathbf{r}$ of G such that*

– *every two edges that are in a common orbit of the automorphism group have the same length*
– $\rho(G, \mathbf{r}) = \frac{\lambda_2}{2} \cdot v(G)$.

Proof. Let $\mathbf{r}_0$ be a λ_2 eigenvector; it is orthogonal to $\mathbf{1}$. Let $Aut(G) = \{\sigma_1, \ldots, \sigma_t\}$ and for $i = 1, \ldots, t$ let r_i be the vector obtained from r_0 by applying the permutation σ_i to the indices. It follows from vertex transitivity that the matrix with rows $r_1, \ldots, r_t$ has all columns of the same norm, since

$$\|\mathbf{r}(u)\|^2 = \sum_{k=1}^{t}((r_0)_{\sigma_k(u)})^2 = \sum_{k=1}^{t}((r_0)_{\sigma_k(\sigma(u))})^2 = \|\mathbf{r}(\sigma(u))\|^2, \tag{5}$$

for any $\sigma \in Aut(G)$. So an appropriate scaling of the matrix with columns $r_1, \ldots, r_t$ is a representation matrix of a unit barycentre $\mathbf{0}$ representation of G and Proposition 1 shows that it is optimal. Further, taking an edge (u, v) and letting $(r_0)_u, (r_0)_w$ be the value of r_0 in position u, v, respectively. It follows that for every $\sigma \in Aut(G)$

$$\|\mathbf{r}(u) - \mathbf{r}(w)\|^2 = \sum_{k=1}^{t}((r_0)_{\sigma_k(u)} - (r_0)_{\sigma_k(w)})^2 = \sum_{k=1}^{t}((r_0)_{\sigma_k(\sigma(u))} - (r_0)_{\sigma_k(\sigma(w))})^2$$
$$= \|\mathbf{r}(\sigma(u)) - \mathbf{r}(\sigma(w))\|^2,$$

which shows that edges in the same orbit of the automorphism group have the same length. $\qquad\square$

The nice property here is that edges from the same orbit have the same length, which can not be obtained as easily by the spectral graph representation algorithm.

5 Random Regular Graphs

5.1 High Girth

Here we shall turn our attention to regular graphs with large girth and show that such graphs have ρ asymptotically close to the upper bound. Our argument is based on a lovely theorem of Nilli [12] who showed how to construct vectors orthogonal to $\mathbf{1}$ that are close to λ_2 eigenvectors (in the sense of Rayleigh quotient). In particular, the multiplicative factor that appears in our theorem is the same as that from the paper of Nilli.

Next we introduce the vectors that will be used here and in the forthcoming subsection on random regular graphs. The *distance* $dist(u, v)$ between vertices u and v is the length of the shortest path in G connecting them. If $A, B \subseteq V$ then $dist(A, B) = \min_{x \in A, y \in B} dist(x, y)$. This defines distance between two edges or between a vertex and an edge. For every edge e and nonnegative integer s we define $V_s(e)$ to be the set of vertices of distance s to e. Now let $e, \bar{e}$ be edges of distance at least $2k + 2$ in a d-regular graph G and construct the row vector $\mathbf{w}_{e,\bar{e}}$ (indexed by $V(G)$) as follows:

$$(\mathbf{w}_{e,\bar{e}})_v = \begin{cases} (d-1)^{\frac{-s}{2}}, & \text{if } v \in V_s(e) \text{ for some } s \leq k \\ -(d-1)^{\frac{-s}{2}}, & \text{if } v \in V_s(\bar{e}) \text{ for some } s \leq k \\ 0, & \text{else} \end{cases} \tag{6}$$

The key feature of the vector $\mathbf{w}_{e,\bar{e}}$ is the following bound.

Lemma 3. *Let G be a d-regular graph, let $e, \bar{e}$ be edges with distance greater than $2k + 2$ and assume that the subgraph induced by all vertices of distance at most k to e or $\bar{e}$ has no cycle. Then using A for the adjacency matrix of G we have*

$$||\mathbf{w}_{e,\bar{e}}||^2 = 4(k+1) \qquad and \qquad \mathbf{w}_{e,\bar{e}} A \mathbf{w}_{e,\bar{e}}^\top = 4 + 8k\sqrt{d-1}$$

Proof. First we note that $|V_s(e)| = |V_s(\bar{e})| = 2(d-1)^s$ holds for all $0 \leq s \leq k$ since there are no cycles in the subgraph induced by all vertices of distance at most k to e or $\bar{e}$. Therefore $||\mathbf{w}_{e,\bar{e}}||^2 = 2\sum_{s=0}^{k} |V_s(e)|(d-1)^{-s} = 4(k+1)$ as claimed. For two disjoint subsets, say U, W of $V(G)$, we let $e(U, W)$ denote the number of edges with one end in U and the other in W. Similar to the above, $e(V_s(e), V_{s+1}(e)) = e(V_s(\bar{e}), V_{s+1}(\bar{e})) = 2(d-1)^{s+1}$ holds for all $0 \leq s \leq k - 1$.

The only edges $uv \in E(G)$ for which $\mathbf{w}_{e,\bar{e}}$ assigns both u and v nonzero weight are as follows: $e, \bar{e}$, and those edges with one end in $V_s(e)$ ($V_s(\bar{e})$) and the other in $V_{s+1}(e)$ ($V_{s+1}(\bar{e})$) for some $0 \leq s \leq k - 1$. Our result now follows from the calculation below:

$$\mathbf{w}_{e,\bar{e}} A \mathbf{w}_{e,\bar{e}}^\top = 2 \sum_{uv \in E(G)} (\mathbf{w}_{e,\bar{e}})_u (\mathbf{w}_{e,\bar{e}})_v$$

$$= 2\left(2 + 2\sum_{s=0}^{k-1}(d-1)^{-s-\frac{1}{2}}e(V_s(e), V_{s+1}(e))\right)$$

$$= 4 + 8k\sqrt{d-1}.$$

$\square$

Next we use these vectors to find good representations for regular graphs of high girth.

Theorem 3. *Let G be d-regular with girth $g(G) > 2k + 2$, then*

$$\left(2\sqrt{d-1} - \frac{2\sqrt{d-1}-1}{k+1}\right)\frac{v(G)}{2} \leq \rho(G).$$

Proof. We begin by claiming there exist two enumerations of the edges of G, $e_1, \ldots, e_{e(G)}$ and $\bar{e}_1, \ldots, \bar{e}_{e(G)}$ with the property that e_i and $\bar{e}_i$ have distance at least $2k + 2$ for $1 \leq i \leq e(G)$. To see why these enumerations exist consider the bipartite graph H with bipartition $(E(G) \times \{1\}, E(G) \times \{2\})$ with $(e, 1) \sim (f, 2)$ if e and f are distance at least $2k + 2$ in G. It follows from the assumptions that G is regular and of girth at least $2k + 3$ that H is regular. Since every regular bipartite graph has a perfect matching, the desired enumerations exist.

Now define W to be the matrix with rows $\mathbf{w}_1, \ldots, \mathbf{w}_{e(G)}$ where $\mathbf{w}_i = \mathbf{w}_{e_i, \bar{e}_i}$ and $\mathbf{r}_W$ be the representation with representation matrix W. Note that $\mathbf{r}_W$ is barycentre $\mathbf{0}$ as the entries of $\mathbf{w}_{e_i, \bar{e}_i}$ sum up to 0 for every i.

Any column of W can be transformed into any other column of W by permuting entries and changing signs: Given a vertex v, the number of edges e such that $v \in V_s(e)$ and $s \leq k$ is the same for every vertex v because of the girth assumption. In particular, any two columns of W have the same norm as the number of entries with value $(d-1)^{-s/2}$ (or $-(d-1)^{-s/2}$) is the same.

Lemma 3 implies that the sum of the squares of the norms of the rows of W is $e(G)4(k+1) = \frac{d}{2}v(G)4(k+1)$ so defining $t = \sqrt{2d \cdot (k+1)}$ the matrix $W' = \frac{1}{t}W$ is a representation matrix of a unit barycentre $\mathbf{0}$ representation $\mathbf{r}'$ of G. The result is then given by the following calculation:

$$\rho(G) \geq \rho(G, \mathbf{r}')$$
$$= \frac{1}{2} \sum_{k=1}^{e(G)} \frac{1}{t^2} \mathbf{w_i} A \mathbf{w_i}^\top$$
$$= \frac{e(G)}{2t^2} \left(4 + 8k\sqrt{d-1}\right)$$
$$= \frac{v(G)}{2} \left(2\sqrt{d-1} - \frac{2\sqrt{d-1}-1}{k+1}\right).$$

$\square$

5.2   Random Graphs

Before proving our final result on the behaviour of ρ for random regular graphs we require one straightforward result. We define $\mathbb{R}^+ = \{x \in \mathbb{R} \mid x \geq 0\}$ and for a function $f : S \to \mathbb{R}^+$ and a subset $A \subseteq S$ we let $f(A) = \sum_{x \in A} f(x)$.

Lemma 4. *Let $G = (V, E)$ be a complete graph and let $f : V \to \mathbb{R}^+$. If*

1. $(\star)$ $f(v) \leq \frac{1}{2}f(V)$ *for every $v \in V$,*

then there exists $g : E \to \mathbb{R}^+$ satisfying $\sum_{e : v \in e}(g(e))^2 = f(v)$ for every $v \in V$.

Proof. We proceed by induction on $|V|$. As a base case, when $|V| = 1$ condition $(\star)$ implies that $f = 0$ and the results holds trivially since there are no edges (i.e. the sum $\sum_{e : v \in e}(g(e))^2$ is an empty sum). Suppose $|V| \geq 2$. First we consider the case when there exists a vertex v which achieves equality in $(\star)$. Then setting $g(uv) = \sqrt{f(u)}$ for every $u \sim v$ and $g(e) = 0$ for every edge e not incident to v yields the desired function.

Next we consider when there does not exist a vertex which achieves equality in $(\star)$. Choose distinct vertices $x, y \in V$ with $0 < f(x) \leq f(y)$. Consider the complete graph $G' = G - x$ with weight function $f' : V(G') \to \mathbb{R}^+$ given by $f'(y) = f(y) - f(x)$ and $f'(z) = f(z)$ for all $z \in V \setminus \{x, y\}$. If G' (and f') satisfy $(\star)$ then the result follows by applying induction and then modifying this solution by giving the edge xy the value $\sqrt{f(x)}$ and all other edges incident to x the value 0. Otherwise there exists $v \in V \setminus \{x, y\}$ violating $(\star)$ (in the graph G'), hence $f(v) > \frac{f(y)-f(x)}{2} + \frac{1}{2}\sum_{z \in V \setminus \{x,y\}} f(z)$. In particular, $\frac{f(v)}{2} > \frac{f(z)}{2}$ for every $z \in V \setminus \{x, y, v\}$, hence z does not violate $(\star)$ since $f(z) < \frac{f(z)}{2} + \frac{f(v)}{2} \leq \frac{f(y)-f(x)}{2} + \frac{1}{2} \sum_{z \in V \setminus \{x,y\}} f(z)$. In this case we let $f'' : V \to \mathbb{R}^+$ be given by $f''(z) = f(z) - \frac{1}{2} f(V) + f(v)$ for $z = x, y$ and $f''(z) = f(z)$ for every $z \in V \setminus \{x, y\}$. Now the graph G and the weight function f'' satisfy $(\star)$. However, $(\star)$ is tight for v so the result holds for G and f'' (as shown above). Modifying this solution to change the value on the edge xy from 0 to $\sqrt{\frac{1}{2} f(V) - f(v)}$ gives a solution in the original graph. $\square$

We also require a result on the number of short cycles in random regular graphs of Wormald.

Proposition 2. *[14, Corollary 4] Let $j \geq 3$. The expected number of j-cycles in a random d-regular graph on n vertices, with n even if d is odd, is asymptotic to*

$$\frac{(d-1)^j}{2j}$$

as $n \to \infty$.

Theorem 4. *Let G be a random regular graph with degree d. For every $\epsilon > 0$ the inequality*

$$\left(2\sqrt{d-1} - \epsilon\right) \frac{v(G)}{2} \leq \rho(G) \leq \left(2\sqrt{d-1} + \epsilon\right) \frac{v(G)}{2}$$

holds asymptotically almost surely.

Proof. The upper bound follows from Proposition 1 and a remarkable result by J. Friedmann [3] showing that the second largest eigenvalue of a d-regular graph is asymptotically almost surely smaller than $2\sqrt{d-1} + \epsilon$. We now move on to the lower bound. Let $n := v(G)$. We begin by fixing k sufficiently large such that $\frac{2\sqrt{d-1}-1}{k+1} \leq \frac{\epsilon}{2}$ and fixing a constant $B > \sum_{j=3}^{2k+2} \frac{(d-1)^j}{2j}$. We may assume that n is sufficiently large so that the expected number of cycles of length at most $2k+2$ is at most B by Proposition 2. By Markov's inequality the probability that G has more than $\log(n)$ cycles of length at most $2k+2$ is at most $\frac{B}{\log(n)}$ which goes to 0. Therefore it will suffice to show our bound under the assumption that G has at most $\log(n)$ cycles of length at most $2k+2$. We proceed in the same fashion as in the proof of Theorem 3. We construct a matrix W^*, show that it has the desired properties and calculate $\rho(G)$ for this representation.

We begin by claiming that there exist two enumerations of the edges of G given by $e_1, \cdots, e_m$ and $\overline{e}_1, \cdots, \overline{e}_m$ such that e_i and $\overline{e}_i$ have distance at least $2k+2$. As in the proof, we define a bipartite graph H with bipartition $(E(G) \times \{1\}, E(G) \times \{2\})$ with $(e, 1) \sim (f, 2)$ if e and f are distance at least $2k+2$ in G. Since the number of edges at distance less than $2k+2$ is bounded above by a function of d and k, for G suitably large the graph H will have m vertices on each side of the bipartition and minimum degree greater than $\frac{m}{2}$, thus implying the existence of a perfect matching by Hall's Matching Theorem. (A set of size less than $\frac{m}{2}$ will have neighbourhood size at least $\frac{m}{2}$. A set of size greater than $\frac{m}{2}$ will have neighbourhood of size m, in order to meet the minimum degree requirement on the vertices outside of the neighbourhood.)

Define e to be a *bad edge* if there is a cycle C of length at most $2k+2$ such that $dist(e, V(C)) \leq k$. We call an edge *good* if it is not bad. The number of edges of distance at most k from a given vertex is at most d^{k+1}. Since there are at most $\log(n)$ cycles of length $\leq 2k+2$, the number of bad edges is at most $(2k+2)(d^{k+1})\log(n)$.

Let I be the set of indices i for which both e_i and $\bar{e}_i$ are good, noting that $|I| \geq m - 2(2k+2)(d^{k+1})\log(n)$. Now define a matrix W where the rows are indexed by $i \in I$ and the i-th row is $\mathbf{w}_i = \mathbf{w}_{e_i,\bar{e}_i}$. As we did in the previous subsection, we set $t = \sqrt{2d(k+1)}$ and define $W' = \frac{1}{t}W$. If there are no bad edges in G then W' is a unit barycentre $\mathbf{0}$ representation and we are done as in Theorem 3. So we may assume that there exists a cycle of length $\leq 2k+2$ in G.

For every vertex v let $\mathbf{u}_v$ be the column of W' associated with v and observe that $\|\mathbf{u}_v\|^2 \leq 1$. Further, $\|\mathbf{u}_v\|^2 = 0$ for every vertex v on a cycle of length $\leq 2k+2$ since there are no good edges e such that v is of distance at most k to e. For every vertex v define $f(v) = 1 - \|\mathbf{u}_v\|^2$ and note that $f(v) \leq 1$ for every vertex v (Also note that $f(v) = 1$ for every vertex v on a cycle of length $\leq 2k+2$, so in particular there are at least two vertices with $u \neq v$ such that $f(u) = f(v) = 1$.) Hence condition $(\star)$ in the statement of Lemma 4 is satisfied and we may apply Lemma 4 to choose weights w_{uv} for every pair of distinct vertices, u, v, so that $f(v) = \sum_{u \in V(G)\setminus\{v\}} w_{uv}^2$ holds for every vertex v. Now for every $w_{uv} > 0$ add a row vector to W' with w_{uv} at vertex u and $-w_{uv}$ at vertex v, and all other entries 0. The matrix W'' obtained at the end of this process has all columns of norm 1 and all rows summing to zero so W'' is the representation matrix of a suitable representation $\mathbf{r}''$ of G and

$$\rho(G) \geq \rho(G, \mathbf{r}'') \geq \frac{1}{2} \sum_{i \in I} \frac{1}{t^2} \mathbf{w}_i A \mathbf{w}_i^\top = \frac{|I|}{2t^2}\left(4 + 8k\sqrt{d-1}\right)$$

$$= \frac{1}{d}|I|\left(2\sqrt{d-1} - \frac{2\sqrt{d-1}-1}{k+1}\right).$$

Recall that $|I| \geq m - 2(2k+2)(d^{k+1})\log(n)$. We have

$$\frac{1}{d}|I|\left(2\sqrt{d-1} - \frac{2\sqrt{d-1}-1}{k+1}\right)$$

$$\geq \frac{1}{d}\left(m - 2(2k+2)(d^{k+1})\log(n)\right)\left(2\sqrt{d-1} - \frac{2\sqrt{d-1}-1}{k+1}\right)$$

$$\geq \frac{n}{2}2\sqrt{d-1} - \frac{n}{2}\left(\frac{2\sqrt{d-1}-1}{k+1}\right)$$

$$- 4(2k+2)d^k\sqrt{d-1}\log(n).$$

Recall that we have chosen k sufficiently large such that $\frac{2\sqrt{d-1}-1}{k+1} \leq \frac{\epsilon}{2}$. As well because k is fixed, $4(2k+2)d^k\sqrt{d-1}\log(n)$ is $\leq n \cdot (\frac{\epsilon}{2})$ for sufficiently large n. Hence we have

$$\rho(G) \geq \frac{n}{2}\left(2\sqrt{d-1} - \epsilon\right).$$

This completes the proof.   □

Acknowledgements. We thank the referees for helpful comments which improved the presentation of the paper, in particular for the shortened proof of Theorem 5 in the full version of the paper [1].

References

1. DeVos, M., Rogers, D., Wesolek, A.: On minimizing the energy of a spherical graph representation. arXiv preprint arXiv:2309.02817 (2023)
2. Fiedler, M.: A property of eigenvectors of nonnegative symmetric matrices and its application to graph theory. Czechoslovak Math. J. **25**(100)(4), 619–633 (1975)
3. Friedman, J.: A Proof of Alon's Second Eigenvalue Conjecture and Related Problems. American Mathematical Society, Providence (2008)
4. Gärtner, B., Matousek, J.: Approximation Algorithms and Semidefinite Programming. Springer, Heidelberg (2012). https://doi.org/10.1007/978-3-642-22015-9
5. Godsil, C.D.: Algebraic Combinatorics. Chapman and Hall Mathematics Series, Chapman & Hall, New York (1993)
6. Godsil, C., Royle, G.: Algebraic Graph Theory. Graduate Texts in Mathematics, vol. 207. Springer, New York (2001). https://doi.org/10.1007/978-1-4613-0163-9
7. Goemans, M.X., Williamson, D.P.: Improved approximation algorithms for maximum cut and satisfiability problems using semidefinite programming. J. Assoc. Comput. Mach. **42**(6), 1115–1145 (1995). https://doi.org/10.1145/227683.227684
8. Gotsman, C., Gu, X., Sheffer, A.: Fundamentals of spherical parameterization for 3D meshes. ACM Trans. Graph. **22**(3), 358–363 (2003)
9. Hall, K.M.: An r-dimensional quadratic placement algorithm. Manag. Sci. **17**(3), 219–229 (1970)
10. Koren, Y.: Drawing graphs by eigenvectors: theory and practice. Comput. Math. Appl. **49**(11–12), 1867–1888 (2005)
11. Lovász, L., Schrijver, A.: On the null space of a Colin de Verdière matrix. Ann. Inst. Fourier (Grenoble) **49**(3), 1017–1026 (1999)
12. Nilli, A.: On the second eigenvalue of a graph. Discret. Math. **91**(2), 207–210 (1991)
13. Pisanski, T., Shawe-Taylor, J.: Characterizing graph drawing with eigenvectors. J. Chem. Inf. Comput. Sci. **40**(3), 567–571 (2000)
14. Wormald, N.C.: The asymptotic distribution of short cycles in random regular graphs. J. Comb. Theory, Ser. B **31**(2), 168–182 (1981)

Posters

One-Bend Drawing of K_n in 3D, Revisited

Olivier Devillers[✉][iD] and Sylvain Lazard

Université de Lorraine, CNRS, Inria, LORIA,Nancy, France
{Olivier.Devillers,Sylvain.Lazard}@inria.fr

Although graphs are usually drawn in 2D, some authors have considered 3D-drawings with vertices placed on an integer grid and try to minimize the volume of the enclosing box. If the edges are constrained to be straight line segments, Cohen et al. [2] proposed a crossing free drawing of the complete graph K_n within an optimal bounding box of volume $\Theta(n^3)$. Some authors consider the possibility of allowing some bends per edges. Bose, et al. [1] showed a $\Omega(n^2)$ lower bound on the volume of a k-bend drawing of K_n (for any k). Dyck et al. [4] achieved a $O(n^2)$ construction with two bends per edge. If we restrict the number of bends to at most one per edge, Morin and Wood [5] get a volume of $O(\frac{n^3}{\log^2 n})$, later improved by Devillers et al. to $O(n^{2.5})$ [3]. Both constructions split the set of vertices in k groups of $\frac{n}{k}$ vertices and use two basic tools: the collinear drawing of the complete graph of each group (all vertices of a group are collinear) and the bi-collinear drawing of the complete bi-partite graph for each pair of groups (vertices are on two parallel lines). The trick is to arrange these subgraphs so that their hulls are disjoints.

In this paper, we allow the hulls of some bi-partite graphs to intersect provided that the bunches of edges do not intersect. Edges are drawn iteratively, checking the possible conflicts with already drawn edges. We draw the complete graph in a box of size $[0, \sqrt{n}] \times [0, Y_{\max}] \times [-n, n \log n]$, where $Y_{\max}$ is computed for each n. We conjecture and check experimentally that $Y_n = O(\sqrt{n} \log n)$ reaching a volume of $O(n^2 \log^2 n)$ close to the quadratic lower bound.

Collinear Drawing of Complete Graphs. (See Fig. 1)
Morin and Wood proposed to draw the complete graph K_k with all vertices on the Y-axis and $O(k^3)$ volume (with various aspect ratio for the enclosing box). We will use it with a box $[0, 1] \times [0, k] \times [-\frac{k^2}{2}, 0]$ by placing the vertices at $(0, i, 0)$ for $i \in [0, k)$ and bend at $(1, y, j)$ where y is any integer in $[0, k]$ and j is used to number the $\frac{k(k-1)}{2}$ edges of K_k. Each edge is drawn in its own page (plane containing the Y-axis), and thus there are no crossings.

Bi-Collinear Drawing of Complete Bi-Partite Graphs. (See Fig. 2)
The construction proposed by Morin and Wood [5, §3] uses, as building blocks, drawings of complete bi-partite graphs with exactly one bend point per edge. These drawings are called one-bend bi-collinear drawings because the two sets of vertices of $K_{n,m}$ are furthermore constrained to lie on two different lines parallel to the Y-axis.

Vertices are placed at $(0, i, 0)$ for $0 \leqslant i < n$ and $(x, y + j, z)$ for $0 \leqslant j < m$ with $(x, y, z) \in \mathbb{Z}^3$. The bend point is placed at $(x', j, z' + i)$ $0 < x' < x$ and $z' \in \mathbb{Z}$

© The Author(s), under exclusive license to Springer Nature Switzerland AG 2023
M. A. Bekos and M. Chimani (Eds.): GD 2023, LNCS 14466, pp. 235–237, 2023.
https://doi.org/10.1007/978-3-031-49275-4

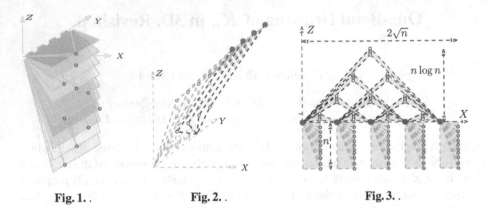

Fig. 1. .   **Fig. 2.** .   **Fig. 3.** .

Choosing $x' = 1$, $z' = 0$, $(x, y, z) = (2, 0, 0)$, the volume is $3nm$. Notice that this construction is contained in two convex bodies, called *wedges*, which are the convex hulls of, respectively, the bends and the n vertices, and the bends and the m vertices.

Disjointness of edges is granted by considering the different planes parallel to the Y-axis they belong to.

Notice that this construction still works if we place the bends of the edges from $(0, i, 0)$ to $(x, y + j, z)$ at $(x', y_{ij}, z' + i)$ (instead of $(x', j, z' + i)$) where $(y_{i,j})_j$ are sequences of integers that are strictly increasing with j (sequences may depend on i).

Our construction. (See Fig. 3)

We follow the idea of previous constructions to (i) split the set of n vertices in k groups of size $\frac{n}{k}$, (ii) place the groups on some lines parallel to the Y-axis in the XY-plane, (iii) draw for each group its complete graph in a box below the XY-plane, and (iv) draw for each pair of groups its complete bi-partite graph in two wedges above the XY-plane. However, unlike previous constructions, we allow the possibility for some wedges to intersect, although the different line segments in the wedges remain crossing free. We thus propose the following strategy:

– We split the vertices in $k = \sqrt{n}$ groups of k vertices.
– Vertex $v_{i,j}$, $0 \leqslant i, j < k$ is at coordinates $(2i, j, 0)$.
– The complete graph of group $v_{i,\star}$ is drawn in the box $[2i, 2i+1] \times [0, k] \times [-k^2, 0]$.
– The bend for the edge from $v_{i,j}$ to $v_{i',j'}$ with $i < i'$ is placed at $e_{i,j,i',j'} = (i' + i, y_{i,j,i',j'}, z_{i'-i}k - j)$

where

$z_1 = 1$, $z_i = \left\lceil \frac{i}{i-1} z_{i-1} \right\rceil + 1$, and $y_{i,j,i',j'}$ is determined algorithmically. Such a placement guarantees disjoint up-going (resp.down-going) wedges and a maximal height of $O(n \log n)$. To avoid intersections between up-going and down-going segments the bends are placed iteratively at the smallest y that preserves the monotonicity and does not create crossing with already drawn edges. We achieve a drawing within a box

$[0, 2\sqrt{n}] \times [0, Y_{max}] \times [-n, n \log n]$, where $Y_{max} = O(n^2)$ is determined by the algorithm. Based on the the computation of Y_{max} for many values of n (see below), we conjecture that $Y_{max} = O(\sqrt{n} \log n)$, yielding a volume of $O(n^2 \log^2 n)$.

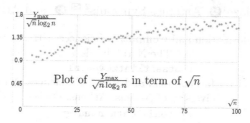

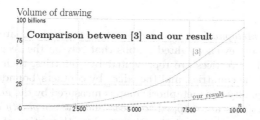

References

1. Bose, P., Czyzowicz, J., Morin, P., Wood, D.R.: The maximum number of edges in a three-dimensional grid-drawing. J. Graph Algorithms Appl. **8**(1), 21–26 (2004). https://doi.org/10.7155/jgaa.00079
2. Cohen, R.F., Eades, P., Lin, T., Ruskey, F.: Three-dimensional graph drawing. Algorithmica **17**(2), 199–208 (1997). https://doi.org/10.1007/BF02522826
3. Devillers, O., Everett, H., Lazard, S., Pentcheva, M., Wismath, S.: Drawing K_n in Three Dimensions with One Bend per Edge. J. Graph AlgorithmsbAppl. **10**(2), 287–295 (2006). https://doi.org/10.7155/jgaa.00128
4. Dyck, B., Joevenazzo, J., Nickle, E., Wilsdon, J., Wismath, S.: Drawing K_n in Three Dimensions with Two Bends Per Edge. Technical report, TR-CS-01-04, University of Lethbridge (2004). https://citeseerx.ist.psu.edu/document?repid=rep1&type=pdf&doi=359925ca9da74cceef62c31a9f8461889e7e592c
5. Morin, P., Wood, D.R.: Three-dimensional 1-bend graph drawings. J. Graph Algorithms Appl. **8**(3), 357–366 (2004). https://doi.org/10.7155/jgaa.00095

Accurate and Simple Metaphoric Maps

Tamara Mchedlidze[1] , Antonios Symvonis[2]([✉]) , and Athanasios Tolias[2]

[1] Department of Information and Computing Sciences, Utrecht University, Utrecht,
The Netherlands
t.mtsentlintze@uu.nl
[2] School of Applied Mathematical and Physical Sciences, National Technical
University of Athens, Athens, Greece
symvonis@math.ntua.gr

Abstract. *Metaphoric maps* or *contact representations* are visual representations of vertex-weighted graphs that rely on the geographic map metaphor. The vertices are represented by countries, their weights by the areas of the countries, and the edges by contacts/boundaries among them. The quality of a metaphoric map is measured by the *accuracy* with which the weights are mapped to areas and the *simplicity* of the polygons representing the countries. We introduce the notion of *region stiffness* and we incorporate it in the force-directed algorithm of Mchedlidze & Schnorr [8]. The new algorithm results in simple metaphoric maps of nearly perfect area accuracy with a little sacrifice in the areas' simplicity. Our algorithm is also able to accommodate non-triangulated graphs.

Introduction. Map-like graph visualization is an alternative way to represent graphs where vertices are represented by polygonal regions and edges by contacts among them; see Fig. 1. Metaphoric maps can display the vertex weights by resizing the corresponding map's regions. They are also known as *area-proportional contact representations* [1] and are closely related to *cartograms* [9, 11].
Metaphoric maps have been studied, mostly in the form of area-universal planar drawings, in [2, 3, 4, 5, 7, 10]. A force-directed algorithm for the creation of metaphoric maps has been provided by Mchedlidze and Schnorr [8]. The quality of a metaphoric map is typically measured in terms of its *cartographic error* which expresses how much the actual area of a region is away from the desired, and its *polygon complexity* which quantifies the complexity of the polygonal shapes of the map's regions. Mchedlidze and Schnorr [8] adopted (after slightly modifying it) the polygon complexity metric proposed by Brinkhoff [6].

The Mchedlidze-Schnorr [8] Algorithm (MS-algorithm, for Short). The MS-algorithm is a typical force-directed algorithm that employs antagonistic forces applied to the vertices of the metaphoric map. The algorithm employs three forces (vertex-vertex repulsion, vertex-edge repulsion and angular resolution) targeted towards producing metaphoric maps of low polygon complexity and one force (air-pressure) working towards reducing the cartographic error.

© The Author(s), under exclusive license to Springer Nature Switzerland AG 2023
M. A. Bekos and M. Chimani (Eds.): GD 2023, LNCS 14466, pp. 238–240, 2023.
https://doi.org/10.1007/978-3-031-49275-4

Fig. 1. (a-b) An internally triangulated plane graph and its corresponding metaphoric map. (c-f) A non-triangulated graph, its layout after applying Tutte's algorithm, and the corresponding metaphoric maps with and without holes, resp.

Our Metaphoric Map Generation Algorithm. We revise the air-pressure force utilized in the `MS-algorithm` by (a) incorporating in it a *stiffness coefficient* and (b) by adopting it in order to accommodate non-triangulated graphs. *The Stiffness of each Map Region.* For every region, we introduce a variable, which we refer to as its *stiffness coefficient*, that accounts for its resilience to respond to the air-pressure force. Given a region g, we define the *stiffness coefficient of g at the i-th iteration of the algorithm*, denoted by $s_i(g)$, as:

$$s_i(g) = \begin{cases} 1, & i = 0 \\ \min(s_{high}, \max(s_{low}, s_{i-1}(g) + \alpha \cdot step)), & i \geq 1 \end{cases}$$

where s_{low}, s_{high} and *step* are constants that are determined experimentally and α is equal to -1, 0, or 1 depending on whether the pressure $P_{i-1}(g)$ at the previous iteration is smaller, equal, or greater than 1, respectively. Then, the *revised air-pressure force* of region g on its boundary edge e during the i-th iteration of the algorithm, $i \geq 1$, denoted by $F_i'(e, g)$, is defined as

$$F_i'(e, g) = s_i(g) \cdot F_i(e, g)$$

where $F_i(e, g)$ is the air-pressure force computed at the i-th iteration of the `MS-algorithm`. Note that the "stiffness" attribute is different for each region and demonstrates an adaptive behaviour over time.

Dealing with Non-triangulated Graphs. In order to deal with non-triangulated plane graphs, it is sufficient to ensure that the initial layout of the given graph satisfies the *barycenter visibility property*, that is, for any interior face g and any vertex v on its boundary, the open segment connecting the barycenter of f with v lies entirely inside f. The barycenter visibility property can be established by (a) placing inside any non-triangulated face an auxiliary vertex and connecting it with the vertices in its boundary, and (b) running Tutte's barycentric algorithm [12]. Then, we can either remove the auxiliary vertices or keep them in the graph and assign a weight to them and, subsequently, run our force-directed algorithm. In the former case the resulting metaphoric map where more than three regions meet at a point while, in the later case, the resulting map has holes.

Experimental Evaluation. Compared to the metaphoric maps produced by the `MS-algorithm` having average cartographic error up to 30%, our experiments showed that our algorithm achieves cartographic error close to zero. This comes at the cost of a small increase (at most 5%) in the polygon complexity.

Demo Link: Visit http://aarg.math.ntua.gr/demos/metaphoric_maps/

References

1. Alam, M.J.: Contact representations of graphs in 2D and 3D. Doctoral thesis, The University of Arizona (2015). https://www.proquest.com/openview/5bf2b2dc7f21bf9f8759bdc89632b50b/1?pq-origsite=gscholar&cbl=18750
2. Alam, M.J., Biedl, T., Felsner, S., Gerasch, A., Kaufmann, M., Kobourov, S.G.: Linear-time algorithms for hole-free rectilinear proportional contact graph representations. Algorithmica **67**(1), 3–22 (2013)
3. Alam, M.J., Biedl, T., Felsner, S., Kaufmann, M., Kobourov, S.G., Ueckerdt, T.: Computing cartograms with optimal complexity. Discret. Comput. Geom. **50**(3), 784–810 (2013). https://doi.org/10.1007/s00454-013-9521-1
4. Alam, M.J., Kobourov, S.G., Veeramoni, S.: Quantitative measures for cartogram generation techniques. Comput. Graph. Forum **34**(3), 351–360 (2015). https://doi.org/10.1111/cgf.12647
5. de Berg, M., Mumford, E., Speckmann, B.: On rectilinear duals for vertex-weighted plane graphs. Discret. Math. **309**(7), 1794–1812 (2009). https://doi.org/10.1016/j.disc.2007.12.087
6. Brinkhoff, T., Kriegel, H., Schneider, R., Braun, A.: Measuring the complexity of polygonal objects. In: Bergougnoux, P., Makki, K., Pissinou, N. (eds.) Proceedings of the 3rd ACM International Workshop on Advances in Geographic Information Systems, Baltimore, Maryland, USA, 1–2 December 1995, in conjunction with CIKM 1995, p. 109. ACM (1995)
7. Kleist, L.: Drawing planar graphs with prescribed face areas. J. Comput. Geom. **9**(1), 290–311 (2018). https://doi.org/10.20382/jocg.v9i1a9
8. Mchedlidze, T., Schnorr, C.: Metaphoric maps for dynamic vertex-weighted graphs. In: Agus, M., Aigner, W., Hoellt, T. (eds.) EuroVis 2022 - Short Papers. The Eurographics Association (2022). https://doi.org/10.2312/evs.20221090
9. Nusrat, S., Kobourov, S.G.: The state of the art in cartograms. Comput. Graph. Forum **35**(3), 619–642 (2016). https://doi.org/10.1111/cgf.12932
10. Thomassen, C.: Plane cubic graphs with prescribed face areas. Comb. Probab. Comput. **1**, 371–381 (1992). https://doi.org/10.1017/S0963548300000407
11. Tobler, W.: Thirty five years of computer cartograms. Ann. Assoc. Am. Geogr. **94**(1), 58–73 (2004). https://doi.org/10.1111/j.1467-8306.2004.09401004.x
12. Tutte, W.T.: How to draw a graph. Proc. Lond. Math. Soc. **3**(1), 743–767 (1963)

Quantum Tutte Embeddings

Shion Fukuzawa[1]([✉])[iD], Michael T. Goodrich[1][iD], and Sandy Irani[1,2][iD]

[1] Dept. of Computer Science, Univ. of California, Irvine, USA
{fukuzaws,goodrich}@uci.edu, irani@ics.uci.edu
[2] Simons Institute, Berkeley, USA

Abstract. Using the framework of Tutte embeddings, we begin an exploration of *quantum graph drawing*, which uses quantum computers to visualize graphs. We discuss how to construct a graph-drawing quantum circuit from a given graph, and show how to calculate a Tutte embedding as a quantum state that can then be sampled to extract the embedding. To evaluate the complexity of our quantum Tutte embedding circuits, we compare them to theoretical bounds established in the classical computing setting derived from a well-known classical algorithm for solving the types of linear systems that arise from Tutte embeddings. We also present empirical results obtained from experimental quantum simulations.

Keywords: Tutte embeddings · Quantum computing · Linear systems

1 Introduction

In this work, we begin an exploration of *quantum graph drawing*, which studies how to use quantum computers to visualize graphs. As a first step in this exploration, we focus on quantum circuits for what is arguably the first graph drawing algorithm—*Tutte embeddings* [8]. A Tutte embedding of a graph G is a crossing-free straight-line embedding of G such that the outer face, f, is a convex polygon and such that each interior vertex (not on f) is at the average (or barycenter) of its neighbors' positions; see, e.g., [2,5]. It is well known now, due to Tutte, that such a configuration can be found by minimizing the energy of a physical system in which edges are represented as springs connecting massless particles representing the vertices. This system of vertices is described by a linear system defined by combining the following pair of equations for every vertex u:

$$\sum_{v \in N(u)} (v_x - u_x) = 0, \qquad \sum_{v \in N(u)} (v_y - u_y) = 0 \qquad (1)$$

Thus the Tutte embedding problem is reduced to solving the above set of linear equations.

A quantum linear systems solver [4] computes a unit vector x that satisfies $Ax = b$ for input matrix, A, and unit vector, b. The leading classical linear systems solver is due to Spielman and Teng [6] and can solve a system of N

equations in time $\tilde{O}(N\log(\kappa/\epsilon))$ where the $\tilde{O}$ suppresses sublinear terms in N. In contrast, the quantum linear systems solver can prepare the solution vector b in time $\tilde{O}(\log(N)s^2\kappa^2/\epsilon)$, where s is the sparsity of the matrix A, κ is the condition number (the ratio between the largest and smallest eigenvalues), and ϵ is the additive approximation error of the solution vector.

There are three challenges that must be handled to port the linear system in Eq. 1 into a quantum linear systems solver: preparing the input state $|b\rangle$, constructing the circuit representing A, and reading out the output vector $|x\rangle$.

In order to preprocess the graph as an input to the quantum linear system solver, we add three additional vertices at the coordinates $(0,0)$, $(1,0)$, and $(0,1)$ and connect our external face to these vertices. We chose to add additional vertices rather than pinning the original outer face because this allows more flexibility for what polygons are allowed as the outer face. A simple analysis shows that the vector $|b\rangle$ is now fixed with a single 1 in a fixed coordinate and 0's everywhere else, unifying the input across all possible graphs of the same size.

The quantum linear system solver requires implementing a unitary circuit approximating e^{-iAt} for Hermitian A. This problem is of large interest in the quantum computing community, and is referred to as Hamiltonian simulation (see [1] for a survey). Briefly summarized, the graph Laplacian A is decomposed into a sum of matrices with sparsity 1, which are known to be easily simulated. This decomposition can be accomplished using a simple greedy edge coloring algorithm using 2Δ colors, where Δ is the maximum degree of G. The solution is approximated by using the Lie-Trotter theorem [7] which can be used to approximate products of non-commuting matrices up to arbitrary precision.

Given the above setting, the quantum linear systems solver will terminate storing the solution vector $|x\rangle$. Fully sampling the coordinates for all N indices will take at least $\Omega(N)$ time, which leads to losing the quantum advantage. However, the advantage can be maintained for applications where the coordinates for only a constant or logarithmic subset of the vectors are important, perhaps a windowed graph drawing of an extremely large graph.

In our main paper [3] we provide more in-depth descriptions of the procedures introduced above, as well as an analysis of the condition number of the linear systems that appear in Tutte embeddings. Our hope is that this work initiates a bridging between the quantum computing and graph drawing communities, leading to insights and algorithms for both areas.

Acknowledgements. We would like to thank David Eppstein for helpful discussions regarding the topics of this paper. This work was supported in part by NSF grant 2212129.

References

1. Dervovic, D., Herbster, M., Mountney, P., Severini, S., Usher, N., Wossnig, L.: Quantum linear systems algorithms: a primer (2018)

2. Di Battista, G., Eades, P., Tamassia, R., Tollis, I.G.: Graph drawing: algorithms for the visualization of graphs. Prentice Hall (1999)
3. Fukuzawa, S., Goodrich, M.T., Irani, S.: Quantum Tutte Embeddings, arxiv.org/abs/2307.08851
4. Harrow, A.W., Hassidim, A., Lloyd, S.: Quantum algorithm for linear systems of equations. Phys. Rev. Lett. **103**, 150502 (2009), https://link.aps.org/doi/10.1103/PhysRevLett.103.150502
5. Kobourov, S.G.: Force-directed drawing algorithms. In: Tamassia, R. (ed.) Handbook of Graph Drawing and Visualization, chap. 12, pp. 383–408. CRC Press (2013)
6. Spielman, D.A., Teng, S.H.: Nearly-linear time algorithms for graph partitioning, graph sparsification, and solving linear systems. In: 36th ACM Symposium on Theory of Computing (STOC), pp. 81–90 (2004)
7. Trotter, H.F.: On the product of semi-groups of operators. Proc. Am. Math. Soc. **10**(4), 545–551 (1959)
8. Tutte, W.T.: How to draw a graph. Proc. Lond. Math. Soc. **3**(1), 743–767 (1963)

The Widths of Strict Outerconfluent Graphs

David Eppstein[✉]

Computer Science Department, University of California, Irvine, USA
eppstein@uci.edu

Introduction. *Confluent drawing* permits certain non-planar graphs to be drawn without crossings [5–8, 12, 14, 16]. In these drawings curved *tracks* meet at vertices and *junctions*; vertices are adjacent when connected by a smooth curve connects in the union of tracks. Applications include syntax diagrams [1] and the Hasse diagrams of posets [10]. *Strict confluent drawing* forbids multiple connections between vertices, or smooth curves from a vertex to itself [9, 11]. *Outerconfluent drawings* place tracks in a disk with the vertices on its boundary. *Strict outerconfluent graphs* are somewhat mysterious: they can be recognized efficiently given the vertex ordering around the disk but without it this remains open [9].

Strict outerconfluent graphs can be dense, with unbounded treewidth. They include distance-hereditary graphs [11] and *tree-like* strict outerconfluent graphs [7], both of bounded clique-width. In this work, we show that strict outerconfluent graphs have unbounded clique-width but bounded twin-width.

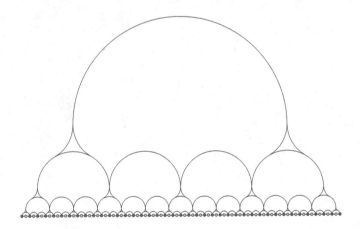

Fig. 1. Construction for strict outerconfluent graphs of unbounded clique-width

Unbounded Clique-Width. We prove that the recursively-constructed graphs of Fig. 1 have unbounded clique-width. It is convenient to use *rank-width*, derived from ternary trees with the graph vertices as leaves. Each tree edge defines a partition of the vertices into two subsets; the rank-width is the maximum rank of

This research was supported in part by NSF grant CCF-2212129.

M. A. Bekos and M. Chimani (Eds.): GD 2023, LNCS 14466, pp. 244–247, 2023.
https://doi.org/10.1007/978-3-031-49275-4

the biadjacency matrix of such a partition, for a ternary tree that minimizes this maximum. Rank-width r and clique-width c are related by $r \leq c \leq 2^{r+1} - 1$ [15].

Proof sketch: Some edge of any ternary tree partitions the vertices roughly evenly; we show that this partition has high rank, through the following steps.

- The partition splits the vertices into contiguous intervals; selected interval endpoints induce a matching of rank proportional to the number of intervals.
- If there are only $O(1)$ intervals, then some two intervals on opposite sides of the partition are both much longer than the gap between them.
- For these two intervals, $\omega(1)$ nested semicircles in the drawing each connect a vertex in one interval to a vertex in the other.
- From these nested semicircles we construct a square submatrix of the biadjacency matrix, of size equal to the number of semicircles, of full rank.

Bounded Twin-Width. In contrast, we prove that strict outerconfluent graphs have bounded *twin-width*. Graphs of bounded twin-width include planar graphs and k-planar graphs, important in graph drawing [2, 13]. Twin-width is defined by merging clusters of vertices in pairs, starting with one-vertex clusters, until only one cluster is left. At each step, two clusters are connected by a *red edge* if some but not all adjacencies exist between their vertices. The twin-width is the minimum d so that this merging process can limit the degree of the graph of red edges to at most d [4]. Twin-width can also be characterized imprecisely by counting *ordered graphs*, graphs assigned a linear order on vertices. A family of ordered graphs is *hereditary* if induced subgraphs with induced orders remain in the family. It is *small* if it has singly-exponentially many n-vertex ordered graphs. Every hereditary family of graphs of bounded twin-width can ordered as a small family of ordered graphs, and every small family of ordered graphs has bounded twin-width [3]. We prove that (with their natural vertex orderings around the disk on which they are drawn) strict outerconfluent graphs form a small hereditary family of ordered graphs, and therefore have bounded twin-width.

Proof Sketch: This family is hereditary: removing any vertices and unused tracks produces another strict outerconfluent graph. From any n-vertex strict outerconfluent drawing of a graph G, form a plane graph D, by reinterpreting the junctions of the drawing as vertices of D and adding one more vertex o, outside the disk of the drawing, connected to the vertices of G by edges external to the disk. An equivalent ordered drawing to the one we started with can be recovered from the plane embedding of D by specifying which vertex is o, which neighbor of o is the start of the linear vertex ordering, and (for each non-neighbor of o) how to partition the incoming edges into tracks meeting smoothly to form a junction. Thus, the number of ordered strict outerconfluent graphs is at most the number of possible combinations of a diagram and this specification of extra information. There are $O(n)$ junctions in the drawing [9], from which it follows that D has $O(n)$ vertices. The number of maximal planar graphs with n vertices is singly exponential [17], from which it follows from the uniqueness of their plane embeddings (up to choice of outer face) and by counting subgraphs that the number of plane graphs is also singly exponential. The numbers of ways to specify o and the start of the linear ordering are $O(n)$, and the number of ways to turn vertices of D into smooth junctions is singly exponential in n. Multiplying these choices gives a singly-exponential bound on strict outerconfluent drawings, showing that the family of strict outerconfluent graphs is small.

For details, see the full version of this paper, arXiv:2308.03967.

References

1. Bannister, M.J., Brown, D.A., Eppstein, D.: Confluent orthogonal drawings of syntax diagrams. In: Di Giacomo, E., Lubiw, A. (eds.) GD 2015. LNCS, vol. 9411, pp. 260–271. Springer, Cham (2015). https://doi.org/10.1007/978-3-319-27261-0_22

2. Bonnet, É., Geniet, C., Kim, E.J., Thomassé, S., Watrigant, R.: Twin-width II: small classes. In: Marx, D. (ed.) Proceedings of the 2021 ACM-SIAM Symposium on Discrete Algorithms, SODA 2021, Virtual Conference, 10–13 January 2021, pp. 1977–1996. Society for Industrial and Applied Mathematics (2021). https://doi.org/10.1137/1.9781611976465.118

3. Bonnet, É., Giocanti, U., Ossona de Mendez, P., Simon, P., Thomassé, S., Toruńczyk, S.: Twin-width IV: ordered graphs and matrices. In: Leonardi, S., Gupta, A. (eds.) STOC '22: 54th Annual ACM SIGACT Symposium on Theory of Computing, Rome, Italy, June 20–24, 2022, pp. 924–937, New York, NY, USA (2022). Association for Computing Machinery. https://doi.org/10.1145/3519935.3520037

4. Bonnet, É., Kim, E.J., Thomassé, S., Watrigant, R.: Twin-width I: tractable FO model checking. J. ACM **69**(1), A3:1–A3:46 (2022). https://doi.org/10.1145/3486655

5. Cornelsen, S., Diatzko, G.: Planar confluent orthogonal drawings of 4-modal digraphs. In: Angelini, P., von Hanxleden, R. (eds.) GD 2022. LNCS, vol. 13764, pp. 111–126. Springer, Cham (2022). https://doi.org/10.1007/978-3-031-22203-0_9

6. Dickerson, M., Eppstein, D., Goodrich, M.T., Meng, J.Y.: Confluent drawings: visualizing non-planar diagrams in a planar way. J. Graph Algorithms Appl. **9**(1), 31–52 (2005). https://doi.org/10.7155/jgaa.00099

7. Eppstein, D., Goodrich, M.T., Meng, J.Y.: Delta-confluent drawings. In: Healy, P., Nikolov, N.S. (eds.) GD 2005. LNCS, vol. 3843, pp. 165–176. Springer, Heidelberg (2006). https://doi.org/10.1007/11618058_16

8. Eppstein, D., Goodrich, M.T., Meng, J.Y.: Confluent layered drawings. Algorithmica **47**(4), 439–452 (2007). https://doi.org/10.1007/s00453-006-0159-8

9. Eppstein, D., Holten, D., Löffler, M., Nöllenburg, M., Speckmann, B., Verbeek, K.: Strict confluent drawing. J. Comput. Geom. **7**(1), 22–46 (2016). https://doi.org/10.20382/jocg.v7i1a2

10. Eppstein, D., Simons, J.A.: Confluent Hasse diagrams. J. Graph Algorithms Appl. **17**(7), 689–710 (2013). https://doi.org/10.7155/jgaa.00312

11. Förster, H., Ganian, R., Klute, F., Nöllenburg, M.: On strict (outer-)confluent graphs. J. Graph Algorithms Appl. **25**(1), 481–512 (2021). https://doi.org/10.7155/jgaa.00568

12. Hirsch, M., Meijer, H., Rappaport, D.: Biclique edge cover graphs and confluent drawings. In: Kaufmann, M., Wagner, D. (eds.) GD 2006. LNCS, vol. 4372, pp. 405–416. Springer, Heidelberg (2007). https://doi.org/10.1007/978-3-540-70904-6_39

13. Hliněný, P., Jedelský, J.: Twin-width of planar graphs is at most 8, and at most 6 when bipartite planar. Electronic preprint arxiv:2210.08620 (2023)
14. Hui, P., Pelsmajer, M.J., Schaefer, M., Štefankovič, D.: Train tracks and confluent drawings. Algorithmica **47**(4), 465–479 (2007). https://doi.org/10.1007/s00453-006-0165-x
15. Oum, S., Seymour, P.: Approximating clique-width and branch-width. J. Comb. Theory Ser. B **96**(4), 514–528 (2006). https://doi.org/10.1016/j.jctb.2005.10.006
16. Quercini, G., Ancona, M.: Confluent drawing algorithms using rectangular dualization. In: Brandes, U., Cornelsen, S. (eds.) GD 2010. LNCS, vol. 6502, pp. 341–352. Springer, Heidelberg (2011). https://doi.org/10.1007/978-3-642-18469-7_31
17. Turán, G.: On the succinct representation of graphs. Discret. Appl. Math. **8**(3), 289–294 (1984). https://doi.org/10.1016/0166-218X(84)90126-4

Convex-Geometric k-Planar Graphs Are Convex-Geometric $(k{+}1)$-Quasiplanar

Todor Antić[✉]

Faculty of Mathematics and Physics, Charles University, Prague, Czech Republic
todorantic29@gmail.com

A *topological* graph is a graph drawn in the plane with vertices as distinct points and edges as simple curves between the points. We say that a topological graph is *simple* if each pair of edges intersects at most once. A topological graph is called a *geometric graph* if its edges are drawn as straight-line segments between the vertices. A geometric graph is called *convex-geometric* graph if its vertices are in convex position.

We say that a graph drawing is k-*quasiplanar* if there is no set of k pairwise crossing edges. We say that an abstract graph is (convex-geometric) k-quasiplanar if it admits a (convex-geometric) k-quasiplanar drawing.

A graph drawing is said to be k-*planar* if every edge has at most k crossings with other edges in the graph. Moreover, an abstract graph is said to be (convex-geometric) k-planar if it admits a (convex-geometric) k-planar drawing. A closely related notion to k-planar graphs is the *local crossing number*, the maximal number of edges crossing a single edge in any drawing of the graph. In particular, a graph is k-planar if it has local crossing number at most k. Notice that both k-planar and k-quasiplanar graphs are defined by forbidden configurations; for an example see Fig. 1. Notice the following connection between these two classes of graphs:

Observation 1. *If G is a simple k-planar graph then G is simple $(k + 2)$-quasiplanar.*

To see that this is true, notice that in a set of $k + 2$ pairwise crossing edges, each edge crosses $k + 1$ other edges, so the local crossing number cannot be k.

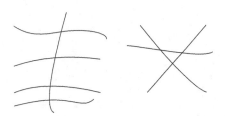

Fig. 1. Forbidden configurations in 3-planar and 3-quasiplanar graphs.

Further, every k-planar (k-quasiplanar) graph is $(k + 1)$-planar ($(k + 1)$-quasiplanar) so they form a natural hierarchy. It is natural to ask what is the

M. A. Bekos and M. Chimani (Eds.): GD 2023, LNCS 14466, pp. 248–250, 2023.
https://doi.org/10.1007/978-3-031-49275-4

growth of a function f such that every k-planar graph is $f(k)$-quasiplanar. For a long time the best bound for f was the obvious one from Observation 1. The first improvement was made by Angelini et al. [3], who proved that every simple k-planar graph is simple $(k + 1)$-quasiplanar. For now, there is no better bound for this problem nor an analogous result for non-simple drawings.

Both k-planar and k-quasiplanar graphs are a part of the family of so-called "beyond planar" graphs. The study of beyond-planar graphs has emerged as an area of interest in graph drawing community recently. This is in part due to practical applications in data visualisation, since it has been shown that topological and geometrical properties of graph drawings impact humans understanding on the graph. For more information about beyond planar graphs see the survey by Didimo, Liotta, Montecchiani [5].

Convex-geometric, k-quasiplanar graphs have been a particular area of interest and have proven to be easier to work with than general topological and geometric counterparts. For example, the following conjecture is still open for general k-quasiplanar graphs:

Conjecture 1. Every k-quasiplanar graph on n vertices has at most $O_k(n)$ edges.

Conjecture 1 is solved for $k = 4$ [1, 2] and is still open for $k > 4$ [6]. While it is known to be true for convex-geometric k-quasiplanar graphs [4]. For more information about convex-geometric graphs see [7].

1   Contribution

We focus on the class of convex-geometric graphs. Inspired by the main result from the paper by Angelini et al. [3] we use variations of the authors techniques and prove that convex-geometric k-planar graphs are convex-geometric $(k + 1)$-quasiplanar for all sufficiently large k. It is important to note that we don't guarantee that the resulting $(k + 1)$-quasiplanar drawing is still k-planar. Our proof relies on the fact that crossings in convex-geometric graphs are completely determined by the ordering of the vertices of the graph along the circle. This allows us to define a very natural flipping operation which "untangles" sets of $k + 1$ pairwise crossing edges. For simplicity, we will call a set of $k + 1$ pairwise crossing edges a $(k + 1)$-crossing. By *untangling* of a $(k + 1)$-crossing we understand the process of redrawing the graph in such a way that at least two edges of a given crossing do not cross anymore. We also introduce a special ordering on the set of $(k + 1)$-crossings in a convex-geometric graph which might be useful in the general study of beyond-planar convex-geometric graphs.

Ackgnowledgments. This work is supported by project 23-04949X of the Czech Science Foundation (GAČR).

References

1. Ackerman, E.: On the maximum number of edges in topological graphs with no four pairwise crossing edges. Discr. Comput. Geom. **41**(3), 365–375 (2009). https://doi.org/10.1007/s00454-009-9143-9

2. Ackerman, E.: Quasi-planar graphs. In: Hong, S.H., Tokuyama, T. (eds.) Beyond Planar Graphs, pp. 31–45. Springer, Singapore (2020). https://doi.org/10.1007/978-981-15-6533-5-3
3. Angelini, P., et al.: Simple k-planar graphs are simple $(k + 1)$-quasiplanar. J. Combin. Theory Ser. B **142**, 1–35 (2020). https://doi.org/10.1016/j.jctb.2019.08.006
4. Capoyleas, V., Pach, J.: A Turán-type theorem on chords of a convex polygon. J. Combin. Theory Ser. B **56**(1), 9–15 (1992). https://doi.org/10.1016/0095-8956(92)90003-G
5. Didimo, W., Liotta, G., Montecchiani, F.: A survey on graph drawing beyond planarity. ACM Comput. Surv. **52**, 1–37 (2019)
6. Fox, J., Pach, J., Suk, A.: Quasiplanar graphs, string graphs, and the Erdos-Gallai problem. In: Graph Drawing and Network Visualization. LNCS, vol. 13764, pp. 219–231. Springer, Cham (2023). https://doi.org/10.1007/978-3-031-22203-0-16
7. Pach, J.: Geometric graph theory. In: Handbook of Discrete and Computational Geometry, pp. 257–279. CRC Press, Boca Raton (2018)

A Collection of Benchmark Datasets
for Evaluating Graph Layout Algorithms

Sara Di Bartolomeo[1,2]([✉]) [iD], Eduardo Puerta[1] [iD], Connor Wilson[1] [iD],
Tarik Cronvrsanin[1] [iD], and Cody Dunne[1] [iD]

[1] Northeastern University, Boston, USA
{puerta.e,wilson.conn,t.crnovrsanin,c.dunne}@northeastern.edu
[2] Universität Konstanz, Konstanz, Germany
di-bartolomeo@dbvis.inf.uni-konstanz.de

Abstract. We built a website to help graph drawing researchers find benchmark datasets to use for evaluating graph layout algorithms. Find it here: https://visdunneright.github.io/gdbenchmarksets/. The datasets and supplemental materials are also available at https://osf.io/j7ucv/.

Benchmarking is a crucial aspect of computer science, as it allows researchers, developers, and engineers to compare the performance of various systems, algorithms, or hardware. A benchmark is a standardized test or set of tests used to measure and compare the performance of hardware, software, or systems under specific conditions. Benchmarking aims to provide objective and consistent metrics that allow for fair comparisons and informed decision-making. Benchmarks are widely used in various fields, including computer hardware evaluation, software optimization, and system performance analysis. In all these fields, benchmarking provides a standardized and objective way to compare and assess the performance of different systems, algorithms, or software implementations. It aids in making informed decisions about which solution best suits a specific use case or requirement.

The same is true for the field of graph drawing, and in particular, for studying the performance and results of graph layout algorithms. Benchmark datasets can provide a standardized set of graphs with known properties and characteristics. These graphs can vary in size, density, connectivity, and structure. Researchers can objectively compare their performance or the quality of their results by applying various graph layout algorithms to the same benchmark dataset.

Because of our own challenges in finding appropriate benchmark sets to evaluate layout algorithms that we developed, we built a collection of benchmark datasets used in previous graph layout algorithm papers and a website to peruse the collection we put together.

The key objective of the work we are doing with benchmark datasets is not only aimed at improving the discoverability of these datasets and easing the

Supported in part by the U.S. National Science Foundation (NSF) under award number IIS-2145382.

running of benchmarks for graph layout algorithms of a vast amount of types and categories, but we also want to place a strong emphasis on the replicability of the experiments that are run. Indeed, reliable access to datasets is fundamental for replicability.

The Collection: the information we collected is a by-product of a larger systematic review we conducted related to graph layout algorithms, which included 206 papers—the core of them being the last 7 years of Graph Drawing proceedings, filtering out the papers with no computational evaluations. Our research expanded to include papers from TVCG and CGF, sourced from the IEEE, ACM, and Wiley digital libraries. For each paper, we noted the algorithm features and datasets used. To locate datasets, we checked supplemental materials, searched online, or contacted authors. We sought permissions for dataset redistribution and stored unclaimed or approved datasets for preservation. We respect ownership rights and will remove any dataset upon the owner's request.

The website is accessible at https://visdunneright.github.io/gd_benchma rk_sets/. Every dataset is accompanied by:

- Labels describing what graph features can be found in the dataset. Additionally, we offer a summary analysis of the contents of the dataset, including information about the distribution of node degrees, or how many graphs are contained in a given dataset, or how many nodes and edges they have.
- A link to where to find the dataset and what paper was associated with its initial publication (if any).
- A list of papers that have used the dataset that exemplify its use in previous research. Moreover, we include representative images of how the dataset has been used in previous research to provide an immediate impression of how the dataset would look. Additionally, we collected the text descriptions of the dataset in these previous papers, which reports useful information in addition to, in some instances, additional insights obtained by the authors.
- A link to the storage location on OSF, which includes converting the original data format to four common formats: GEXF, GraphML, GML, and JSON.

Conclusion: Benchmarking is an important tool in computer science, especially in graph drawing, where consistent datasets are essential for evaluating algorithms. Addressing the challenge of sourcing these datasets, we have curated a collection from prior studies and created a user-friendly website for accessibility. This endeavor not only streamlines the benchmarking process but also emphasizes the replicability of experiments. The website offers a concise overview of each dataset, its features, and associated research. We hope to assist researchers in efficiently finding the right datasets for their work.

Editing Graph Visualizations by Prompting Large Language Models

Evmorfia Argyriou, Jens Böhm, Anne Eberle, Julius Gonser,
Anna-Lena Lumpp, Benjamin Niedermann(✉), and Fabian Schwarzkopf

yWorks GmbH, Tübingen, Germany
{evmorfia.argyriou,jens.bohm,anne.eberle,julius.gonser,anna-lena.lumpp,
benjamin.niedermann,fabian.schwarzkopf}@yworks.com

Abstract. Today's Large Language Models (LLM) provide the possibility of translating user requests given in natural language into executable code. Based on this, we present an approach for interactively modifying graph visualizations. We explain how to prompt such LLMs and how to tackle technical restrictions as limitations of the LLMs on input sizes.

Keywords: Large language model · Code generation · Graph interaction

1 Introduction

Graph editors allow users to interactively modify the layout and the style of graph visualizations. For instance, typically users create graph elements like nodes and edges, move them individually, or change the entire arrangement based on graph drawing algorithms. For usability, the user interface (UI) must be well-balanced. It should provide understandable tools that can be customized to solve specific tasks. However, to keep the UI simple, their generality is highly limited. For example, consider the specific request "create a binary rooted tree of depth 5, layout it radially and color the nodes from blue to red depending on their Euclidean distances to the root". Such requests either require detailed manual work or a sophisticated application programming interface (API), which additionally requires basic programming skills. On the other hand, formulating a request in natural language is intuitive.

With the highly increasing power of large language models (LLM), it now becomes possible to translate natural language into formal language with low effort. Based on this development, we present an approach that allows users to modify the graph visualization based on natural language. It is placed alongside other approaches recently popping up for querying (graph-based) databases [2] and for code generation [1, 4]. The approach translates the user's requests into executable code using the API of the graph editor. To that end, the LLM is prompted both with the user's request and the API. We explain how to formulate the prompts as well as how to tackle technical limitations such as the restricted input size of the LLM, which is far too small for the size of a full-grown graph library. Concerning data privacy, we emphasize that the approach does not send the graph data to the LLM, which is often hosted by external providers.

M. A. Bekos and M. Chimani (Eds.): GD 2023, LNCS 14466, pp. 253–254, 2023.
https://doi.org/10.1007/978-3-031-49275-4

2   Approach

The user initiates communication with the LLM by transmitting a prompt that encapsulates the desired inquiry. Based on the prompt, an *assistant component* of the system formulates explicit instructions for the LLM serving as a roadmap to effectively address the user's query. The primary task of the LLM is the generation of JavaScript code through the utilization of a predefined API. The directives given to the LLM delineate a structured set of prerequisites, serving as imperatives for the creation of executable JavaScript code while minimizing the potential for hallucinations on the part of the LLM. Subsequently, the assistant component transmits the user request along with the instructions and the designated API to the LLM. Following the code generation process, the resultant LLM-generated code is employed to realize the desired graph visualization.

The application of LLMs introduces considerations of prolonged processing times associated with complex prompts and token limitations that potentially impact real-time responsiveness. Therefore, it is crucial to effectively design the prompts. In our approach, the prompt transmitted to the LLM consists also of the predefined API. Instead of using the entire API, we introduce a simplified version using a facade pattern. However, due to the number of methods, this can be still long and complex. To address this, the API methods are partitioned based on their thematic content and tagged with categories. Hence, the assistant component instructs the LLM to select only the API categories corresponding to the user's request, facilitating the transmission of exclusively relevant API segments. This significantly minimizes the overall size of the input prompt.

We have implemented a prototype based on the graph editor yEd Live and ChatGPT [3]. It supports requests in the form of text and speech; for a demo see https://www.yworks.com/yed-live/ and a video https://vimeo.com/yworks/gd2023. We tried requests to modify both the styling and layout of the graph. In many cases, we received correct and executable code. However, there is no guarantee that LLMs provide a correct solution. In future work, we plan to introduce a "self-healing" process to correct defective code parts.

References

1. López Espejel, J., Yahaya Alassan, M.S., Chouham, E.M., Dahhane, W., Ettifouri, E.H.: A comprehensive review of state-of-the-art methods for java code generation from natural language text. Nat. Lang. Process. J. **3**, 100013 (2023)
2. Neo4j: Generating cypher queries with ChatGPT 4 on any graph schema (2023). https://neo4j.com/developer-blog/generating-cypher-queries-with-chatgpt-4-on-any-graph-schema/
3. OpenAI: ChatGPT: Conversational AI model (2021). https://chat.openai.com/
4. Zhang, S., Chen, Z., Shen, Y., Ding, M., Tenenbaum, J.B., Gan, C.: Planning with large language models for code generation. In: The Eleventh International Conference on Learning Representations (2023). https://openreview.net/forum?id=Lr8cOOtYbfL

On Layered Area-Proportional Rectangle Contact Representations

Carolina Haase$^{(\boxtimes)}$ and Philipp Kindermann

Universität Trier, Trier, Germany
{haasec,kindermann}@uni-trier.de

Word clouds can be used to visualize the importance of (key-)words in a given text. Usually, words will be scaled according to their frequency. Motivated by a survey by Viegas et al. [5] are *semantic word clouds*, which are arranged in such a way that closely related words are placed closer together than words that are unrelated. Semantic relatedness can be measured by how often two words occur together in the same sentence [2].

To formalize the problem of drawing semantic word clouds, Barth et al. [1] introduced the problem CONTACT REPRESENTATION OF WORD NETWORKS (CROWN). Given a graph $G = (V, E)$, where every vertex v_i of G corresponds to a word of width w_i and height h_i, and every (weighted) edge between two vertices indicates the level of semantic relatedness between the corresponding words, the goal is to draw a contact representation where each vertex v_i is drawn as an axis-aligned rectangle of width w_i and height h_i such that bounding boxes of semantically related words touch.

Nöllenburg et al. [4] introduced the more restricted variant (MAX-)LAYERED-CROWN. Let $G = (V, E)$ be a planar *layered graph* with L layers, i.e., each vertex is assigned to one of L layers, the order of vertices within a layer is fixed, and edges can only exist between neighboring vertices on the same layer and between vertices on adjacent layers; see Fig. 1. Every vertex v_i is assigned to an axis-aligned rectangle R_i of with w_i and height 1. The goal is to place all rectangles on their layers such that their interiors do not overlap. An edge $\{v_i, v_j\}$ is *realized* if R_i and R_j are in *contact*, i.e., if they share a line segment of length $\varepsilon > 0$. Contacts between rectangles of non-adjacent vertices are forbidden.

The maximization problem MAX-LAYEREDCROWN is to find a valid representation for a given graph G such that the number of realized contacts is maximized. The respective decision problem k-LAYEREDCROWN is to decide whether there exists a valid contact representation that realizes at least k contacts. Many fonts are monospaced, i.e., all letters and characters occupy the same amount of horizontal space. Thus, we also consider the further restriction that rectangles may only be of integer width and may only be placed with their lower left corner on integer coordinates. This implies that two rectangles are in contact if and only if the intersection of their boundaries is a line segment of positive integer length. We call those problems MAX-INTLAYEREDCROWN and k-INTLAYEREDCROWN.

Related Work. Barth et al. [1] have shown that CROWN is strongly NP-hard even when restricted to trees and weakly NP-hard even when restricted to stars, but

© The Author(s), under exclusive license to Springer Nature Switzerland AG 2023
M. A. Bekos and M. Chimani (Eds.): GD 2023, LNCS 14466, pp. 255–257, 2023.
https://doi.org/10.1007/978-3-031-49275-4

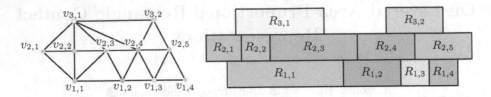

Fig. 1. An internally triangulated graph (left) and a contact representation (right).

can be solved in linear time on irreducible triangulations. They also provided constant-factor approximation algorithms for several graph classes like stars, trees, and planar graphs. These were improved by Bekos et al. [3] and partially implemented and compared to other algorithms by Barth et al. [2].

HIER-CROWN restricts the input to a directed acyclic graph with a single source and a plane embedding [1]. It can be solved in polynomial time, but can be shown to become weakly NP-complete if rectangles are allowed to be rotated.

Nöllenburg et al. [4] introduced MAX-LAYEREDCROWN, but they only considered internally triangulated graphs. They gave a linear-time algorithm for triangulated graphs with only 2 layers and proposed an ILP-formulation for triangulated graphs with more than 2 layers. They further showed how to minimize the area while realizing all contacts in polynomial time with a flow formulation.

Our Contribution. We study the computational complexity of (INT)LAYERED-CROWN and algorithms for MAX-(INT)LAYEREDCROWN. We obtain the following using a reduction from PLANAR MONOTONE 3-SAT.

Theorem 1. *k-INTLAYEREDCROWN is NP-complete for triangulated graphs and k-LAYEREDCROWN is NP-complete for planar graphs.*

On the algorithmic side, we present a $1/2$-approximation for MAX-LAYERED-CROWN on triangulated graphs. We then focus on MAX-INTLAYEREDCROWN with the additional constraint that the maximum rectangle width is at most polynomial in n. Note that practical instances of MAX-INTLAYEREDCROWN will always have bounded maximum rectangle width, as each rectangle corresponds to a word, and words have an upper limit of letters in most languages (in fact, the longest word in an English dictionary, has 45 letters: *pneumonoultra-microscopicsilicovolcanoconiosis*). We find an XP-algorithm based on a dynamic program, which we then use to obtain a PTAS.

Theorem 2. *MAX-INTLAYEREDCROWN is solvable in time $\mathcal{O}(nW)^L$, where W is the maximum rectangle width and L is the number of layers of the input graph. Furthermore, for every $\varepsilon > 0$, MAX-INTLAYEREDCROWN admits a $(1-\varepsilon)$-approximation in $\mathcal{O}(nW)^{1+\frac{1}{\varepsilon}}$ time.*

Open Problems. Several interesting problems remain open, e.g.: (1) Is there an FPT-algorithm parameterized by the number of layers for MAX-INTLAYERED-

CROWN? (2) Is there a PTAS for MAX-INTLAYEREDCROWN for which the running time does not depend on the maximum rectangle width? (3) What can we do if rectangles can have different (integer) heights, thus spanning more layers?

References

1. Barth, L., et al.: Semantic word cloud representations: hardness and approximation algorithms. In: Pardo, A., Viola, A. (eds.) LATIN 2014. LNCS, vol. 8392, pp. 514–525. Springer, Heidelberg (2014). https://doi.org/10.1007/978-3-642-54423-1_45
2. Barth, L., Kobourov, S.G., Pupyrev, S.: Experimental comparison of semantic word clouds. In: Gudmundsson, J., Katajainen, J. (eds.) SEA 2014. LNCS, vol. 8504, pp. 247–258. Springer, Cham (2014). https://doi.org/10.1007/978-3-319-07959-2_21
3. Bekos, M.A., et al.: Improved approximation algorithms for box contact representations. Algorithmica **77**(3), 902–920 (2017). https://doi.org/10.1007/s00453-016-0121-3
4. Nöllenburg, M., Villedieu, A., Wulms, J.: Layered area-proportional rectangle contact representations. In: Purchase, H.C., Rutter, I. (eds.) GD 2021. LNCS, vol. 12868, pp. 318–326. Springer, Cham (2021). https://doi.org/10.1007/978-3-030-92931-2_23
5. Viegas, F., Wattenberg, M., Feinberg, J.: Participatory visualization with wordle. IEEE Trans. Vis. Comput. Graph. **15**(6), 1137–1144 (2009). https://doi.org/10.1109/tvcg.2009.171

`odgf-python` – A Python Interface for the Open Graph Drawing Framework

Simon D. Fink[(✉)] [iD] and Andreas Strobl

Faculty for Mathematics and Computer Science, University of Passau, Passau,
Germany
{finksim,strobland}@fim.uni-passau.de

Abstract. The Open Graph Drawing Framework (OGDF) is a C++
library that contains a vast amount of algorithms and data structures for
automatic graph drawing. While the library is powerful, it is also not easily accessible for new users and the nature of C++ makes implementing
even simple algorithms cumbersome to non-experts. The `odgf-python`
project remedies these problems by making the full functionality of
the OGDF available from Python, providing a visual way to iteratively
develop graph algorithms.

Keywords: OGDF · Graph algorithms · Python · Jupyter notebooks

The Open Graph Drawing Framework (OGDF) [1] is a C++ library that contains
a vast amount of algorithms and data structures for automatic graph drawing.
While the library is powerful, it is also not easily accessible for new users and the
nature of C++ makes implementing even simple algorithms cumbersome to non-
experts. New C++ projects require non-trivial set-up and distinct code-compile-
execute iterations make it hard to incrementally develop a new algorithm. This
is because they separate the code from its results and there is no easy way to
debug and visually analyze the current state of the program.

Many of these issues have been resolved within the Python ecosystem [7]. The
`odgf-python`[1] project remedies these problems also for the OGDF by making its
full functionality available from Python. A similar approach has recently been
taken to make the Computational Geometry Algorithms Library (CGAL) more
accessible [2]. A Python interface greatly reduces the overhead and complications
when using a library for the first time and unlocks a large ecosystem of other
libraries and tools, for example the interactive computing Notebooks provided
by Project Jupyter [3]. These Notebooks allow iteratively developing algorithms
and visualize their results inline right next to their code [7]. They are widely
used in different scientific contexts and constitute the de facto standard in data
science [6].

---

[1] https://github.com/ogdf/ogdf-python.

Funded by the Deutsche Forschungsgemeinschaft (German Research Foundation, DFG)
under grant RU-1903/3-1.

A key component of the `odgf-python` library is its graph display widget for Jupyter Notebooks [8]. It can be used as-is to quickly explore and modify the current graph in-memory. At the same time, it can be used as a building block for more complex user interfaces, where its interactivity can be fully customized to suit different applications. Combining this with the UI components of Jupyter's `ipywidgets` allows to easily build platform-independent user interfaces for interacting with graphs and algorithms. Example use cases include the iterative development of new graph algorithms, step-by-step debugging of implemented algorithms, visual editing of in-memory graphs and variables, interactive visualization of algorithms for teaching, and flexible user interfaces for domain-specific applications.

Internally, `ogdf-python` uses the `cppyy` library [4, 5] to automatically generate python bindings for the C++ OGDF library. One of the main advantages of `cppyy` is that there is no need for explicitly declared bindings or interfaces. This makes `ogdf-python` future-proof by being automatically compatible with future versions of the OGDF without the need for adaptions or manual updates to declarations. The `cppyy` library also allows for arbitrary C++ code to be loaded and called from Python. This allows for variables, functions and even classes to be created and used in either language.

Features

- **No C++ skills needed**: The full OGDF API is available from Python.
- **Rapid prototyping**: Python needs less boilerplate and allows more idiomatic constructions, without the need to configure and compile anything.
- **Iterative execution**: Jupyter Notebooks allow individual lines of code to be adapted and re-run, retaining all previous variable values.
- **Inline results**: Graphs are visually displayed right next to the code that generates them.
- **Interactive graph exploration**: The inline display allows interactive zooming and panning to easily explore the graph.
- **Extensible building block**: The library can be easily combined with other projects from the Python ecosystem, for example with `ipywidgets` to build portable user interfaces for graphs.

Installation

The `ogdf-python` library comes with a prebuilt version of the OGDF called `ogdf-wheel` that works on Linux, macOS and in the Windows Subsystem for Linux. Both components can easily be installed using the Python package manager pip:

```
pip3 install ogdf−python ogdf−wheel notebook
jupyter notebook # start an interactive notebook
```

References

1. Chimani, M., Gutwenger, C., Jünger, M., Klau, G.W., Klein, K., Mutzel, P.: The open graph drawing framework (OGDF). In: Tamassia, R. (ed.) Handbook on Graph Drawing and Visualization, pp. 543–569. Chapman and Hall/CRC, Boca Raton (2013)
2. Goren, N., Fogel, E., Halperin, D.: CGAL made more accessible. CoRR (2022). https://doi.org/10.48550/arXiv.2202.13889
3. Kluyver, T., et al.: Jupyter notebooks - a publishing format for reproducible computational workflows. In: Loizides, F., Scmidt, B. (eds.) Proceedings of the 20th International Conference on Electronic Publishing, pp. 87–90. IOS Press (2016). https://eprints.soton.ac.uk/403913/
4. Kundu, B., Vassilev, V., Lavrijsen, W.: Efficient and accurate automatic python bindings with cppyy & cling. CoRR (2023). https://doi.org/10.48550/arXiv.2304.02712
5. Lavrijsen, W.T., Dutta, A.: High-performance python-c++ bindings with PyPy and cling. In: Proceedings of the 6th Workshop on Python for High-Performance and Scientific Computing (PyHPC 2016). IEEE (2016). https://doi.org/10.1109/pyhpc.2016.008
6. Perkel, J.M.: Why Jupyter is data scientists' computational notebook of choice. Nature **563**(7729), 145–146 (2018). https://doi.org/10.1038/d41586-018-07196-1
7. Perkel, J.M.: Ten computer codes that transformed science. Nature **589**(7842), 344–348 (2021). https://doi.org/10.1038/d41586-021-00075-2
8. Strobl, A.: A generic widget library for rapid-prototyping of graph algorithms in Jupyter notebooks (2020). https://www.fim.uni-passau.de/fileadmin/dokumente/fakultaeten/fim/lehrstuhl/rutter/abschlussarbeiten/2022-Andreas_Strobl-BA.pdf

Drawing Reeb Graphs

Erin Chambers[1] , Brittany Terese Fasy[2] , Erfan Hosseini Sereshgi[3](✉) ,
Maarten Löffler[3,4] , and Sarah Percival[5]

[1] Saint Louis University, Saint Louis, MO, USA
`erin.chambers@slu.edu`
[2] Montana State University, Bozeman, MT, USA
`brittany.fasy@montana.edu`
[3] Tulane University, New Orleans, LA, USA
`{shosseinisereshgi,mloeffler}@tulane.edu`
[4] Utrecht University, Utrecht, The Netherlands
`m.loffler@uu.nl`
[5] Michigan State University, East Lansing, MI, USA
`perciva9@msu.edu`

1   Introduction

Reeb Graphs. Reeb graphs have become an important tool in computational
topology for the purpose of visualizing continuous functions on complex spaces
as a simplified discrete structure. Essentially, these graphs capture how level sets
of a function evolve and behave in a topological space; the components of each
level set become a vertex, and edges connect vertices on adjacent level sets that
are connected.[1] To define the Reeb graph more precisely, let $\mathcal{M}$ be a compact,
connected, and orientable manifold of dimension n, and let $f : \mathcal{M} \to \mathbb{R}$ be a
smooth function. Define an equivalence relation on the points of M by setting
$y \sim y'$ if $y, y' \in g^{-1}(a)$, and y and y' are in the same path component of $g^{-1}(a)$.
Then the Reeb graph is the resulting quotient space $G_f = M/\sim$. Originally
introduced in [11], recent work on computing Reeb graph efficiently [7, 10] has
led to their increasing use in areas such as graphics and shape analysis; see [2,
13] for recent surveys on some of these applications, particularly in graphics and
visualization.

Despite how prevalent the use of Reeb graphs is, surprisingly little work has
been done on generating nice drawings of these structures. The only work in
this area we are aware of considers book embeddings of these graphs [9]; while
of combinatorial interest, the algorithms seem less practical for easy viewing of
larger Reeb graphs. So, this leads to a very natural question, both theoretical
and practical: how difficult is it to draw Reeb graphs?

---

[1] We direct the reader to the poster for illustrations of Reeb graph and other concepts
throughout this abstract.

The authors wish to thank the organizers of the 2023 Annotated Signatures Workshop
in Dauphin Island, Alabama, where this work began (NSF grant 2107434). Research
supported in part by NSF grants 1907612, 2106672 1664858, 2046730.

M. A. Bekos and M. Chimani (Eds.): GD 2023, LNCS 14466, pp. 261–263, 2023.
https://doi.org/10.1007/978-3-031-49275-4

Given the structural properties of Reeb graphs, there is an obvious connection to level drawings of graphs, since Reeb functions are simply real-valued labels on the vertices.

Connection to Layered Graph Drawing. *Layered graph drawing*, also known as *hierarchical* or *Sugiyama-style* graph drawing is a type of graph drawing in which the vertices of a directed graph are drawn in horizontal rows, or *layers*, and in which edges are *y*-monotone curves [1, 3, 12]; depending on the application, vertices may or may not be pre-assigned to levels. As with many drawing styles, layered drawings tend to be more readable when the number of crossings is low; hence a significant amount of effort has been directed towards crossing minimization in layered drawings of graphs. When a drawing without any crossings is possible, the graph is called *upward planar* or *level planar*. While it is possible to test whether a given directed acyclic graph with a single source and a single sink admits an upward planar drawing [5], the problem becomes NP-complete when there can be multiple sources or sinks [6]. When the layers are preassigned, the problem is easier, and testing whether a graph admits a *level planar* embedding is possible in linear time [8]. However, when a planar embedding is not possible, the problem of minimizing crossings is still of interest, and this problem is NP-complete, even when there are only two levels [4].

In Reeb graphs, vertices are preassigned to layers, since the *y*-coordinate carries information that should not be lost.

Challenge. Reeb graphs may be seen as a special case of layered graph drawing with preassigned layers; hence, existing algorithms can be used. Unfortunately, because the problem is NP-hard, existing approaches are mostly heuristic in nature. On the other hand, Reeb graphs have specific structural properties, which may make drawing them easier than the more general layered graph drawing problem. For example, the hardness proof by Garey and Johnson [4] crucially uses high-degree vertices, while Reeb graphs of *generic* (i.e. tame and constructible) manifolds only have vertices of degree 1 or 3. In addition, Reeb graphs have a graph-theoretical *genus* that is directly related to the topological genus of the underlying manifold, and thus it is of interest to design algorithms that can exploit a small number of (undirected) cycles in the graph.

Contribution. We present the following results.

– The problem of deciding whether a given (generic) Reeb graph admits a drawing with at most k crossings, is NP-complete. Our proof uses only vertices of constant degree, but does require a large number of cycles.
– We conjecture that the problem of drawing Reeb graphs may be fixed-parameter tractable (FPT) in the number of (undirected) cycles in the graph.
– As a first step towards establishing our conjecture, we show that when the graph *is* a cycle, we can find a crossing-minimal embedding in polynomial time.

References

1. Bastert, O., Matuszewski, C.: Layered drawings of digraphs. In: Kaufmann, M., Wagner, D. (eds.) Drawing Graphs. LNCS, vol. 2025, pp. 87–120. Springer, Heidelberg (2001). https://doi.org/10.1007/3-540-44969-8_5
2. Biasotti, S., Giorgi, D., Spagnuolo, M., Falcidieno, B.: Reeb graphs for shape analysis and applications. Theoret. Comput. Sci. Comput. Algebraic Geom. Appl. **392**(13), 5–22 (2008)
3. Di Battista, G., Eades, P., Tamassia, R., Tollis, I.G.: Layered Drawings of Digraphs, pp. 265–302. Prentice Hall, Hoboken (1998)
4. Garey, M.R., Johnson, D.S.: Crossing number is NP-complete. SIAM J. Algebraic Discre. Methods **4**(3), 312–316 (1983)
5. Garg, A., Tamassia, R.: Upward planarity testing. Order **12**(2), 109–133 (1995)
6. Garg, A., Tamassia, R.: On the computational complexity of upward and rectilinear planarity testing. SIAM J. Comput. **31**(2), 601–625 (2001)
7. Harvey, W., Wang, Y., Wenger, R.: A randomized $O(m \log m)$ time algorithm for computing Reeb graphs of arbitrary simplicial complexes. In: Proceedings of the 2010 Annual Symposium on Computational Geometry, SoCG '10, pp. 267–276, New York, NY, USA. ACM (2010)
8. Jünger, M., Leipert, S.: Level planar embedding in linear time. In: Kratochvíyl, J. (ed.) GD 1999. LNCS, vol. 1731, pp. 72–81. Springer, Heidelberg (1999). https://doi.org/10.1007/3-540-46648-7_7
9. Kurlin, V.: Book embeddings of Reeb graphs (2013)
10. Parsa, S.: A deterministic $O(m \log m)$ time algorithm for the Reeb graph. In: Proceedings of the 28th Annual ACM Symposium on Computational Geometry, SoCG '12. ACM (2012)
11. Reeb, G.: Sur les points singuliers d'une forme de pfaff completement integrable ou d'une fonction numerique [on the singular points of a completely integrable pfaff form or of a numerical function]. Comptes Rendus Acad. Sci. Paris **222**, 847–849 (1946)
12. Sugiyama, K., Tagawa, S., Toda, M.: Methods for visual understanding of hierarchical system structures. IEEE Trans. Syst. Man Cybernet. SMC **11**(2), 109–125 (1981)
13. Yan, L., et al.: Scalar field comparison with topological descriptors: properties and applications for scientific visualization (2021)

What Happens at Dagstuhl? Uncovering Patterns Through Visualization

Felix Klesen[1], Jacob Miller[2], Fabrizio Montecchiani[3],
Martin Nöllenburg[4], and Markus Wallinger[4](✉)

[1] University of Würzburg, Würzburg, Germany
`felix.klesen@uni-wuerzburg.de`
[2] University of Arizona, Tucson, USA
`jacobmiller1@arizona.edu`
[3] University of Perugia, Perugia, Italy
`fabrizio.montecchiani@unipg.it`
[4] TU Wien, Vienna, Austria
`{noellenburg,mwallinger}@ac.tuwien.ac.at`

1 Introduction

Schloss Dagstuhl is a well-known venue for computer science meetings. In particular, Dagstuhl Seminars[1] cover any topic that is related to computer science. These seminars focus on the exchange and development of ideas, they typically lasts for a week, and they are led by at most four organizers, who are established leaders in their field and represent the different communities invited to the seminar. While most seminars discuss an established field within computer science, Dagstuhl Seminars are also known for establishing new directions by bringing together separate fields or even scientific disciplines.

Analyzing the publicly available data about seminar topics and participants over the years has (at least) two potential benefits: (1) revealing important patterns and trends in computer science; (2) witnessing the impact of such meetings in terms of productivity and cross contamination among different scientific communities. Driven by these motivations, we describe a prototype web-based implementation of a tool for the visual analysis and exploration of the network of Dagstuhl seminars. As a first step towards a richer system, our prototype focuses on the first of the two goals above. It follows the classic overview-first and details-on-demand mantra, namely, individual Dagstuhl seminars are embedded based on the covered topics, which allows for a natural topic-based clustering with multiple levels of details; see Fig. 1 for an illustration. Also, the system shows how participants who attended multiple seminars connect each other, so to reveal connectivity among communities. Finally, users can select and move a time window to gain insight about how seminar topics changed over time.

---

[1] https://www.dagstuhl.de/en/seminars/dagstuhl-seminars.

This research was partially funded by the Vienna Science and Technology Fund (WWTF) [10.47379/ICT19035].

M. A. Bekos and M. Chimani (Eds.): GD 2023, LNCS 14466, pp. 264–266, 2023.
https://doi.org/10.1007/978-3-031-49275-4

2   The Visualization System

Dataset Crawling and Network Creation. We queried all publicly available information from Dagstuhl's website and modelled it as a bipartite graph with two vertex types. We considered events as the first vertex type and stored additional information such as event type, seminar descriptions, key phrases, or associated publications as attributes. We modelled participants as a second vertex type. This information was queried from individual events and we added edges between seminars and their respective participants. Each participant vertex contains the name and if available a DBLP page and id. Edges contain additional information such as role of a participant, or their affiliation. To focus on Dagstuhl seminars we extracted a subgraph by filtering for event vertices of type 'seminar' and all their connected participants. We excluded seminars before the year 2005 as participant information and key phrases are generally not available. In total, the graph contains 980 seminars, 18876 participants and 34854 edges.

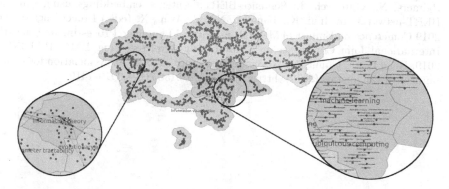

Fig. 1. The Dagstuhl seminar network visualized with a map metaphor. The left circle shows embedded participants while the right circle highlights finer clusters.

Layout Process. The layout is driven by the key phrases associated with each seminar. We first use a pretrained sentence transformer to assign each key phrase a position in high dimensional space based on its semantic meaning [3]. Each seminar is then placed in this space as the barycenter of its key phrases. From this high dimensional representation, we apply the UMAP dimension reduction algorithm [2] as UMAP will generally emphasize clusters and is robust to parameter changes. Though initially hidden in the visualization, individual's (people's) positions in the drawing are computed as the barycenter of their attended seminars. We then find a two-level hierarchical clustering of the embedded seminars. The higher order clustering partitions the data into five distinct top-level clusters (such as 'information visualization' and 'machine learning'). Each of these five clusters are then further partitioned into sub-clusters which represent more narrow topics (such as 'computational geometry' or 'parameterized complexity'). These clusters are represented visually using a map metaphor similar to GMap [1]. Following this map metaphor, we treat each seminar like a city; as one zooms into a particular part of the map, details such as the seminar title appear. One can click on a seminar to reveal the individuals who attended that seminar and the 'roads' (edges) between them.

References

1. Gansner, E.R., Hu, Y., Kobourov, S.G.: GMap: visualizing graphs and clusters as maps. In: IEEE Pacific Visualization Symposium PacificVis 2010, Taipei, Taiwan, 2–5 March 2010, pp. 201–208. IEEE Computer Society (2010). https://doi.org/10.1109/PACIFICVIS.2010.5429590
2. McInnes, L., Healy, J., Saul, N., Großberger, L.: UMAP: uniform manifold approximation and projection. J. Open Source Softw. **3**(29), 861 (2018). https://doi.org/10.21105/joss.00861
3. Reimers, N., Gurevych, I.: Sentence-BERT: sentence embeddings using siamese BERT-networks. In: Inui, K., Jiang, J., Ng, V., Wan, X. (eds.) Proceedings of the 2019 Conference on Empirical Methods in Natural Language Processing and the 9th International Joint Conference on Natural Language Processing, EMNLP-IJCNLP 2019, Hong Kong, China, 3–7 November 2019, pp. 3980–3990. Association for Computational Linguistics (2019). https://doi.org/10.18653/v1/D19-1410

Author Index

M. A. Bekos and M. Chimani (Eds.): GD 2023, LNCS 14466, pp. 267–269, 2023.
https://doi.org/10.1007/978-3-031-49275-4

Printed in the United States
by Baker & Taylor Publisher Services